v. 2

International Study of Achievement in Mathematics II

Associate Editors:

Benjamin S. Bloom
Maurice L. Hartung
Gilbert F. Peaker
Douglas A. Pidgeon
Robert L. Thorndike
David A. Walker

INTERNATIONAL PROJECT FOR THE EVALUATION
OF EDUCATIONAL ACHIEVEMENT (IEA)

Phase I

International Study of Achievement in Mathematics

A Comparison of Twelve Countries

VOLUME
II

EDITED BY

TORSTEN HUSÉN

Chairman of the IEA

ALMQVIST & WIKSELL
STOCKHOLM

JOHN WILEY & SONS
NEW YORK · LONDON · SIDNEY

© *1967*
International Project for the
Evaluation of Educational Achievement, Hamburg
Almqvist & Wiksell/Gebers Förlag AB, Stockholm

This book or any part thereof
must not be reproduced in any form
whatsoever without the written
permission of the publisher

PRINTED IN THE UNITED STATES OF AMERICA

Contents

Preface 17

CHAPTER 1. Mathematics Tests and Attitude Inventory Scores 21
Total Scores on Mathematics Tests 21
Scores on Separate Parts of the Mathematics Tests 31
Correlations with Mathematics Achievement 36
    School Characteristics 37
    Teacher Characteristics 37
    Student Characteristics 39
Analysis of Single Item Results 42
Student Attitudes 43

CHAPTER 2. Correlations Between and Within Countries 49

CHAPTER 3. The Relation of School Organisation to Attainments in Mathematics 56
Introduction 56
The Testing of Hypotheses 61
    Age of Entry 61
    Age at which School Education Ceases 68
    Size of School 74
    Size of Class 79
    Specialisation 85
    Achievement in Selective and Comprehensive Systems 87
    Interest in Mathematics in Selective and Comprehensive Systems 102
    Socio-economic Status in Selective Systems 107
    Mathematical Achievement and School Retentivity 116
    An Attempt to Construct a Model of the Effects of Selection 135
Summary and Conclusion 139

CHAPTER 4. Problems Related to the Curriculum and Instructional Methods 143
Introduction 143
Achievement in Relation to Interests and Attitudes 144
    Formulation of Hypotheses 144
    Findings 148

Inquiry Centered Methods 148
   Interests and Attitudes 152
   Interests and other Factors 157
   Achievement in Relation to Teachers' Perceptions and Training 162
      Formulation of Hypotheses 162
      Findings 167
      Achievement and Opportunity to Learn 167
      Achievement and National Emphasis 170
      Teacher perceptions of Freedom 175
      Achievement, Interests and Attitudes in Relation to In-Service Training of Teachers 177
      Amount and Type of Pre-Service Training of Teachers 178
   School Decisions Relative to Curriculum 182
      Formulation of Hypotheses 182
      Findings 184
      Time Spent on Schooling 184
      Participation in Special Opportunities 189
      Courses in "New" Mathematics 190
   Summary and Conclusions 194
      Summary of Major Findings 194
      Some General Implications of the Findings 197

CHAPTER 5. Social Factors in Education 199
   Introduction 199
   Home Background 204
      Socio-economic Background and Educational Selectivity 204
      Socio-economic Background and Mathematics Achievement 205
      Socio-economic Status, Level of Mathematics Instruction, and Further Plans for Education 209
      Socio-economic Status, Interest in Mathematics and Amount of Homework 211
      Socio-economic Differences in Mathematics Achievement Within School Programs 212
      School Socio-economic Variability and Mathematics Achievement 219
   Urban-Rural Differences in Mathematics Achievement 223
   Financial Support for Education and Mathematics Achievement 232
   Sex Differences in Mathematics Achievement and Attitudes Towards Mathematics 233
      Review of Previous Research 237
      Differences Between the Sexes in Achievement 239
      Interest in and Attitude Towards Mathematics 243
      Single Sex and Coeducational Schools 247
   Mathematics Achievement and Educational and Vocational Aspirations 250
   Summary and Conclusions 253

CHAPTER 6. A Regression Analysis 260
   Introduction 260
   Parental Variables 265

Teacher Variables 269
School Variables—I 274
School Variables—II 278
Student Variables 281
Summary and Conclusions 284

CHAPTER 7. Summary of Major Findings 287
Introduction 287
Pattern of Test Results 288
Problems Related to School Organisation 290
Studies of Certain Features of School Organisation 295
Problems Related to Instruction and Curriculum 298
   Student Achievements and Non-cognitive Outcomes 298
   Achievements as Related to Opportunity to Learn and to Teacher Competence 299
   Mathematics Achievement as Related to Opportunities Provided by the School 300
Mathematics Achievement and Societal Factors 301
   Students' Social Background and Mathematics Performance 302
   Sex Differences in Performance and Attitudes 304
Further Research 307
   Objectives for the Next IEA Stage 307
   Particular Problems to be Considered 308
   Developing New Subject Matter 309
   Case Studies of Particular Countries 309
   Cognitive Styles 310

Appendix I 311

Appendix II 312

Appendix III 360

References 362

Subject Index 366

Tables

CHAPTER 1

1.1 Total Mathematics Test Score Distributions—13-Year-Olds: Population 1a 22
1.2 Total Mathematics Test Score Distributions—Grade Level Containing Most 13-Year-Olds: Population 1b 23
1.3 Total Mathematics Test Score Distributions—Mathematics Students in Final Secondary Year: Population 3a 24
1.4 Total Mathematics Test Score Distributions—Nonmathematics Students in Final Secondary Year: Population 3b 25
1.5 Estimate of Increase in Mathematics Performance from 13-Year-Olds to Terminal Mathematics Students 30
1.6 Increment in Mathematics Achievement in Relation to Two Indices of Educational Selectivity 30
1.7 Subscores for National Groups Expressed as Standard Scores of 13-Year-Old Students: Population 1a 32
1.8 Subscores for National Groups Expressed as Standard Scores of Grade Group Corresponding to 13-Year-Olds: Population 1b 32
1.9 Subscores for National Groups Expressed as Standard Scores of Mathematics Students in Terminal Secondary Year: Population 3a 33
1.10 Subscores for National Groups Expressed as Standard Scores of Nonmathematics Students in Terminal Secondary Year: Population 3b 34
1.11 Intercorrelations of Mathematics Total Score and Subscores 37
1.12 Median Correlations of School, Teacher, and Student Variables with Mathematics Achievement 38
1.13 Correlation of Mathematics Score with Teacher's Rating of Opportunity to Learn 40
1.14 Correlations of Mathematics Score with Parental Education and Occupation 41
1.15 National Means and Standard Deviations on Five Attitude Dimensions (13-Year-Old Students) 46
1.16 National Means and Standard Deviations on Five Attitude Dimensions (Grade Containing Most 13-Year-Olds) 46
1.17 National Means and Standard Deviations on Five Attitude Dimensions (Math. Students in Final Secondary Year) 47
1.18 National Means and Standard Deviations on Five Attitude Dimension (Nonmath. Students in Final Secondary Year) 47
1.19 Correlations of National Mean Achievement and National Mean Attitude 48

CHAPTER 2

2.1 Correlations with Total Corrected Mathematics Scores 50
2.2 Country Means in Code for Variables in Table 2.1 52–53

CHAPTER 3

3.1 Mean Ages and Standard Deviations of Age for Populations 1a and 1b 64
3.2 Mean Scores and Standard Deviations of Scores in Mathematics for Different Ages of Entry 65
3.3 Differences Between Mean Scores—Population 1a 66
3.4 Differences Between Mean Scores—Population 1b 66
3.5 Mean Score in Mathematics by Socio-Economic Status 67
3.6 Mean Ages and Dispersions of Students by Country and Population (in Months) 69
3.7 Relationship Between 1b Score, 1a Score, and Age 72
3.8 Relationship Between 3a Score, 1a Score, Age, and Retentivity 73
3.9 Relationship Between 3b Score, 1a Score, Age, and Retentivity 73
3.10 Mean Enrollments per School by Country and Population 76
3.11 Weighted Mean Scores by Population, Type of School and Size of School 77
3.12 Mean Scores Showing Significant Differences Within Countries by Type of School and Size (Population 1a) 77
3.13 Mean Scores Showing Significant Differences Within Countries by Type of School and Size (Population 3a) 78
3.14 Mean Mathematics Scores for Different Sizes of Class (Population 1a) 80
3.15 Mean Mathematics Scores for Different Sizes of Class (Population 1b) 81
3.16 Mean Mathematics Scores for Different Sizes of Class (Population 3a) 82
3.17 Mean Mathematics Scores for Different Sizes of Class (Population 3b) 83
3.18 Significant Differences in Score by Different Sizes of Class 84
3.19 Variation of Score with Number of Subjects Studied (Population 3a) 86
3.20 Mean Corrected Mathematics Test Scores for Students Following Academic, Vocational, and General Programs in both Comprehensive and Specialized Schools in Population 1a 93
3.21 Mean Corrected Mathematics Test Scores for Students Following Academic Programs in Comprehensive and Specialized Schools in Population 3a 94
3.22 Mean Corrected Mathematics Test Scores for Students Following Academic, Vocational, and General Programs in Comprehensive Schools in Various Countries in Population 1a 95
3.23 Mean Corrected Mathematics Test Scores for Students Following Academic, Vocational, and General Programs in Specialized Schools in Various Countries in Population 1a 95
3.24 Mean Corrected Mathematics Test Scores for Students Following Academic Programs in Comprehensive and Specialized Schools in Various Countries in Population 3a 96
3.25 Mean Corrected Mathematics Test Scores for Students Following Academic, Vocational, and General Programs in Either Comprehensive or Specialized Schools in England 97
3.26 Mean Corrected Mathematics Test Scores for Students Following Academic and General Programs in Either Comprehensive or Specialized Schools in Scotland 98

3.27 Mean Corrected Mathematics Test Scores for Students Following Academic and General Programs in Either Comprehensive or Specialized Schools in Australia 98
3.28 Number of Students Classified by Type of School and Degree of Differentiation of Courses—Populations 1a and 1b 104
3.29 Mean Scores on "Interest in Mathematics" for Different Types of School and Degrees of Differentiation—Populations 1a and 1b 105
3.30 Relation Between Interest Score and Emphasis Within a Country on Comprehensive and Undifferentiated Education—Population 1a and 1b 106
3.31 Percentages of Students by Country, Population, and Occupational Category of Parent 111
3.32 Relationship Between Retentivity and Social Background of Enrollment at the Preuniversity Level (Population 3a and 3b Combined) 113
3.33 Indices of Social Bias 114
3.34 Percentage of an Age Group in the Preuniversity Year Enrolled in Terminal Mathematics-Science and in Full-Time Schooling 116
3.35 Indices of Retentivity, Income, Industrialization, and Comprehensive Education 117
3.36 Mean Mathematics Score and Percent of Age Group in Population for Populations 3a and 3b 118
3.37 Means and Standard Deviations of Mathematics Test Scores for Total Sample and Equivalent Proportion of the Age Group in Each Country at the Terminal Mathematics Level 122
3.38 Means and Standard Deviations of Mathematics Test Scores of Total Sample and Equivalent Proportion of the Age Group in Each Country at the Terminal Nonmathematics Level 125
3.39 Rank Orders of Mean Mathematics Scores for Countries with Comprehensive School Systems 127
3.40 Percentage of Preuniversity Mathematics Students Reaching Given Standards (Population 3a) 129
3.41 Percentage of Age Group Reaching Given Standards (Population 3a) 131
3.42 Percentage of Preuniversity Nonmathematics Students Reaching Given Standards (Population 3b) 133
3.43 Percentage of Age Group Reaching Given Standards (Population 3b) 133
3.44 Data for Constructing a Model (Population 3a) 135

CHAPTER 4

4.1 Means, Standard Deviations, and Number of Cases of the Scale: Description of Mathematics Teaching and School Learning for Populations 1a and 3a 150
4.2 Mean Total Mathematics Scores, Adjusted for Level of Instruction with S.D.'s and N.'s for Students in "Little" and "Much" Inquiry Groups, for Populations 1a, 1b, and 3a 151
4.3 Mean Total Mathematics Scores (Adjusted for Level of Instruction) Related to Inquiry-Centered Approaches to Learning for Various Countries—Populations 1a and 3a 152
4.4 Mean Total Mathematics Scores, Adjusted for Student Opportunity to Learn, with S.D.'s and N.'s for Students in "Little" and "Much" Inquiry Groups, for Populations 1a, 1b, and 3a 153

4.5 Mean Total Scores (Adjusted for Student Opportunity to Learn) Related to Inquiry-Centered Approaches to Learning for Various Countries—Populations 1a and 3a 153
4.6 Correlations with Total Mathematics Scores for Interest and a Number of Attitudes Over All Countries, Populations 1a, 1b, 3a, and 3b 154
4.7 Means, Standard Deviations, and Number of Cases of Scores on the Scale of Interest in Mathematics for Populations 1a, 3a, and 3b 155
4.8 Correlations Between National Mean Scores in Mathematics and Means of Interest and Various Attitude Scales for Populations 1a, 1b, 3a and 3b 156
4.9 Means, Standard Deviations, and Number of Cases of Interest Related to Extent that Learning Is Viewed as Inquiry Centered for Population 1a 158
4.10 Means, Standard Deviations, and Number of Cases of Interest in Mathematics in Relation to Scientific or Nonscientific Nature of Father's Occupation for Population 1a 161
4.11 Rank Correlation Coefficients of National Mean Scores on Interest and Selected Attitudes Between Different Populations 162
4.12 National Means of Teachers' Ratings of Opportunity to Learn and of Mathematics Scores 170
4.13 National Means, Standard Deviations, and Number of Cases for Scores on Degree of Freedom Given Teachers, for Populations 1b and 3a 176
4.14 Correlation Coefficients Between National Mean on In-Service Training and Five Other Variables for Populations 1a and 3a 178
4.15 Means of Mathematics Scores, Standard Deviations, and Number of Cases Related to the Type of Training of Teachers and to Whether the Teachers had Little (L) or Much (M) Training—Population 1a 180
4.16 Means of Mathematics Scores, Standard Deviations, and Number of Cases Related to the Type of Training of Teachers and to Whether the Teachers had Little (L) or Much (M) Training—Population 3a 181
4.17 Means, Standard Deviations, and Number of Cases of Time Given Mathematics Instruction and of Total Mathematics Scores, by Countries, for Population 3a 185
4.18 Partial Regression Coefficients for All Homework with Mathematics Scores on Lower Mental Process (LMP) and on Higher Mental Process (HMP), with Level of Instruction Held Constant 186
4.19 Means, Standard Deviations, and Number of Cases of Hours per Week for Mathematics Homework (MH) and All Homework (AH), and Percent of MH to AH by Populations—Populations 1a and 3a 187
4.20 Coefficients of Correlation Between National Mean Mathematics Scores and National Means of Several Time Variables for Populations 1a and 3a 188
4.21 Percent of Participation in Special Opportunities Related to Difference Between Mean Achievement of Students Who Participated and Students Who Did Not—Population 3a 189
4.22 Means, Standard Deviations, and Differences of Means of Scores on Items of Traditional Type by Students Who Have Had and Have Not Had Courses in New Mathematics—Populations 1a and 1b 193

CHAPTER 5

5.1 Percentages of the Three Higher Occupational Status Groups in the Different Samples 205

5.2 Mean Level of Father's Education 205
5.3 Correlations Between Mathematics Achievement and Father's Educational Level and Father's Occupational Status 206
5.4 Total Mathematics Test Score by Level of Father's Occupational Status, Population 1b 207
5.5 Correlations of Father's Level of Education with the Student's Level of Mathematics Instruction and Number of Years of Additional Education Expected 211
5.6 Correlations of Father's Level of Education with the Student's Interest in Mathematics and Number of Hours of Homework per Week 212
5.7 Percent of High Occupational Status Students in Academic and General School Programs in Selected Countries, Population 1b 213
5.8 Total Mathematics Test Scores (Corrected) as Related to Socio-Economic Background (School Program Being Held Constant)—Population 1a 214
5.9 Total Mathematics Test Scores (Corrected) as Related to Socio-Economic Background (School Program Being Held Constant)—Population 1b 215
5.10 Total Mathematics Test Scores (Corrected) as Related to Socio-Economic Background (School Program Being Held Constant)—Population 3a 216
5.11 Total Mathematics Test Scores (Corrected) as Related to Socio-Economic Background (School Program Being Held Constant)—Population 3b 217
5.12 Total Mathematics Test Score as Related to Father's Occupational Status and School Variability—Population 1a 221
5.13 Total Mathematics Test Score as Related to Father's Occupational Status and School Variability—Population 1b 221
5.14 Total Mathematics Test Score as Related to Father's Occupational Status and School Variability—Population 3a 221
5.15 Total Mathematics Test Score as Related to Father's Occupational Status and School Variability—Population 3b 221
5.16 Place of Parents' Residence, Mean Corrected Scores, Standard Deviations, and Sample Size. Population 1a 226
5.17 Place of Parents' Residence, Mean Corrected Scores, Standard Deviations, and Sample Size. Population 1b 226
5.18 Place of Parents' Residence, Mean Corrected Scores, Standard Deviations, and Sample Size. Population 3a 227
5.19 Place of Parents' Residence, Mean Corrected Scores, Standard Deviations, and Sample Sizes. Population 3b 227
5.20 Differences in Mathematics Scores (Corrected for Level of Instruction) Between Students Grouped by Place of Parents' Residence 228
5.21 Summary of Significant and Non-significant Differences Between Urban, Town, and Rural Groups 229
5.22 Correlation of Total Mathematics Score with Total Per-Student Expenditure and Expenditure for Teachers' Salaries 233
5.23 Ratio of Male to Female Students 234
5.24 Ratio of Students in Single-Sex Schools to Those in Coeducational Schools 236
5.25 Sex Differences in Total Mathematics Scores 240
5.26 Sex Differences in Total Mathematics Score by Country 240
5.27 Comparison of Boys and Girls on Verbal and Computational Problems 241
5.28 Sex Differences in Verbal and Computational Problem Scores by Country 242
5.29 Correlations Between Sex of Student and (A) Interest in Mathematics,

(B) Plans to Take Further Mathematics and (C) Attitudes Toward the Difficulty of Learning Mathematics 243
5.30 Sex Differences in Interest in Mathematics—All Countries 244
5.31 Sex Differences in Interest in Mathematics by Country 245
5.32 Interest in Mathematics: Analysis of Variance 245
5.33 Sex Difference in Plans to Take Further Mathematics by Country 246
5.34 Sex Differences in Attitude Toward the Difficulty of Learning Mathematics by Country 247
5.35 Comparison of Mathematics Scores of Boys and Girls in Single-Sex and Coeducational Schools 248
5.36 Sex Differences in Total Mathematics Score (Corrected) in Single-Sex and Coeducational Schools by Country 248
5.37 Comparison of Boys and Girls in Interest in Mathematics in Single-Sex and Coeducational Schools 250
5.38 Partial Correlation Coefficients Between Mathematics Scores, Educational Plans, and Educational Aspirations—Holding Level of Mathematics Instruction Constant 251
5.39 Partial Correlation Coefficients Between Mathematics Scores, Vocational Plans, and Vocational Aspirations—Holding Level of Mathematics Instruction Constant 252

CHAPTER 6
6.1 Percentage of Variance Accounted For: Populations 1a, 1b, 3a, 3b 264
6.2 Parental Group of Variables: Population 1a 266
6.3 Parental Group of Variables: Population 1b 266
6.4 Parental Group of Variables: Population 3a 267
6.5 Parental Group of Variables: Population 3b 267
6.6 Teacher Group of Variables: Population 1a 270
6.7 Teacher Group of Variables: Population 1b 270
6.8 Teacher Group of Variables: Population 3a 271
6.9 Teacher Group of Variables: Population 3b 271
6.10 First Group of School Variables: Population 1a 275
6.11 First Group of School Variables: Population 1b 275
6.12 First Group of School Variables: Population 3a 276
6.13 First Group of School Variables: Population 3b 276
6.14 Second Group of School Variables: Population 1a 279
6.15 Second Group of School Variables: Population 1b 279
6.16 Second Group of School Variables: Population 3a 280
6.17 Second Group of School Variables: Population 3b 280
6.18 Student Group of Variables: Population 1a 282
6.19 Student Group of Variables: Population 1b 282
6.20 Student Group of Variables: Population 3a 283
6.21 Student Group of Variables: Population 3b 283

Figures

CHAPTER 1

1.1 National means expressed as deviations from the grand mean (in standard deviation units) 27
1.2 National standard deviations divided by grand overall standard deviations 28
1.3 Increase in level of mathematics achievements from 13-year-old to terminal math. students 29
1.4 Sample item analysis report 43

CHAPTER 3

3.1 Mean mathematics score at the age of 13 for different ages of entry 66
3.2 Relation of mathematics score to age by country (Population 1b) 70
3.3 Relation of mathematics score to age by country (Population 3a) 70
3.4 Relation of mathematics score to age by country (Population 3b) 71
3.5 Relation of mathematics score to percentage of age group in population, by country (Population 3a) 119
3.6 Relation of mathematics score to percentage of age group in population, by country (Population 3b) 120
3.7 Mean mathematics test scores (1) for the total sample and (2) for equal proportions of the age group in each country for terminal mathematics population 124
3.8 Mean mathematics test scores (1) for the total sample and (2) for equal proportions of the age group in each country for terminal nonmathematics population 126
3.9 Cumulative percentile frequencies (smoothed)—Population 3a 132
3.10 Mean score in mathematics made by population 3a in different countries plotted against the score expected on basis of the model 137
3.11 Ratio of variances in populations 3a and 1a in different countries plotted against the variances expected on basis of the model 138
3.12 (Model A.) Hypothetical distribution of mathematics scores for whole age group with score k cutting off population 3a to form proportion q of the age group 138
3.13 (Model B.) The correlation surface relating mathematics score to selection variable, with k cutting off proportion q from the rest 138

CHAPTER 4

4.1 Correlations between the scale Descriptions of mathematics teaching and school learning and Total mathematics scores for Populations 1a, 1b, 3a and 3b 149

4.2 Correlations between Interest in mathematics and Student's view about mathematics teaching for Populations 1a and 3a 158
4.3 Correlations between Interest in mathematics and Student's views about mathematics as an open system for Populations 1a and 3a 159
4.4 Correlations between Interest in mathematics and Status of father's occupation for Populations 1a and 3a 160
4.5 Correlations between Total mathematics score and Teacher's ratings of opportunities to Learn 169
4.6 National profiles of percent emphasis on selected topics in academic and nonacademic courses, Population 1b 171
4.7 National profiles of percent emphasis on selected topics in academic and nonacademic courses, Population 3a 172
4.8 National profiles of percent emphasis on selected topics in academic and nonacademic courses, Population 3b 173

CHAPTER 5

5.1 Total mathematics score as related to level of father's occupational status, Population 1b 208

Erratum

It was discovered, at the last moment when this volume was already in page proof, that certain major errors had occurred in the coding and weighting of the data from Finland for Populations 1 a and 1 b. There are in Finland two main categories of schools in which 13-year-olds appear, elementary and secondary. Pupils were tested in both, but in addition to forgetting 400 elementary school pupils the coding and weighting of the two groups were distorted. All the materials relating to Populations 1 a and 1 b should therefore be ignored.

It has, however, been possible to calculate from unweighted data univariate statistics for the Finnish Populations 1 a and 1 b. These provide reasonably accurate estimates of the national statistics. They are to be found in Appendix III in this volume (p. 360).

Preface

The present volume reports the findings of the IEA project. Chapter 1 has been written by Professor Robert L. Thorndike and aims at giving a descriptive picture of the mathematics test and attitude inventory scores in the different countries and populations within countries in terms of means, standard deviations, and correlations. The increment in mathematics performance from the age of 13 to the preuniversity year and the correlations between the mathematics scores and the various student, teacher, and school variables have been calculated. Finally, the statistics for the individual test items by nation and population have been explained in Chapter 1. Appendix II contains all the item statistics.

Mr. Gilbert Peaker, C.B.E., has written Chapter 2 in which he describes the between-country correlations on some of the major variables.

Chapters 3, 4, and 5 report the outcomes of the testing of the hypotheses. Chapter 3, which has been edited by Dr. David A. Walker, deals primarily with problems related to school organisation, such as the role of age of entry to school, selectivity, retentivity, and in that connection with the question whether "more means worse". Finally, the relation between selection and social bias has been elucidated.

Chapter 4, which has been edited by Professor Maurice Hartung, reports the outcomes of the testing of hypotheses dealing with the relationship of various curriculum and instruction factors to mathematics achievement. For example, the relation between mathematics performance and the general approach in instruction ("inquiry" versus "drill") has been studied together with the relationship of the relative emphases given to various topics and student performance on these topics. Such factors as hours of homework, training of teachers, the extent to which they are free to plan curricula and choose methods, and the opportunity to learn the "New Mathematics", have also received attention.

Chapter 5, which has been edited by Professor Benjamin S. Bloom, deals primarily with how societal factors are related to mathematics performance. The socio-economic background of the students in various countries has been related to both the cognitive and noncognitive outcomes of mathematics instruction. The differential performance of students from urban and rural areas is examined cross-nationally. In an attempt to understand the way in which the expectations are associated with school performance, sex differences in mathematics achievement, interest, and attitudes have been considered at some length. Furthermore, the relationship between the social class variability of schools and the performance of students from various social classes within them has been studied.

Chapters 3, 4, and 5 are the products of many authors. The hypotheses were assigned to individual members of the IEA group according to expressed preferences. Each author's write-up was then reviewed by one of the other members of the group. The first drafts of most of the individual contributions dealing with single hypotheses were prepared in Chicago in February, 1965, after the first tables had been produced by the computer. In some cases, a new set of tables had to be made because of changes in the form of analysis. The authors then revised their drafts. At the following council meeting in Hamburg in August, 1965, the editors worked with the authors in subgroups and structured the material in clusters of hypotheses. From this, the present Chapters 3–5 emerged.

In Chapter 6, Mr. Peaker reports the results of a 26-variable multiple regression analysis which he also planned.

In Chapter 7, finally, the general editor has tried to summarise the main conclusions of this study, but in so doing he has necessarily omitted many of the interesting details reported in each of the preceding chapters.

The final editing of all chapters was carried out at a session of the Editorial Committee at the Center for Advanced Study in the Behavioral Sciences at Stanford, California, in December, 1965. The general editor, who spent the academic year 1965–66 as a Fellow at the Center, devoted a considerable part of his time to the editing of the present volume.

At this point, some general comments on the statistical analyses which appear in this volume are appropriate. Mathematics scores are weighted and corrected for guessing; furthermore, the standard deviations and product moment correlations are also weighted. The weighting procedures have already been described in Chapter 12 of Volume I. In some cases (which are always indicated), tables report mathematics scores which have been regressed on particular variables. The purpose of this was to remove an undesired source of variance from the data and to bring out the results more clearly. Occasionally, certain tables are incomplete; there are various reasons for this—countries were omitted because the data for a particular variable were not available or, if available, were not in a form suitable for the type of analysis employed. This was occasionally the case where trichotomies were used.

It should also be pointed out that in some instances missing data made it impossible to include certain countries in the analysis relating to hypotheses where regressed scores were used. One example is Scotland, which did not have adequate data on the level of mathematics instruction. Thus, throughout Volume II this country has been omitted from the analyses of all hypotheses utilizing data on level of mathematics instruction.

Bold type has been used in the tables to indicate significance. Standard errors have not been printed, but can easily be estimated in most tables by using the standard deviation per student and the number of students, which are also given in the tables, to obtain the simple random sampling estimate of standard error, and then multiplying the latter by the appropriate factor (see Appendix I). These factors are needed because the primary sampling unit was not the individual student, but the school, or in one case, the area. (The method of obtaining factors is given in Chapter 9 of Volume I.) The factors vary somewhat from country to country and from population to population, but for most purposes it is enough to note that to obtain the complex standard error it would be prudent to double the simple random sampling standard error, and it would be excessively cautious to treble it.

Correlations have been used as summarizing statistics in many parts of this study. They have some notable advantages. First, they are exceedingly compact summaries. Second, they include more of the available information than many alternatives. Third, they make it possible to partial out the effects of other variables by simple operations. And finally, the standard errors do not vary widely. In our study these errors range from .02 to .08, according to the size and constitution of the samples from the various populations in the countries included in the study. (See Tables 9.2 through 9.7 in Volume I, summarized at the end of this volume in Appendix I.) This makes it easy to read a column of correlation coefficients, whether in tabular or diagrammatic form, with due regard to the sampling uncertainty. The more exact values should be used when necessary, but the standard errors can be taken uniformly as .04 without giving a seriously misleading impression of the whole table.

The correlations for students within countries may often seem small, but this is because they in fact measured only a fairly small fraction of the total variation between students. Taken together, the variables that we have studied account for about a third of the whole variation, leaving two thirds for individual differences outside our field of study. The correlations between countries are often much larger; this must be because much of the individual variation cancels out over whole countries. It should be noted that the twelve countries in the study are not to be thought of as a random sample from a larger population of countries. They are themselves the universe of discourse for the study. To think of the correlations between countries as subject to the sampling variation appropriate to a random sample of 12 units would be to misconceive the situation, which is not that of making estimates from a sample about a population, but that of describing a particular universe.

The variables we have studied are not only correlated with our criteria, but are also correlated, sometimes heavily, with each other. In the tables of Chapter 6, the simple correlations of 26 variables with the total corrected mathematics score are set down side by side with the standardized regression coefficients for each of the four populations. These tables show that in some cases a substantial simple correlation with the criterion becomes negligible when the other variables are held constant. How far, and in what circumstances, this means that the simple correlation is illusory is a difficult question, which is discussed in Chapter 6.

All the members of the IEA council and the consultants performed a dedicated work in preparing the drafts for the present volume. Thanks are due also to the members of the Editorial Committee, which had to carry out the heavy task of structuring the material provided by their colleagues. Professor C. Arnold Anderson rendered invaluable help by reading the various drafts and suggesting improvements. Mr. Neville Postlethwaite successfully dealt with the complicated and arduous coordination which inevitably is connected with an international cooperative enterprise like the present one.

TORSTEN HUSÉN
Chairman and Technical Director of IEA

Center for Advanced Study in the Behavioral Sciences,
Stanford, California. 1966.

Chapter 1

Mathematics Test and Attitude Inventory Scores

It is our purpose in this chapter to present a description of the pattern of test results. We shall not be concerned with hypotheses regarding factors producing or associated with certain outcomes; that analysis will occupy subsequent chapters. We will be content to look at mathematics test scores and subscores, at the performance on different measures of attitude, and at some of the intercorrelations among the variables studied. We will also provide an introduction to certain appendices in which material on specific test items is presented so that it may be available to the interested mathematics educator.

Total Scores on Mathematics Tests

The first set of tables (Tables 1.1–1.4, pages 22–25) shows the distribution of mathematics total score by country[1] and by level. The results for 13-year-olds are shown in Table 1.1, where, as in the other tables of the set, frequencies have been reduced to percentages to make them directly comparable from country to country. All scores have been corrected for the possibility of getting multiple-choice questions right by guessing, using the conventional correction formula.

The first general conclusion that emerges from Table 1.1 is that the test was hard for 13-year-olds in almost all countries. On a 70-item test, the typical corrected score was approximately 20. This conclusion applies not merely to the 13-year-old group, but also to a considerable extent to the other target populations as well, as shown by Tables 1.2–1.4. Thus, the group in mathematics courses in the terminal year of secondary education average only 26 out of 69 items right. The tests had been

This Chapter was written by Professor R. L. Thorndike.
[1] The terminal nonmathematics populations (3 b) from France and Netherlands have been deleted from all analyses for reasons stated in Chapter 9 of Volume I. However, in this descriptive chapter, they have been included.

TABLE 1.1. *Total Mathematics Test Score Distributions[a] 13-Year-Olds.*

Population 1a.

Score	Australia	Belgium	England	Finland	France	Japan	The Netherlands	Scotland	Sweden	United States	Total
0	1.0	0.6	3.0	0.1	1.0	0.6	0.5	1.0	1.0	3.0	1.0
1–5	13.0	8.0	21.0	1.0	13.0	7.0	10.0	17.0	15.0	19.0	16.0
6–10	14.0	7.0	13.0	7.0	15.0	7.0	15.0	14.0	19.0	16.0	13.0
11–15	13.0	9.0	12.0	11.0	17.0	7.0	11.0	14.0	19.0	14.0	14.0
16–20	12.0	8.0	10.0	16.0	16.0	8.0	10.0	12.0	15.0	13.0	12.0
21–25	12.0	11.0	7.0	21.0	12.0	8.0	12.0	10.0	11.0	11.0	10.0
26–30	10.0	11.0	7.0	17.0	8.0	8.0	9.0	9.0	9.0	8.0	9.0
31–35	9.0	11.0	6.0	12.0	6.0	11.0	8.0	7.0	5.0	6.0	8.0
36–40	7.0	12.0	6.0	9.0	5.0	11.0	7.0	6.0	3.0	5.0	6.0
41–45	4.0	9.0	5.0	4.0	3.0	9.0	5.0	4.0	2.0	2.0	4.0
46–50	2.0	7.0	5.0	1.0	2.0	9.0	5.0	3.0	0.8	1.0	3.0
51–55	2.0	4.0	2.0	0.6	1.0	8.0	2.0	1.0	0.1	1.0	2.0
56–60	0.6	2.0	2.0	0.0	0.3	5.0	2.0	1.0	0.1	0.3	1.0
61–65	0.3	0.4	0.6	0.0	0.3	1.0	4.0	0.3	0.0	0.1	0.4
66–70	0.1	0.0	0.1	0.0	0.0	0.4	0.0	0.0	0.0	0.0	0.1
Mean	20.2	27.7	19.3	24.1	18.3	31.2	23.9	19.1	15.7	16.2	19.8
S.D.	14.0	15.0	17.0	9.9	12.4	16.9	15.9	14.6	10.8	13.3	14.9
Number of cases	2,917	1,686	2,949	747	2,409	2,050	429	5,256	2,554	6,231	27,228

[a] All scores have been corrected for guessing. Entries are percentages of the total group. Values greater than 1.0 are reported to the nearest whole percent.

TABLE 1.2. *Total Mathematics Test Score Distributions Grade Level Containing Most 13-Year-Olds.*

Population 1 b.

Score	Australia	Belgium	England	Finland	France	Germany	Israel	Japan	The Netherlands	Scotland	Sweden	United States	Total
0	1.0	0.5	4.0	0.1	1.0	0.4	0.6	0.6	0.3	1.0	1.0	2.0	1.0
1-5	12.0	4.0	15.0	2.0	9.0	4.0	4.0	7.0	7.0	13.0	16.0	16.0	10.0
6-10	13.0	6.0	10.0	3.0	12.0	6.0	4.0	7.0	12.0	12.0	20.0	14.0	11.0
11-15	15.0	6.0	11.0	7.0	16.0	11.0	6.0	7.0	14.0	12.0	18.0	14.0	12.0
16-20	15.0	8.0	9.0	12.0	16.0	13.0	8.0	8.0	17.0	12.0	15.0	14.0	12.0
21-25	14.0	10.0	6.0	22.0	13.0	15.0	9.0	8.0	16.0	10.0	11.0	12.0	12.0
26-30	11.0	13.0	7.0	22.0	9.0	16.0	11.0	8.0	11.0	9.0	9.0	10.0	10.0
31-35	9.0	13.0	8.0	14.0	8.0	15.0	12.0	11.0	9.0	8.0	5.0	7.0	10.0
36-40	5.0	13.0	7.0	10.0	5.0	8.0	13.0	11.0	6.0	7.0	3.0	5.0	7.0
41-45	3.0	12.0	6.0	4.0	5.0	6.0	12.0	9.0	3.0	6.0	1.0	3.0	6.0
46-50	1.0	8.0	7.0	3.0	2.0	3.0	9.0	9.0	3.0	5.0	0.4	1.0	4.0
51-55	0.3	4.0	5.0	0.5	3.0	2.0	7.0	8.0	1.0	3.0	0.1	1.0	3.0
56-60	0.1	2.0	2.0	0.0	0.5	0.3	3.0	5.0	0.1	1.0	0.0	0.4	1.0
61-65	0.1	0.5	2.0	0.0	0.2	0.0	2.0	1.0	0.1	0.4	0.0	0.1	0.5
66-70	0.0	0.0	0.4	0.0	0.0	0.0	0.0	0.4	0.0	0.0	0.0	0.0	0.1
Mean	18.9	30.4	23.8	26.4	21.0	25.4	32.3	31.2	21.4	22.3	15.3	17.8	23.0
S.D.	12.3	13.7	18.5	9.6	13.2	11.7	14.7	16.9	12.1	15.7	10.8	13.3	15.0
Number of Cases	3,078	2,645	3,089	841	3,449	4,475	3,232	2,050	1,443	5,718	2,828	6,544	39,392

TABLE 1.3. *Total Mathematics Test Score Distributions Mathematics Students in Final Secondary Year.*

Population 3 a.

Score	Australia	Belgium	England	Finland	France	Germany	Israel	Japan	The Netherlands	Scotland	Sweden	United States	Total
0	0.3	0.0	2.0	0.0	0.5	0.0	0.0	0.1	0.0	0.1	0.0	2.0	0.7
1–5	4.0	1.0	0.0	0.8	0.5	0.6	0.0	2.0	0.0	0.6	1.0	23.0	4.0
6–10	11.0	1.0	1.0	5.0	0.9	3.0	0.0	6.0	0.0	2.0	5.0	21.0	7.0
11–15	15.0	6.0	2.0	11.0	4.0	6.0	0.0	8.0	2.0	11.0	10.0	16.0	10.0
16–20	17.0	8.0	6.0	16.0	12.0	11.0	6.0	11.0	7.0	20.0	15.0	12.0	13.0
21–25	16.0	14.0	10.0	19.0	13.0	17.0	8.0	10.0	14.0	22.0	16.0	8.0	14.0
26–30	17.0	13.0	17.0	18.0	15.0	19.0	13.0	11.0	16.0	18.0	14.0	6.0	14.0
31–35	10.0	13.0	12.0	12.0	15.0	18.0	18.0	9.0	26.0	10.0	13.0	5.0	12.0
36–40	5.0	12.0	14.0	10.0	15.0	13.0	19.0	12.0	18.0	6.0	11.0	3.0	9.0
41–45	2.0	9.0	11.0	6.0	12.0	8.0	18.0	10.0	13.0	4.0	8.0	1.0	6.0
46–50	2.0	6.0	13.0	2.0	6.0	2.0	9.0	9.0	4.0	2.0	4.0	1.0	5.0
51–55	0.4	4.0	6.0	0.2	4.0	2.0	4.0	7.0	0.0	2.0	1.0	1.0	2.0
56–60	0.1	10.0	4.0	0.0	1.0	0.2	5.0	4.0	0.0	1.0	1.0	0.6	2.0
61–65	0.0	3.0	1.0	0.0	0.5	0.0	0.0	1.0	0.0	0.3	0.4	0.3	0.5
66–69	0.0	0.0	0.3	0.0	0.0	0.0	0.0	0.0	0.0	0.0	0.1	0.1	0.0
Mean	21.6	34.6	35.2	25.3	33.4	28.8	36.4	31.4	31.9	25.5	27.3	13.8	26.1
S.D.	10.5	12.6	12.6	9.6	10.8	9.8	8.6	14.8	8.1	10.4	11.9	12.6	13.8
Number of Cases	1,089	519	967	369	222	649	146	818	462	1,422	776	1,568	9,007

TABLE 1.4. *Total Mathematics Test Score Distributions Nonmathematics Students in Final Secondary Year.*

Population 3b.

Score	Belgium	England	Finland	France	Germany	Japan	The Netherlands	Scotland	Sweden	United States	Total
0	0.4	3.0	0.2	3.0	0.0	0.4	0.0	0.5	0.0	5.0	1.0
1–5	2.0	2.0	1.0	1.0	0.0	7.0	0.0	4.0	11.0	38.0	10.0
6–10	4.0	9.0	7.0	3.0	1.0	10.0	10.0	11.0	24.0	24.0	11.0
11–15	9.0	12.0	14.0	9.0	6.0	10.0	15.0	14.0	32.0	13.0	12.0
16–20	19.0	19.0	18.0	14.0	11.0	11.0	15.0	20.0	22.0	8.0	14.0
21–25	22.0	21.0	25.0	18.0	19.0	12.0	21.0	17.0	8.0	5.0	15.0
26–30	17.0	14.0	18.0	20.0	26.0	12.0	17.0	16.0	1.0	4.0	13.0
31–35	14.0	11.0	8.0	16.0	20.0	11.0	11.0	12.0	2.0	1.0	10.0
36–40	8.0	5.0	6.0	11.0	12.0	9.0	11.0	4.0	0.0	1.0	6.0
41–45	3.0	2.0	2.0	5.0	5.0	7.0	0.0	1.0	0.0	0.2	4.0
46–50	1.0	1.0	0.7	0.0	0.0	6.0	0.0	0.3	0.0	0.0	2.0
51–55	0.3	0.3	0.0	0.0	0.0	4.0	0.0	0.0	0.0	0.0	1.0
56–58	0.0	0.0	0.0	0.0	0.0	0.8	0.0	0.0	0.0	0.0	0.3
Mean	24.2	21.4	22.5	26.2	27.7	25.3	24.7	20.7	12.6	8.3	21.0
S.D.	9.5	10.0	8.3	9.5	7.6	14.3	9.8	9.5	6.2	9.0	12.8
Number of Cases	1,004	1,782	399	192	643	4,372	50	2,123	222	2,042	12,828

designed to be inclusive, rather than being tailored to fit the curriculum of any one country, and this may have accounted for the result. Thus, items on the elements of algebra and geometry were included in the test for 13-years-olds, even though in a number of the participating countries most students do not encounter a formal treatment of these topics until a later age. The relation of difficulty to topic is seen in the fact that the 13-year-olds, over all countries, got 48 percent of the possible score on the basic arithmetic items, on the average, but only 21 percent of the possible score on algebra items and 20 percent on the geometry items.

The tests were not devised primarily in order to make total score comparisons between countries possible and certainly not as yard sticks for an "international contest". The mere fact that algebra and geometry items were included in the tests for the 13-year level in spite of the fact that these topics were not dealt with in some countries should discourage national comparisons. The construction of the tests was guided by the hypotheses advanced and the tests are to be used primarily for comparisons between school systems both within and between countries in relation to these hypotheses. Thereby, subscores play as important a role as do variables like student's opportunity to learn the items and emphasis put upon the topics in the instruction.

Though they are considered to be of minor significance for the project, one can hardly avoid being interested in national differences in average score and in variability.

Differences between countries in average score are quite marked; at the 13-year level, the range is from 31.2 for Japan to 15.7 for Sweden. This difference of 15.5 points is more than one standard deviation of the combined distribution of all scores. The range of national means is even greater in the other target populations. Among students in the grade typical for 13-year-olds (Population 1 b), the range is from 32.3 for Israel to 15.3 for Sweden, a difference of 17.0 points, as against a standard deviation of 15.0. In the terminal mathematics group (3 a), the range is from 36.4 in Israel to 13.8 in the United States, a difference of 22.6 points in comparison with a standard deviation of 13.8. Finally, in the terminal nonmathematics group (3 b) the range is from 27.7 in Germany to 8.3 in the United States, a 19.4 point difference in comparison with a standard deviation of 12.8.

The national differences are summarized in Figure 1.1. Here, each national mean has been expressed as a deviation from the grand mean of all cases, using the grand standard deviation for all cases as a unit of measure. An examination of the figure will permit the reader in-

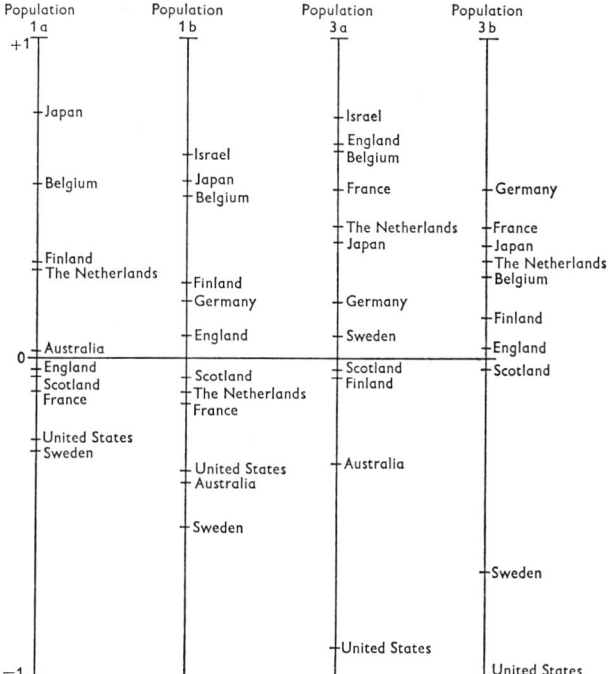

Figure 1.1. National means expressed as deviations from the grand mean (in standard deviation units).

terested in any specific country to determine how that country stands in each of the four target populations. There are several cases in which the number of schools and pupils is quite small, and the results should very possibly be discounted. In the terminal mathematics group, there were only 222 pupils from France and 146 from Israel, two of the four countries with highest means.

Countries differ not only in average score, but also in spread of scores within the country. Thus, in the 13-year-old group, the standard deviation ranges from 9.9 for Finland to 17.0 for England. Finland's standard deviation is below the average in all four groups, while those for Japan and England are uniformly high. The relative spread of scores for each target population in every country can be seen in Figure 1.2, in which each national standard deviation is expressed as a ratio of that for the aggregate of all students tested. There is clearly some consistency in variability over the four target populations, high and low variability tending to be a persisting national characteristic in this study.

It is of some interest to examine the increase in mathematical com-

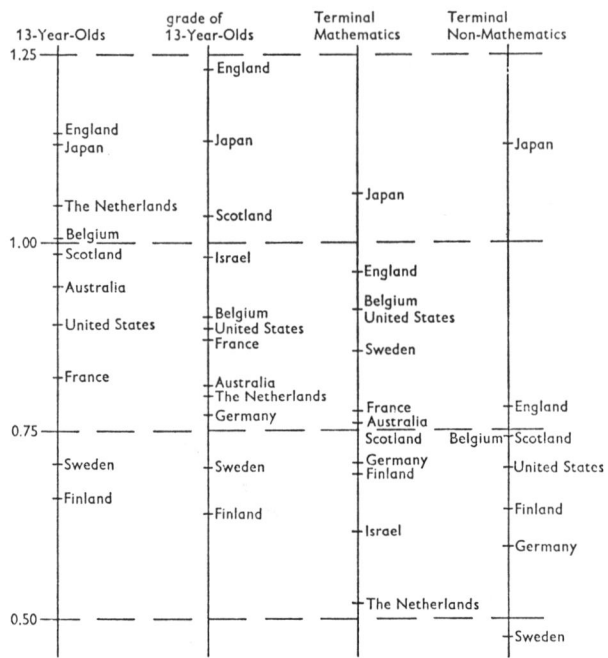

Figure 1.2. National standard deviations divided by grand overall standard deviations.

petence as one goes from the 13-year-olds to the students enrolled in mathematics courses in the final year of secondary education. Unfortunately, no direct comparison of these two groups is possible, because it did not seem appropriate to give groups as different as this a common test. However, an indirect and somewhat hazardous comparison can be made using the terminal nonmathematics group to provide a bridge. Since this group and the 13-year-olds both took test Number 3 of the series of 9 one-hour tests that were prepared for the study, the superiority of the terminal nonmathematics group can be estimated by the difference in mean performance of the two groups on this 23-item test, using the standard deviation of the aggregate terminal nonmathematics group as a common unit. Since both the terminal math and nonmathematics groups had taken test Number 5, we can compare these two groups, once again using the standard deviation for the terminal nonmathematics group as a unit. We determined for each country (1) how far its 13-year-olds were below the grand mean of terminal nonmathematics students on test Number 3 and (2) how far its terminal mathematics students were above the grand mean of nonmathematics students on test Number 5. These results are shown

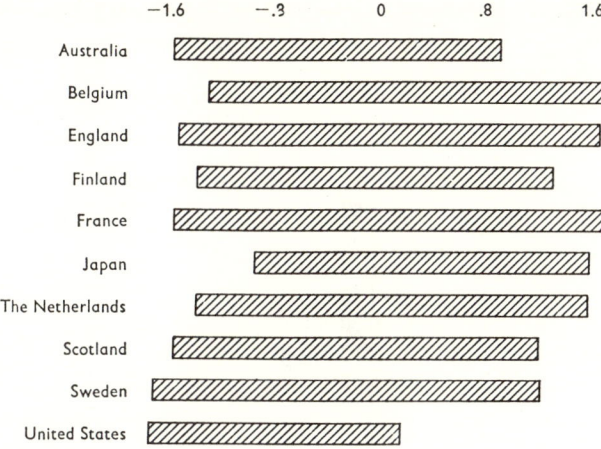

Figure 1.3. Increase in level of mathematics achievements from 13-year-old to terminal math. students.

in Table 1.5 and Figure 1.3. For example, the average 13-year-old from Australia falls at −1.52 standard deviations, in relation to the mean of all terminal nonmathematics students as a zero point, whereas the average terminal mathematics student in Australia falls at +.86 standard deviations. Hence, the increase in score level in Australia from 13-year-olds to terminal mathematics students can be estimated to be 2.38 standard deviation units. The increase ranges from a low of about 1.8 for the United States to a high of about 3.1 for England and France. In Figure 1.3, the left-hand end of each bar represents the level of performance for Population 1 a (13-year-olds), the right-hand end represents the level of performance for terminal mathematics students, and the length of the bar represents the amount of increase from the lower to the higher level.

The size of the increment in performance may be a function of the amount and quality of mathematics instruction and/or of the amount of academic selectivity that has operated upon the group receiving mathematics instruction or of some combination of these and other factors. Some indication of the role of selectivity may be obtained by correlating the above gains with two indices of selectivity, one based on father's education and the other on father's occupation. The education index used here is the difference between average years of education for fathers of the 13-year-old group and that for fathers of the terminal mathematics group. The occupational index is the corresponding ratio of the percentage of fathers with higher level occupations.

TABLE 1.5. *Estimate of Increase in Mathematics Performance from 13-Year-Olds to Terminal Mathematics Students.*

	Level of Performance of 13-Year-Olds[a]	Level of Performance Terminal Math. Students[a]	Increase in Level
Australia	−1.52	0.86	2.38
Belgium	−1.22	1.66	2.88
England	−1.47	1.61	3.08
Finland	−1.33	1.28	2.61
France	−1.53	1.63	3.16
Japan	−0.90	1.55	2.45
The Netherlands	−1.33	1.53	2.86
Scotland	−1.53	1.18	2.71
Sweden	−1.64	1.18	2.82
United States	−1.69	0.14	1.83
Grand Total	−1.49	1.11	2.60

[a] Relative to grand mean and standard deviation of terminal nonmathematics students.

That the variables are correlated as between countries can be determined by inspection of Table 1.6. When rank correlations are computed they turn out to be .54 with the increase in father's education and .62 with the ratio of percentage in higher level occupations. The correlation between the two selectivity indices is .50, and the maximum prediction from the two of them considered jointly is approximately .68. These correlations are quite substantial. Furthermore, it must be re-

TABLE 1.6. *Increment in Mathematics Achievement in Relation to Two Indices of Educational Selectivity.*

	Math. Achievement Increment	Increase in Level of Father's Education	Ratio of Percentage in Higher Occupations
Australia	2.38	0.5	2.1
Belgium	2.88	0.6	1.9
England	3.08	1.3	3.7
Finland	2.61	1.4	2.1
France	3.16	3.6	2.8
Japan	2.45	2.0	2.1
The Netherlands	2.86	2.6	2.6
Scotland	2.71	1.3	3.1
Sweden	2.82	3.1	2.4
United States	1.83	0.4	1.3

membered that these are very imperfect indicators of educational selectivity, since they refer to the father rather than the student himself, and were reported by the student with unknown accuracy. In view of these considerations, one may conclude that differences in degree of selectivity are very important determiners of national differences in gains in mathematics achievement from late in primary to end of secondary school (cf. pp. 107–115).

Scores on Separate Parts of Mathematics Tests

In some ways, it is more instructive to compare national and other groups on part scores for the mathematics test than on a single total score. The items were grouped in several different ways. First, they were classified (by the pooled judgment of several reviewers) into items calling for higher mental processes and those calling for lower mental processes. The latter are judged to call for relatively routine application of previously learned techniques, while the former call for a greater amount of ingenuity and inventiveness in the attack upon novel or complex problems. A second subdivision of the items was between those that were verbally formulated and those that involved primarily computation or solution of a problem expressed in numbers or symbols. A third special group of items consisted of those that were judged by the mathematics educators to represent the "New Mathematics". Fourth, items were grouped by content areas: arithmetic, algebra, geometry, etc.

To express subscores in a form in which different ones could be compared, they were converted into standard-score form. The grand mean of all pupils tested at a given level served as the zero point, and the standard deviation of all pupils tested served as the unit of measure. (To eliminate decimal points, national means are reported in hundredths of a standard deviation.) The patterns of performance are shown in Tables 1.7–1.10 (pp. 32–34).

Let us look first at Table 1.7 and study one country as an example of the way in which the table is to be read. Let us use Finland as an illustration. In total performance, the 13-year-olds of Finland average 28/100 of a standard deviation above the grand mean of all 13-year-olds tested. The Finnish 13-year olds are somewhat better on the higher-mental-process items than on the lower-mental-process items, but the difference is only a fifth of a standard deviation. Again, they perform slightly better on the verbal than on the computational items. They do quite well on the "New Mathematics", a rather surprising result since none of the teachers in Finland *reported* that they were teaching

TABLE 1.7. Subscores for National Groups Expressed as Standard Scores of 13-Year-Old Students.
Population 1 a.

	No. of Items	Australia	Belgium	England	Finland	France	Japan	The Netherlands	Scotland	Sweden	United States
Total Score	70	2	53	-3	28	-10	76	27	-5	-28	-25
Lower Process	49	4	54	-7	21	-9	77	23	-6	-28	-22
Higher Process	21	-2	41	8	43	-12	59	32	-1	-22	-27
Verbal	41	3	50	0	34	-13	76	23	-4	-23	-29
Computational	29	1	52	-10	18	-5	69	29	-6	-31	-16
New Mathematics	13	1	42	-1	50	-28	36	7	-6	-11	-9
Basic Arithmetic	18	-5	37	-23	51	-7	50	34	-4	-15	-10
Advanced Arithmetic	14	0	56	-3	40	-9	82	28	-14	-31	-20
Algebra	17	12	37	12	16	-32	62	11	6	-36	-23
Geometry	17	4	62	7	-16	14	75	19	-8	-20	-37

TABLE 1.8. Subscores for National Groups Expressed as Standard Scores of Grade Group Corresponding to 13-Year-Olds.
Population 1 b.

	No. of Items	Australia	Belgium	England	Finland	France	Germany	Israel	Japan	The Netherlands	Scotland	Sweden	United States
Total Score	70	-28	49	5	22	-14	16	62	54	-11	-5	-52	-35
Lower Process	49	-26	52	1	16	-12	15	58	57	-13	-5	-53	-31
Higher Process	21	-27	34	13	36	-17	17	63	38	-2	-2	-41	-37
Verbal	41	-24	44	7	26	-16	25	58	53	-14	-6	-46	-39
Computational	29	-30	51	1	15	-9	2	60	49	-5	-3	-54	-25
New Mathematics	13	-15	42	10	54	-29	10	35	21	-16	-6	-28	-16
Basic Arithmetic	18	-31	32	-19	45	-12	28	43	32	12	-7	-40	-19
Advanced Arithmetic	14	-30	45	2	26	-16	28	62	56	-2	-12	-56	-32
Algebra	17	-9	47	29	30	-25	-27	40	52	-29	14	-46	-24
Geometry	17	-24	53	8	-29	6	23	75	50	-22	-13	-43	-48

TABLE 1.9. *Subscores for National Groups Expressed as Standard Scores of Mathematics Students in Terminal Secondary Year.*

Population 3a.

	No. of Items	Australia	Belgium	England	Finland	France	Germany	Israel	Japan	The Netherlands	Scotland	Sweden	United States
Total Score	69	−33	65	66	−6	53	19	75	38	42	−4	9	−90
Lower Process	41	−24	61	59	3	51	19	82	36	49	−4	5	−94
Higher Process	28	−43	53	66	−20	49	16	53	36	24	−4	13	−70
Verbal	31	−38	39	65	−17	38	20	59	41	25	0	14	−77
Computational	38	−26	72	60	2	59	17	79	33	50	−7	3	−91
New Mathematics	17	−41	66	45	−27	67	−11	22	34	27	−8	−17	−29
Algebra	19	−35	98	59	2	32	−10	92	30	14	−1	19	−81
Geometry	5	−21	26	42	18	12	14	19	40	27	11	−8	−69
Analytical Geometry	5	−11	36	47	−10	51	46	40	32	65	−7	13	−34
Calculus	9	−6	−32	93	−10	14	66	79	−6	37	2	12	−90
Analysis	13	−32	39	24	15	49	18	89	61	87	−24	29	−79
Sets	5	−31	65	2	−40	89	−13	1	28	16	−8	−18	−5
Logic	6	−25	40	39	−49	33	−5	−20	−1	−18	11	−26	0

33

TABLE 1.10. *Subscores for National Groups Expressed as Standard Scores of Non-mathematics Students in Terminal Secondary Year.*

Population 3 b.

	No. of items	Belgium	England	Finland	Germany	Japan	Scotland	Sweden
Total Score	58	26	3	12	52	34	−3	−66
Lower Process	34	34	−3	17	57	32	−2	−71
Higher Process	24	11	11	3	39	32	−4	−53
Verbal	39	16	12	5	44	32	3	−47
Computational	19	38	−5	21	58	32	−12	−89
New Mathematics	10	12	−1	−10	38	32	−3	−58
Advanced Arithmetic	6	−5	−12	12	39	44	−2	−22
Algebra	15	44	12	12	63	19	−3	−72
Geometry	15	24	14	24	33	24	6	−38
Analytical Geometry	[a]	34	−22	31	78	27	−24	−70
Analysis	6	−4	−19	−3	31	33	−7	−62
Sets	5	−8	9	−15	20	22	4	−51

[a] Less than 5.

"New Mathematics". In the content areas, the Finnish students make the best showing in arithmetic, are intermediate in algebra, and are weakest in geometry. The difference in performance level between basic arithmetic and geometry is quite marked, amounting to two thirds of a standard deviation.

There are several other aspects of the patterning in Table 1.7 that merit comment. In the first place, the differences between the lower-process and the higher-process scores and between the verbal and computational types of problems are quite small in all the participating countries. The differences that were assumed to exist in the tasks are not reflected in differences of behavior in different national groups. There is a suggestion that countries in which French is spoken do relatively well in geometry (largely descriptive and intuitive at this level). Japan, which did best on the total test, is relatively weak on the "New Mathematics", while Sweden and the United States, which are weakest in total performance, did relatively well on the "New Mathematics". Other minor differences can be found by the person interested in searching for them.

Table 1.8 includes many of the same pupils as Table 1.7, the sample being drawn from a single grade group rather than an age group. The contrasts that emerge are much the same, with the addition of one or two from the inclusion of two more national groups. Thus, Germany appears weakest on the algebra items, and Israel does relatively less well on the "New Mathematics".

In Table 1.9 we are dealing with quite a different educational level and quite a different set of content categories. The mental process classification and the verbal-computational one continue to produce only small differences in relative performance within each country. However, as among the content categories, the within-country differences are quite marked. For example, the students in Belgium were more than a standard deviation higher on algebra than on calculus, while those in England were nearly a standard deviation higher on calculus than on "New Mathematics". The Finnish students were half a standard deviation better on geometry than on "New Mathematics", while in France this relationship was reversed. German students scored about three fourths of a standard deviation higher on calculus than on algebra, while in Japan the calculus score fell about this same distance below the analysis score. Other substantial differences will be found in the table. These and other differences are dealt with by some countries in their own IEA *National* Reports.

The results for nonmathematics students in the final year of secondary education are presented in Table 1.10. The material on which to study patterning is less rich for this group, since the only content areas with more than 6 items are algebra and geometry. Subscore differences seem less marked than on other levels, and they must be viewed tentatively because of the small size of some samples and the small number of items in some of the scores. One may note the relatively low score on sets for Belgium and Finland; the strength of Germany on the few analytic geometry items; and the weakness of Belgium and England on the analysis items.

For some countries, at least, the patterns that have been noted in the previous paragraphs are known to parallel curricular emphases in the country. Thus, in the final years of secondary education, England places a good deal of stress on calculus in both pure and applied mathematics, while the English General Certificate Examinations pay little attention to the "New Mathematics". Some analysis of the relationship between score on specific topics and students' opportunity to learn the topics (as rated by their teachers) is given in Chapter 4.

Correlations of Educational and Other Variables with Mathematics Achievement

As part of the analysis of the results of the study, correlations were computed among several of the subscores of the mathematics test and between mathematics test scores and a range of items of information concerning the background, education, and attitudes of the individual students. These were calculated separately for each country and for each of the four target populations. Out of the mass of detail, we have extracted certain results to report here.

Tables 1.11 A–D show the median value of the national intercorrelations between total mathematics scores and several of the subscores. Many of the correlations are spuriously high because of the presence of the same items in more than one of the scores and therefore are difficult to interpret. The correlations between items measuring "lower" and "higher" mental processes, and between verbal and computational items, however, involve no common items. The median correlations of higher with lower mental process scores for the four levels are, respectively, .76, .74, .68, and .66. Any interpretation of these correlations must take account of the unreliability of each score. Our rough estimate, based on data on the reliability of total score for the test, is that if these correlations were corrected for unreliability in the separate scores, they would become .90 to .95. Thus, there appears to be some slight difference in the function being measured by the two scores, but the difference *is* slight, and the two scores are hardly different enough to make their separate analysis promising.

The median correlations between verbal and computational items in the four levels are, respectively, .84, .79, .69, and .63. Once again, appropriate correction for the unreliability of the part scores, it is estimated, would raise these correlations into the .90's, and it would be unprofitable to spend much effort to identify factors associated more with the one than with the other.

We turn now to the correlates of mathematics achievement, limiting our attention to total score as an indicator of achievement. Correlations were computed between achievement and 45 other scores or facts about the individual students. These correlations were available for each country and each educational level. Since it is clearly impractical to present all this detail, we report the median of the within-country correlations for each variable at each level in Table 1.12 (pp. 38–39). We present a numerical value for the correlation only when either the

TABLE 1.11. *Intercorrelations of Mathematics Total Score and Subscores.*

	A. 13-Years-Olds						B. Grade of 13-Year-Olds					
	1	2	3	4	5	6	1	2	3	4	5	6
1. Total Score	—	98	86	96	94	85	—	98	86	96	92	84
2. Lower Process	98	—	76	92	95	82	98	—	74	92	94	81
3. Higher Process	86	76	—	89	73	76	86	74	—	90	71	76
4. Verbal Items	96	92	89	—	84	85	96	92	90	—	79	85
5. Computational Items	94	95	73	84	—	75	92	94	71	79	—	72
6. Test Number 3	85	82	76	85	75	—	84	81	76	85	72	—

	C. Terminal Math						D. Terminal Nonmath						
	1	2	3	4	5	6	1	2	3	4	5	6	7
1. Total Score	—	95	87	89	93	84	—	93	87	94	83	86	85
2. Lower Process	95	—	68	74	96	82	93	—	66	84	87	84	82
3. Higher Process	87	68	—	93	70	70	87	66	—	89	62	71	72
4. Verbal Items	89	74	93	—	69	76	94	84	89	—	63	78	81
5. Computational Items	93	96	70	69	—	79	83	87	62	63	—	80	69
6. Test Number 5	84	82	70	76	79	—	86	84	71	78	80	—	61
7. Test Number 3							85	82	72	81	69	61	—

median value of the correlation was at least .10 or the correlation had the same sign in all or all but one of the participating countries.

School Characteristics

The first ten variables characterize the schools. The first variable is school size, and in Populations 1 a and 1 b size and high performance go together in almost all of the countries, although the modest correlations range from the 20's in Scotland to essentially zero in Sweden. In Populations 3 a and 3 b, this relationship becomes quite inconsistent from country to country, and the median value is near zero. The other school variables give generally small and inconsistent relationships.

Teacher Characteristics

The next seven variables relate to the teachers of the classes tested. Generally speaking, pupil performance tends to be slightly better with teachers who have had more training, although the relationship is admittedly a weak one. There is some indication of a relationship, but quite an inconsistent one across countries, between the teacher's judgment of the opportunities that the pupils have had to learn the topics covered in the test and the pupils' scores. The correlations run as high as .50 to .60 in England and Scotland, but are near zero in a number

TABLE 1.12. *Median Correlations of School, Teacher, and Student Variables with Mathematics Achievement.*

	13-Year-Olds	Grade of 13-Year-Olds	Terminal Math.	Terminal Nonmath.
1. Total Enrollment of School	12	14	a	a
2. Percentage Male Teachers	a	a	a	11[a]
3. Sex of School Children	a	a	a	a
4. Type of School	a	a	a	a
5. Educational Differentiation	a	a	a	a
6. Number Subjects Taken Grade 12	a	04	a	a
7. Number Subjects Taken Grade 8	a	a	a	a
8. Per Expenditures, Teachers Salaries	10[a]	a	a	a
9. Per Student Expenditures, Total	10[a]	a	a	a
10. Variability, Father's Occupational Status	18	—	a	—
11. Sex of Teacher	a	a	a	a
12. Length of Training of Teacher	08	07	a	a
13. Type Professional Training of Teacher	a	a	a	a
14. Recent In-Service Mathematics Training	a	a	a	08
15. Degree Freedom Given Teacher	a	a	a	a
16. Student Opportunity to Learn Higher Process	16[a]	12	a	a
17. Student Opportunity to Learn All Items	19[a]	10	19[a]	a
18. Perception of Mathematics Teaching	10[a]	a	a	a
19. Perception of School as Discovery	a	a	a	a
20. Attitude—Mathematics as Process	a	a	a	a
21. Attitude—Difficulty Learning Mathematics	06	a	a	a
22. Attitude—Place of Mathematics in Society	a	a	a	a
23. Attitude—School and School Learning	−16	−14	−06	a
24. Attitude—Man and His Environment	−18	−12	a	a
25. Sex (1 male, 2 female)	−10	−10	−12	−28
26. Age of Student	08	a	−16	−08
27. School Course or Program	−29	−27	−30[a]	−14[a]
28. Hours of Mathematics Instruction	a	a	13[a]	10[a]
29. Hours in School Week	a	a	a	a
30. Hours Mathematics Homework per Week	a	a	a	a
31. Hours of Homework per Week	22	15	a	a
32. Place of Parents Residence	a	a	a	a
33. Father's Education	18	16	a	a

Tab. 1.12 (cont.)

	13-Year-Olds	Grade of 13-Year-Olds	Terminal Math.	Terminal Nonmath.
34. Mother's Education	15	14	07	a
35. Status of Father's Occupation	24	22	a	a
36. Father's Occupation Technical	15	13	a	a
37. Years Education Expected	42	38	23	18
38. Years Education Desired	38	35	16	14
39. Status Level of Expected Occupation	20	18	10[a]	a
40. Status Level of Desired Occupation	18	22	a	11
41. Expects to Take More Mathematics	18	19	24	15
42. Desires to Take More Mathematics	22	25	24	18
43. Level of Mathematics Instruction	35	30	28	24
44. Interest in Mathematics	30	30	36	30
45. Hours Devoted to School Work per Week	18	16	a	a

[a] Correlations low or erratic from country to country.

of the other countries. Possibly, this could be thought of as an indirect indicator of the amount of differentiation in levels of instruction offered to different students within a given level. The correlations by level and country are shown in Table 1.13 for those countries in which teacher ratings were carried out. If our hypothesis is correct, then we would conclude that England and Scotland make the greatest differentiation in levels of instruction offered to 13-year-olds (and the corresponding grade group), and Japan makes the most to different students in the terminal mathematics program and to those in the terminal secondary year but not in mathematics programs.

Student characteristics

The remaining variables relate directly to students. Variables 18–24 are derived from the opinion and attitude questionnaire filled out by each student. Of the attitude scales, only two showed a consistent relationship to achievement, and this primarily in the less advanced student groups. Those who made high scores on the mathematics tests tended to express somewhat *less* enthusiastic attitudes toward school and school learning and somewhat *less* belief that man can cope with and master his environment.

Sex was related to mathematics achievement in almost all countries, the boys scoring higher than the girls at all levels and especially in the

TABLE 1.13. *Correlation of Mathematics Score with Teacher's Rating of Opportunity to Learn.*

	13-Year-Olds	Grade Group of 13-Year-Olds	Terminal Math.	Terminal Nonmath.
Australia	—	—	40	—
England	55	51	13	—
Finland	04	04	−09	−13
France	26	27	06	30
Germany	—	03	30	−03
Israel	—	10	−10	—
Japan	09	09	44	31
Scotland	56	60	18	—
Sweden	−03	−04	20	—
United States	19	17	29	04

terminal nonmathematics population (3 b). Age showed a consistent relationship for the terminal groups, the younger students tending to score slightly better. Achievement was related to the school program in which the student was enrolled, but due to the nature of this variable, which consists of discrete categories rather than a continuum, about all that one can safely say is that students in academic programs did better than those in other types. At the lower academic level, students reporting spending more time on homework achieved better than those reporting less time.

The education of father and mother showed a low positive relationship to achievement in Populations 1 a and 1 b. The level of father's occupation was also positively related, higher achievement going with having a father in a professional or managerial type of occupation, but once again the relationship was consistently present only at the lower educational level. There is evidence (see Chapter 5) that parental education and occupation are much more homogeneous for pupils at the end of secondary education in most of the participating countries, and this may account for the disappearance of the relationship. National differences in the size of the relationships with family variables may be of some interest, and these are shown in Table 1.14 for two of the target populations. Family variables appear to be substantially associated with achievement at the earlier academic level in England, the Netherlands, Japan, and the United States, but the United States is the only country that continues to show the relationship at the end of the secondary school.

TABLE 1.14. *Correlations of Mathematics Score with Parental Education and Occupation.*

	Grade of 13-Year-Olds			Terminal Mathematics		
	Father's Education	Mother's Education	Status of Father's Occupation	Father's Education	Mother's Education	Status of Father's Occupation
Australia	07	06	20	02	03	05
Belgium	14	07	19	08	09	04
England	25	26	36	09	12	08
Finland	14	10	07	−04	03	−11
France	18	13	22	02	00	−01
Germany	13	13	15	−02	08	01
Israel	18	16	22	−12	09	01
Japan	33	32	25	13	13	08
The Netherlands	26	21	22	05	00	−04
Scotland	15	12	26	04	06	06
Sweden	15	15	17	−02	−04	−03
United States	30	28	29	32	25	24

A group of variables relates to the student's own educational plans, hopes, and aspirations. Especially at the lower levels, but continuing also in the terminal groups, the amount of further education that a student expects and the amount he wishes to have are related to mathematics score. Furthermore, there is also a relationship, but a lesser one, to the level of the expected and of the desired occupation. Those who score better on the test are likely both to wish and to expect to take more mathematics courses. By the same token, mathematics achievement tends to be related to an index of mathematics interest. Finally, the score is related to the level of the mathematics courses the pupil has taken. These indicators of mathematics experience, mathematics interest, and educational plans are the closest correlates of mathematics achievement among those that we have studied. Of course, it is not clear in which direction influences operate. Does a student work hard on his mathematics and do well on it because he wishes to continue his education, or does the fact that he has done well on his studies (including mathematics) lead him to aspire to further schooling? Is he successful in mathematics because he is interested in it, or interested because he is successful? We suspect that both are true, and that these several factors act back and forth upon one another in such a way as to be mutually reinforcing, so that the final resultant is the product of years of interaction.

Analysis of Single Item Results

To the mathematics educator, the most concrete and tangible indicator of the performance of pupils is their degree of success on a specific test item; he can examine the item, see what percentage of students succeeded with it, and note the most frequent errors. However, with 60 to 70 items for each of four different populations in each of 12 participating countries, the amount of detail from this type of analysis tends to become overpowering. We have tried to achieve a reasonable compromise by relegating the material on specific items to Appendix II and by condensing it in certain ways. We will describe and illustrate here the material that is presented in Appendix II.

Figure 1.4 shows the format that has been used for presenting item results, illustrating it with item Number 3 of Test A. On the left, the complete item is displayed, together with the error choices if the item is in multiple choice format. The correct answer for multiple choice questions is marked with an asterisk. On the right are shown the results for that item. The target population is always indicated, using the following code:

Target Population 1 a	Thirteen-year olds.
Target Population 1 b	The grade group containing the largest fraction of the 13-year-olds.
Target Population 3 a	Mathematics students in the final year of secondary education.
Target Population 3 b	Nonmathematics students in the final year of secondary education.

Following the identification of the group is a number showing the difficulty of the item. This number signifies the proportion getting the item right and is an unweighted averaging of the difficulty values for the separate countries. (However, the original proportions were converted to deviation values on the abscissa of the normal curve before averaging, and the averaged normal deviation value was converted back to a proportion.)

Item difficulties in specific countries vary from what would be expected on the basis of the average difficulty over all countries, taken together with the average level of performance in the specific country. For this reason, a notation is made of those countries in which the item appeared either especially easy or especially difficult. A notation appears in Appendix II whenever the obtained value for a given country deviated from the expected value by as much as three tenths of a

Test Item	Target Population	Average Difficulty	Average Discrimination	Order of error frequency	Easier in	Harder in
Test no. A						
3. $(22 \times 18) - (47 + 59)$ is equal to	1a	.85	.29	E, D	Finland	England
*A. 290 B. 300 C. 384 D. 408 E. 502	1b	.87	.29	E, D	Finland	England

Figure 1.4. Sample item analysis report.

standard deviation. The country is listed under "Easier in" if the percent of success is greater than anticipated and under "Harder in" if the percent of success is less than expected. Thus, for our illustrative item, the results showed it to be unexpectedly easy in Finland and unexpectedly difficult in England.

As a service to mathematics educators who are especially interested in the item statistics for single countries, difficulty and discrimination indices have been prepared in typed form and deposited with the American Documentation Institute as Document Number 9189. A copy may be secured by citing the Document number and by remitting $ 6.25 for photoprints, or $ 2.50 for 35 mm microfilm.

One fact of interest about a test item is the degree to which those students who are good performers on the total test surpass those who are poor performers. The statistic that has been used to express this relationship in the present study is the point biserial correlation coefficient, a correlation between success-fail on the item and score on the total test. This index was computed for each country, but only the median of the values for the separate countries is reported in Appendix II, under the designation "Average Discrimination". For item Number A 3, the value is .29.

Mathematics educators and test-makers may be interested in the relative attractiveness of different error choices. The typical popularity sequence is indicated for each multiple-choice item under the caption "Order of error frequency". Thus, for item Number A 3, the most popular error was E, followed by D.

Student Attitudes

In this inquiry an attempt has been made to assess pupils' attitudes as well as their accomplishments. We elicited attitudes toward mathematics, education, and the world. Five dimensions of attitude were assessed. The development of the attitude scales and the nature of the attitude dimensions are discussed more fully in Chapter 6 of Volume I. Here we present descriptive data on attitudes in different countries. The attitude data also enter into certain of the hypotheses that will be tested in later chapters.

Means and standard deviations of the attitude scales for each participating country in the four groups are shown in Tables 1.15–1.18, pages 46–47.

The first scale inquires about the degree to which mathematics is viewed as fixed and given once and for all time (a low score), or as something that is developing, growing, and changing (a high score). On a scale where the possible range of scores is from 0 to 16, the mean of all cases (pooled over all countries) is 7.7 for Population 1 a and 7.6 for the corresponding grade group. For end-of-secondary-education mathematics students, the mean is 6.6 and for nonmathematics students 7.3. It is interesting that in all countries, without exception, the more advanced mathematics students view mathematics as more fixed and "frozen" than do the younger or the less advanced students. Differences between countries are quite marked, amounting to about one full standard deviation of the pooled distribution of student scores. A country that is high at the lower school level tends to be high also at the upper level. There is a suggestion that those countries in which the teachers report teaching "New Mathematics" (See Volume I, Chapter 14) are the countries that see mathematics as more of an open and changing system.

The second scale refers to the perceived ease of learning mathematics. In this case a low score means that mathematics is perceived as difficult and something reserved for the intellectual elite, whereas a high score means that it is perceived as within the reach of most students. The possible range of scores is from 0 to 14, and the mean for all examinees is 9.1 for Population 1 a and 8.2 for terminal mathematics students. Mathematics is perceived as a little more difficult and demanding at the upper level. The range of national differences is slightly less for this scale than for the previous one—from 8.2 to 9.4 for Population 1 a and from 7.3 to 8.6 for terminal mathematics students.

The third scale relates to the role of mathematics in contemporary society. A high score represents an expression of the belief that mathematics has an important and vital role in our society, whereas a low score indicates a judgment that mathematics is of little value. The possible range of scores is from 0 to 16. On this scale, the grand mean is 8.8 for Population 1 a and 8.1 for mathematics students in their terminal year. It is of some interest that the lowest appraisal of mathematics is made by those students with the largest exposure and academic commitment to it. On this scale, there are fairly substantial differences between countries—from 8.2 to 10.2 among Population 1 a and from 7.1 to 9.6 in the terminal year mathematics group. In general, the importance of mathematics is rated lowest in the countries where English is spoken. The consistency within a language grouping may reflect consistency in the national value systems, but we must entertain the possibility that it reflects some subtle difference in the wording of items in translation.

The scale expressing attitude toward school and school learning is basically a dimension of like–dislike. A high score expresses enthusiasm for school and the experiences it provides. The possible range of scores is from 0 to 22, and the means for the four levels are rather similar. It is on this scale that the differences among countries are the greatest, the extreme national groups differing by more than a standard deviation of the total group. Once again, the least favorable attitudes are expressed by students from the countries where English is spoken and the lowest among these are the students from the United States. On the whole, the most positive attitudes are expressed by Japanese students.

The fifth of the attitude scales, dealing with the relationship of man to his environment, concerns the extent to which man is perceived as having effective control of and mastery over his environment. A low score represents an attitude that mankind is helpless in the face of the forces at work in his world, while a high score corresponds to a feeling that he is in some degree master of his own fate. The range of possible scores is from 0 to 18. National differences are of moderate size for this scale, but not as large as for the one just discussed. The greatest optimism is expressed in Finland and Japan, the greatest pessimism in the United States (and, in Populations 3 a and 3 b, in France and Sweden).

We have compared national attitudes with national achievement in Table 1.19. The countries were ranked with respect to mean score on each of the attitude scales and also with respect to mean total score in mathematics. Rank correlations were computed for each of the four levels that we have been studying.

The correlations are consistent with respect to sign across all four populations, and some of them are quite substantial. We may say, in general, that in those countries in which achievement is high pupils have a greater tendency to perceive mathematics as a fixed and closed system, as difficult to learn and for an intellectual elite, and as im-

TABLE 1.15. *National Means and Standard Deviations on Five Attitude Dimensions.*
(13-Year-Old-Students)

	Math. as Process		Difficulty Learning Math.		Place Math. Society		Sch. and Sch. Learning		Man and Environment	
	Mean	S.D.	Mean	S.D.	Mean	S.D.	Mean	S.D.	Mean	S.D.
Australia	7.8	2.2	9.4	1.7	8.2	2.2	9.1	2.1	8.6	2.4
Belgium	7.6	1.9	9.0	1.9	8.6	2.2	10.0	2.0	9.0	2.3
England	7.9	2.2	9.2	1.7	8.2	2.2	9.3	2.0	8.7	2.4
Finland	7.6	1.7	9.0	1.9	9.8	2.1	10.2	2.0	9.6	2.1
France	8.1	1.9	9.2	2.2	10.2	2.3	10.4	2.1	8.8	2.3
Japan	6.2	1.8	8.4	1.5	9.5	2.0	10.5	1.8	9.4	2.0
The Netherlands	6.8	1.9	8.2	2.0	9.1	2.1	10.2	2.1	8.3	2.5
Scotland	7.6	2.1	9.2	1.8	8.6	2.4	9.1	2.1	8.7	2.4
Sweden	7.7	1.9	8.9	1.6	9.0	2.3	10.0	2.1	8.4	2.1
United States	8.0	2.1	9.4	1.7	8.6	2.2	8.4	2.0	7.9	2.4
Total	7.7	2.1	9.1	1.8	8.8	2.3	9.3	2.1	8.6	2.4

TABLE 1.16. *National Means and Standard Deviations on Five Attitude Dimensions.*
(Grade Containing Most 13-Year-Olds)

	Math. as Process		Difficulty Learning Math.		Place Math. Society		Sch. and Sch. Learning		Man and Environment	
	Mean	S.D.	Mean	S.D.	Mean	S.D.	Mean	S.D.	Mean	S.D.
Australia	7.8	2.2	9.5	1.7	8.1	2.2	9.1	2.1	8.6	2.4
Belgium	7.6	1.9	8.9	1.9	8.6	2.1	9.9	2.0	8.9	2.4
England	7.9	2.1	9.1	1.7	8.5	2.4	9.3	2.1	8.5	2.4
Finland	7.5	1.7	9.0	2.0	9.8	2.0	10.2	2.0	9.6	2.2
France	8.1	1.9	9.2	2.1	10.1	2.3	10.2	2.0	8.9	2.3
Germany	6.6	1.8	8.6	1.8	9.9	2.1	11.1	1.9	8.9	2.1
Israel	7.3	2.0	8.8	2.0	9.8	2.3	9.0	2.1	9.0	2.2
Japan	6.2	1.8	8.4	1.5	9.5	2.0	10.5	1.8	9.4	2.0
The Netherlands	7.1	1.8	8.3	2.0	9.1	2.2	10.0	2.1	8.6	2.3
Scotland	7.7	2.2	9.2	1.8	8.6	2.4	9.2	2.1	8.7	2.4
Sweden	7.7	2.0	8.8	1.6	9.1	2.3	10.0	2.0	8.4	2.2
United States	8.0	2.1	9.4	1.7	8.8	2.2	8.4	2.0	7.8	2.4
Total	7.6	2.1	9.0	1.8	9.1	2.3	9.6	2.2	8.6	2.3

TABLE 1.17. *National Means and Standard Deviations on Five Attitude Dimensions.*
(Math. Students in Final Secondary Year)

	Math. as Process		Difficulty Learning Math.		Place Math. Society		Sch. and Sch. Learning		Man and Environment	
	Mean	S.D.	Mean	S.D.	Mean	S.D.	Mean	S.D.	Mean	S.D.
Australia	7.2	2.4	8.3	2.0	8.0	2.4	8.6	2.0	8.0	2.4
Belgium	7.0	2.0	8.3	2.0	8.5	2.1	9.9	2.0	8.3	2.5
England	6.5	2.1	8.0	2.0	7.5	2.4	8.5	1.9	8.1	2.6
Finland	6.3	2.2	8.4	1.9	8.5	2.2	10.5	2.0	9.0	2.3
France	7.1	2.2	8.1	2.1	9.1	2.2	9.7	1.8	7.5	2.5
Germany	5.2	2.2	7.8	1.9	8.2	2.1	10.7	2.0	7.7	2.3
Israel	6.0	2.1	8.3	2.3	9.6	2.2	9.1	2.1	9.0	2.1
Japan	5.8	1.8	8.5	1.5	9.6	2.1	10.6	2.0	9.0	1.9
The Netherlands	5.8	2.2	7.3	1.8	8.4	2.2	10.5	2.1	8.3	2.8
Scotland	7.4	2.4	8.0	2.1	7.8	2.4	8.7	1.9	8.3	2.5
Sweden	5.6	2.2	8.6	1.7	8.8	2.3	10.0	2.0	7.2	2.3
United States	7.4	2.4	8.5	2.0	7.1	2.4	8.0	1.8	7.4	2.4
Total	6.6	2.4	8.2	2.0	8.1	2.5	9.2	2.2	8.0	2.5

TABLE 1.18. *Means and Standard Deviations on Five Attitude Dimensions.*
(Nonmath. Students in Final Secondary Year)

	Math. as process		Difficulty Learning Math.		Place Math. Society		Sch. and Sch. Learning		Man and Environment	
	Mean	S.D.	Mean	S.D.	Mean	S.D.	Mean	S.D.	Mean	S.D.
Belgium	7.6	2.0	8.1	2.1	9.0	2.3	9.7	2.0	8.1	2.4
England	8.2	2.2	7.7	2.0	7.3	2.6	8.5	1.9	7.9	2.6
Finland	6.9	2.2	8.2	2.0	8.3	2.6	10.5	1.9	9.2	2.3
France	7.5	2.1	8.1	2.2	9.3	2.5	9.6	2.0	7.4	2.5
Germany	6.4	2.3	7.4	1.9	9.0	2.4	10.5	1.9	7.7	2.5
Japan	6.2	1.8	8.6	1.5	9.6	2.2	11.4	2.0	9.2	2.0
The Netherlands	7.2	2.6	6.9	1.8	8.4	2.3	9.8	2.3	7.6	2.7
Scotland	8.1	2.3	7.7	2.1	8.2	2.6	8.5	1.8	8.2	2.5
Sweden	7.3	1.7	8.1	1.7	7.6	2.3	9.9	1.9	6.8	2.4
United States	8.0	2.1	8.4	2.0	7.2	2.6	8.3	1.9	7.5	2.3
Total	7.3	2.2	8.2	1.9	8.5	2.6	9.8	2.3	8.4	2.4

TABLE 1.19. *Correlations of National Mean Achievement and National Mean Attitude.*

	Target Population			
	13-Year-Olds	Grade of 13+	Terminal Math.	Terminal Nonmath.
Mathematics as Process	−.70	−.64	−.41	−.61
Difficulty Learning Mathematics	−.32	−.45	−.43	−.23
Place of Mathematics in Society	.15	.27	.45	.88
School and School Learning	.48	.23	.20	.56
Man and His Environment	.65	.81	.38	.15

portant to the future of human society. In these countries pupils tend to be favorably disposed toward school and school learning and to feel that man has some mastery and control over his destiny.

Chapter 2

Correlations Between and Within Countries

Table 1.6 gave the median correlations *within* countries of a number of variables with the total corrected mathematics score. In Table 2.1, some of these are repeated, side by side, with the corresponding correlations *between* countries. The two sets present an interesting contrast. The median correlations within countries are usually small, and often so small that they are indicated by stars. The correlations between countries are usually substantial. Why is this so?

The median correlations within countries may be very small either because the separate correlations are all very small or because, although not all very small, they have an even distribution of sign. Examples can be seen in the 20 tables in Chapter 6. In those tables the simple correlations and the standardized regression coefficients are given for the separate countries, and at the foot of each column the numbers of positive and negative signs are entered, together with the range, the mean, and the corresponding correlations and coefficients over all countries. It will be noted that the correlations and coefficients over all countries sometimes differ substantially from the means of the separate entries for countries. There is a minor reason for this, namely, that in computing the overall figures the countries have been given weight proportional to the size of their samples, whereas the means are simple averages. But there is also a major reason, namely, that the national means of the variables usually differ significantly, and often considerably. It is this fact that is mainly responsible for the moderate differences between the means of the national correlations and the corresponding overall correlations. It is also responsible for the much larger differences between the means (or medians) of the national correlations and the corresponding correlations of the national means. The two sets are entered side by side in Table 2.1. The fact that the correlations

This chapter was written by Mr. G. F. Peaker, C.B.E.

TABLE 2.1. *Correlations with Total Corrected Mathematics Scores.*

Variable	Median Correlations Within Countries		Correlations Between Country Means	
	1 a	3 a	1 a	3 a
1. Total Enrollment of School	12	a	33	−66
2. Percentage Male Teachers	a	a	73	56
5. Educational Differentiation	*	*	60	34
6. Number Subjects Taken in Grade 12	*	04	32	71
7. Number Subjects Taken in Grade 8	*	*	52	86
8. Per Pupil Expenditures, Teacher's Salaries	10^a	*	−87	a
12. Length of Training of Teacher	08	*	−39	−23
14. Recent In-Service Mathematics Training	*	*	−58	−54
15. Degree of Freedom Given Teacher	*	*	−50	−45
17. Student Opportunity to Learn All Items	19	19	62	80
18. Perception of Mathematics Teaching	10	*	70	−51
19. Perception of School as Discovery	*	*	60	10
20. Attitude—Mathematics as Process	*	*	−78	−54
21. Attitude—Difficulty Learning Mathematics	06	*	−64	−39
22. Attitude—Place of Mathematics in Society	*	*	28	57
23. Attitude—School and School Learning	−16	−06	57	65
24. Attitude—Man and His Environment	−18	*	69	49
28. Hours of Mathematics Instruction	*	13	−01	18
29. Hours in School Week	*	*	−18	31
30. Hours Mathematics Homework per Week	*	*	44	35
31. Hours of All Homework per Week	22	*	61	45
35. Status of Father's Occupation	24	*	44	51
42. Desire to Take Additional Mathematics	22	24	−32	47
43. Level of Mathematics Instruction	35	28	64	49
44. Interest in Mathematics	30	36	17	28
45. Hours Devoted to School Work per Week	18	*	39	39

^{a} Correlations low or variable from country to country.

between countries are so much larger than those within countries is evidence that there are substantial differences between countries in respect to these variables, independently of the individual differences between students in the same country. There is nothing surprising about this. It is plain enough that the countries do differ substantially, and it would be a matter for chagrin if our data had provided no evidence of this fact. Nonetheless, some of the correlations disclosed may not be altogether expected. For example, variables 1, 8, 18, and 42 have changes of sign from Population 1 a to Population 3 a. For variables 12, 14, and 15, the signs are all negative, and the correlations are substantial. Since 12 is the length of training of the teacher and 14 the amount of recent in-service mathematical training, this means that the countries that have the lowest scores tend to have the longest training for their teachers and the most in-service training. This could be put in the form that in the countries where the teachers need most training they are given it, and perhaps this is the explanation. At any rate, the relation is substantial.

The reversal of sign between Population 1 a and Population 3 a for the first variable, which is the size of school, mainly reflects the fact that Japan has much the largest schools for Population 1 a, and scores high at that level, whereas for Population 3 a the United States have even larger schools and score low. The reversal for variable 8, which is the cost per student in teacher salaries, arises mainly because Japan, which is a low-cost country, scores high for Population 1 a, whereas the highest scorer for Population 3 a is England, where costs are fairly high. These facts can easily be discovered by plotting the country means for the pair of variables concerned. It would take too much paper to print the scatter diagram corresponding to each of the 52 correlations between countries set out in Table 2.1, but the reader can easily plot it for any case in which he is interested by using the columns of means given in Table 2.2.

It may occur to a reader who plots some of the diagrams that a correlation derived from no more than ten points must be subject to a large amount of sampling fluctuation. But the countries taking part in the study are not to be thought of as a random sample from some larger universe of countries. As pointed out in the preface they together constitute the universe of discourse, so that sampling fluctuation only comes in from the fact that the country means themselves have small standard errors because they are based on samples drawn in each country. That is, the points in the 52 diagrams are to be thought of as having a small amount of play, but not as being subject to replace-

TABLE 2.2. *Country Means in Code (see Appendix, Volume I) for Variables in Table 2.1.*

Variable	Australia	Belgium	England	Finland	France	The Netherlands	Japan	Scotland	Sweden	United States
1. Total Enrollment of School	540	408	501	340	384	330	985	548	446	662
2. Percentage Male Teachers	41	42	39	37	37	43	50	39	39	25
5. Educational Differentiation	26	25	21	40	30	31	39	18	27	22
6. Number of Subjects Taken in Grade 12	62	89	63	90	78	90	—	44	90	52
7. Number of Subjects Taken in Grade 8	87	89	89	90	85	90	90	82	90	73
8. Per Pupil Expenditures, Teacher Salaries	182	—	208	148	197	154	53	294	227	214
12. Length of Training of Teacher	28	24	31	32	21	41	32	40	46	44
14. Recent In-Service Mathematics Training	14	12	15	12	17	12	14	13	16	25
15. Degree Freedom Given Teacher	158	126	163	166	169	179	169	168	190	176
17. Student Opportunity to Learn All Items	—	—	60	47	50	52	63	51	37	48
18. Perception of Mathematics Teaching	126	—	134	115	—	139	140	136	127	135
19. Perception of School as Discovery	107	—	104	110	—	108	109	101	107	103
20. Attitude—Mathematics as Process	78	76	79	76	81	68	62	76	77	80
21. Attitude—Difficulty of Learning Mathematics	94	90	92	90	92	82	84	92	89	94
22. Attitude—Place of Mathematics in Society	82	86	82	99	102	91	95	86	90	86
23. Attitude—School and School Learning	91	100	93	102	104	102	105	91	100	84
24. Attitude—Man and His Environment	86	90	87	96	88	83	94	87	84	79
28. Hours of Mathematics Instruction	51	46	40	30	44	40	45	46	38	46
29. Hours in School Week	38	62	38	32	45	44	39	43	57	47
30. Hours of Mathematics Homework per Week	24	36	17	28	34	26	30	23	19	31
31. Hours of All Homework per Week	61	114	54	110	91	90	83	48	61	69
35. Status of Father's Occupation	56	57	60	53	59	56	50	61	55	53
42. Desires to Take Additional Mathematics	13	13	14	13	16	16	13	15	16	12
43. Level of Mathematics Instruction	22	18	20	18	14	20	30	16	14	21
44. Interest in Mathematics	59	57	57	62	55	54	61	53	58	62
45. Years Devoted to School Work per Week	32	44	30	35	37	35	35	31	36	35

Population 3 a

Variable	Australia	Belgium	England	France	Japan	The Netherlands	Scotland	United States
1. Total Enrollment of School	786	446	645	928	1146	462	768	1434
2. Percentage of Male Teachers	43	48	45	39	56	52	37	36
5. Educational Differentiation	25	25	24	19	29	37	17	20
6. Number of Subjects Taken in Grade 12	61	90	66	88	88	90	44	52
7. Number of Subjects Taken in Grade 8	87	90	89	89	89	90	82	60
8. Per Pupil Expenditures, Teacher Salaries	194	—	237	308	80	357	280	270
12. Length of Training of Teacher	37	40	37	47	37	49	47	47
14. Recent In-Service Mathematics Training	16	18	13	28	13	18	18	33
15. Degree of Freedom Given Teacher	154	139	160	147	169	167	157	172
17. Student Opportunity to Learn All Items	57	—	66	75	63	—	58	50
18. Perception of Mathematics Teaching	127	—	131	—	139	126	135	138
19. Perception of School as Discovery	100	—	99	—	112	119	100	109
20. Attitude—Mathematics as Process	72	70	65	71	58	58	74	74
21. Attitude—Difficulty Learning Mathematics	83	83	80	81	85	73	80	85
22. Attitude—Place of Mathematics In Society	80	85	75	91	96	84	78	71
23. Attitude—School and School Learning	85	99	85	97	106	105	87	80
24. Attitude—Man and His Environment	80	83	81	75	89	83	83	73
28. Hours of Mathematics Instruction	69	74	43	89	54	51	62	50
29. Hours in School Week	38	62	38	63	40	53	43	48
30. Hours Mathematics Homework per Week	61	87	41	95	52	57	41	41
31. Hours of All Homework per Week	197	188	143	220	138	199	120	114
35. Status of Father's Occupation	44	47	41	39	39	38	41	47
42. Desire to Take Additional Mathematics	14	17	14	15	13	14	15	13
43. Level of Mathematics Instruction	67	68	74	70	80	70	82	59
44. Interest in Mathematics	67	60	73	69	63	64	58	64
45. Hours Devoted to School Work per Week	46	52	40	55	40	50	40	41
61. Total Corrected Mathematics Score	216	346	352	334	314	319	255	138

ment by a different set of points representing a different sample of countries from a larger universe.

In how many cases does the between-country component reinforce a substantial median correlation within countries? If we take both populations together, there are only three such cases, namely,

17 Student Opportunity of Learning All Items.
43 Level of Mathematics Instruction.
44 Interest in Mathematics.

There are four more that apply to the younger population only, namely,

18 Perception of Mathematics Teaching.
31 Hours per Week of All Homework.
35 Status of Father's Occupation.
45 Hours Devoted to School Work per Week.

Finally, there are two more that apply to the older population only, namely,

28 Hours of Mathematics Instruction.
42 Desire to Take Additional Mathematics.

Thus, altogether there are 12 cases out of 52 where the component between countries reinforces a general correlation within countries. In the other 40 cases the situation is more complex.

In the ensuing chapters much use is made of correlations both within and between countries. The object of this brief comparison of a number of cases is to draw attention to the fact that it is rare for the same factors to operate at the same strength in different countries, and it is also rare for the same factor to operate at the same strength within and between countries. Frequently, the sense is reversed. This adds to the interest, as it certainly adds to the difficulty, of international studies.

The general position may be seen in more detail by comparing Table 2.1 with Tables 6.2 through 6.21, which give the individual correlations within countries, together with their means, ranges, and distributions of signs. Where there is a sharp contrast in Table 2.1, or where there is a long range, and still more where there is a more or less even distribution of sign in Tables 6.2 through 6.21, hypotheses have to be considered piecemeal, country by country. The same hypothesis may be confirmed by the evidence in one country and confuted by it in another. It may be confirmed (or confuted) within all, or nearly all, countries and confuted (or confirmed) between all countries. Of the variables considered in this chapter only three (interest, opportunity, and level) lend themselves to uniform generalizations needing little qualification. The greater heterogeneity shown between others is responsible for the length, but also for much of the interest of the ensuing chapters where the hypotheses are studied in detail. The scene does not lack variety.

Chapter 3

The Relation of School Organization to Attainments in Mathematics

Introduction

A characteristic of industrially developed countries today is that apart from an *explosion scolaire* there has also been a rapid rate of economic growth; science and technology are bringing about a revolution in both the amount and means of productivity. Science and technology are themselves developing at an enormous rate, and it is often maintained that in 20 years at least one third of present industrial occupations will be obsolete and will be replaced by other and more technical occupations. A highly important factor in economic and technological expansion is manpower educated to fulfill the needs of society at all levels. One of the nations' tasks at the material level must therefore be to identify, develop, and utilize the human intelligence of their societies.

Mathematics is a strategic subject, since science and technology, upon which economic progress depends, require a thorough grounding in mathematics if they are to be successfully studied. The Organization for Economic and Cultural Development, OECD (1961, 1962) has devoted some energy to trying to assess the shortage of mathematics teachers. McIntosh (1962) has indicated the shortage of highly educated manpower in the last two decades in the United Kingdom and particularly in mathematics. As technology develops and reaches more and more into all levels of industry (and commerce), so more mathematics will be needed at all levels.

The relation between technology and mathematics is, however, more complicated than one might assume at first. The applications of mathematics in the everyday life of the majority of the citizens of economically highly developed communities is surprisingly limited (Dahllöf,

The hypotheses in this chapter were dealt with by T. Husén, G. F. Peaker, D. A. Pidgeon, T. N. Postlethwaite, W. Schultze, R. M. Wolf, and D. A. Walker, the last being responsible for the editing and for the mathematical model.

1960). But there is no doubt that well-qualified mathematicians are in short supply.

Although the shortage of mathematicians seems to be a worldwide phenomenon, with no country able to show a system manifestly superior to those in all other countries as far as the production of an adequate supply of well-qualified mathematicians is concerned, those engaged in this international inquiry set out in the faith that each of the countries participating would be able to learn something from the other countries in which methods of school organization sometimes differ markedly. At worst, a study of these differences would indicate which factors appeared to be of little importance in the production of high scores in mathematics; at best, the study could indicate the factors most likely to yield results in any changes a country might make.

This chapter deals with matters of school organization, changes in which are largely in the hands of policy-makers and administrators. Other chapters deal with curriculum, teaching, and sociological and other factors, but in this chapter we study the effects which might result from changes in the organization of schools. We are able to study some of these effects because this has been an international investigation in which the different countries may be regarded as the testing laboratories for some of the treatments which other countries might wish to adopt. Very obvious changes which might be made are in the age of entry to school, the age at which school education ceases, the size of the school, the size of the class, the use of selective schools or of comprehensive schools, the age at which selection is made, and the degree of separation between the academic and nonacademic students once the separation has been made. In these days, when the proportion of an age group continuing to attend school until late adolescence is increasing in almost all countries, the question has been raised whether a necessary consequence of the broadening of opportunities to complete preuniversity courses may be that standards fall. This fear has been crystallized in the phrase, *"Does more mean worse"?* The data provide the opportunity to examine this aspect so far as the outcomes of instruction in mathematics are concerned. All of these topics are discussed in the following pages in the light of the data obtained from the investigation. At the end of the chapter the conclusions reached are drawn together to present a picture of the effects which changes in school organization might produce.

The word "might" is used because it does not follow that procedures which are associated with certain results in some countries will necessarily produce similar results in other countries where the conditions may be very different. At the same time, a procedure which has been found to be associated with certain results in several countries within the group does at least appear to be a promising one to use in another country desiring similar results. In an investigation of this type the findings should probably be stated in the form "y is associated with x". If the writers occasionally lapse into the form "y is an effect of x", it is not that they are unaware of the logic of the situation, but that they are putting forward what to them seems to be a reasonable assumption.

Age of Entry

This is one of the questions which can be dealt with in an international study but which would be extremely difficult to answer from a national study simply because it is difficult to test the value of alternative possible choices within one system. Of the twelve countries in this project, pupils in England and Scotland begin school at the age of 5, in Sweden and Finland at the age of 7, and in the other countries at the age of 6. We are therefore in a position to study the outcomes associated with differing ages of entry between the limits 5 and 7, and this is done in the report on hypothesis 01.

Age at Which School Education Ceases

A second way in which school organizations differ is in the age at which pupils are deemed to have completed their school courses and to be fit to enter university or college courses. Scottish students may enter a university at the age of 17 or 18; the corresponding age in Germany is nearer 20, and other countries fall within that range. It would appear at first sight that the older students would attain a higher standard in mathematics than would the younger students. This aspect is studied in hypothesis 02.

Size of School

Countries also differ among themselves in the size of school they favor. To some extent, the choice of size is limited by geographical factors; in sparsely populated areas there is no choice but to have a number of relatively small schools. In more densely populated areas, however, there is often the choice between providing one large school or several smaller schools. The mathematical attainments of students in larger or smaller schools are studied in the report on hypothesis 03.

Size of Class

The variable which at first sight seems most likely to produce substantial differences in instructional outcomes is size of class. Teachers in several countries have been pressing for some years for smaller classes in order that each pupil might be given more of the teacher's time, and research on this question has already been carried out in several countries. The data obtained in the project gave another opportunity of studying the question and the results are given under hypothesis 04.

Specialization

Countries differ markedly in the degree of specialization in the higher classes, and this might be expected to have a marked effect on the scores made by the students in those classes. We should expect students studying mathematics as one of three or four subjects to make higher scores than do students studying seven or eight subjects. This problem is studied in hypothesis 05.

Selective and Comprehensive Systems

The most striking difference between the educational systems of the participating countries is undoubtedly their use of selective or comprehensive secondary schools. This is sometimes the most controversial aspect of school organization.

The pervading issue in school reform, at least in Western Europe, has been that the broadening of educational opportunities at the secondary level has conflicted with the traditional dualism in school organization. This dualism consists of one elementary school which all children have to attend and an academic secondary school to which transfer takes place at an early age, a school designed for an academic and/or social elite. The need for mass education at the secondary level has raised the problem of the extent to which organizationally different provisions should be made for different "types" of children from the age of 10, 11, or 12 onward. The issue has sometimes been formulated as "the comprehensive versus the selective school". The label "comprehensive" represents different things in different countries. In England and Scotland a comprehensive school is a school covering the age range 11 or 12 to 15 and over and providing a variety of programs for the complete range of ability in the area. In Sweden, a comprehensive school is a 9-year school covering the age range 7 to 16, that is, the entire age range for compulsory schooling, and keeping the children together in the same classes during at least the first eight school years, without any streaming. In the United States, a comprehensive school is a public high school which makes provisions for all the children within a given area. Thus, one might distinguish between three criteria of "comprehensiveness", these being (1) the extent to which a school provides instruction for all children in an area, (2) the extent to which it has a wide variety of programs, and (3) the extent to which it avoids rigid forms of ability grouping.

By its very nature the present investigation is confined to only one or two parts of the field of debate. What attainments in mathematics are achieved by students being educated under the various national forms of the two systems? This is discussed at greater length in the report on hypothesis 06. What differences in interest in mathematics are shown by students in countries and schools using these various national forms? These are studied in the report on hypothesis 07.

Socio-Economic Status in Selective Systems

It is already well known that in countries operating selective systems of secondary education, the socio-economic status of a student's home and the probability of his being allocated to a selective academic course and performing satisfactorily in the course are strongly related (Husén, 1965). The investigation provided a further opportunity of assessing the strength of that association in different countries and of probing more deeply into the factors operating, such as the age at which selection was made. The whole question is discussed in the report on hypothesis 08.

Does More Mean Worse?

Whatever the system of school organization used, it is apparent that in most countries the proportion of students electing to continue at school until a later age is steadily rising. If standards can be maintained, this would appear to herald a solution of the problem of educated manpower. The cry, however, has been raised in several quarters that standards will fall, that "more means worse". This problem is studied in hypothesis 09.

Declining standards might be due to a general lowering over the whole of the larger enrollment or, alternatively, to the combination of high standards maintained by the smaller enrollment with lower standards among the additional students. The loss of the high scoring group would be a disadvantage to any country. This aspect of the problem is studied in hypothesis 10.

It is to be expected that systems with varying degrees of *retentivity* as measured by the proportion of a year group continuing to attend school after the period of compulsory attendance has been completed, will produce different yields at the various levels of competence measured by the tests. This possibility is examined in hypothesis 11.

A Mathematical Model of Retentivity
The results obtained from the analyses described under hypotheses 09 and 10 suggest that a fairly simple mathematical model can be constructed to illustrate the relation between differing degrees of retentivity and the mean scores made by the countries. This model is described at the end of this chapter.

The Testing of Hypotheses

For each of the variables referred to in the introduction to this chapter an appropriate hypothesis has been formed and tested. It may state that there is a particular type of relation between the variable and the scores obtained in the tests and questionnaires used in the investigation. Alternatively, it may postulate that there is no relation between them. In each of the following studies the author states the hypothesis, gives a summary of work which has been done previously on the problem, presents an analysis of the data obtained in the present investigation, and draws his conclusions, stating whether the hypothesis is or is not in accordance with the results obtained.

Age of Entry

The hypothesis to be tested here is: *The level of mathematics achievement at age 13 is not related to the age at which compulsory schooling begins* (Hypothesis 01).

Introduction
In each country there are specific regulations specifying when "normal" children (that is, excluding such children as spastics, extremely mentally retarded, etc.) should at the latest enter school. In some countries (for example, Sweden and Germany) there is a day in the school year on which all children within a year group begin school. In other countries (for example, Scotland and England) there are two or three possible days of entry. In most areas in England, for example, all children who will be five years of age between September and the end of December start school at the beginning of September; those who will be five between January and the end of March start at the beginning of

January; and those who will be five between April and August start in the middle of April.

As with most general regulations, there are exceptions. In certain countries children slightly younger than the mandatory age of entry may begin school if there are sound reasons. It is usually the local school authority which then decides whether the reasons are sound. In other European countries it is possible for children to start school before they reach the mandatory age, if it can be shown that they are "mature" enough for school. This judgment of maturity has up to the present involved physical tests of fitness for school as well as certain group tests of reasoning. Examples of these tests are the *Skolmognadstest* in Sweden and the *Schulreife* test in Germany.

It should be remembered, furthermore, that in all countries preschools are attended to different extents (compare Chapter 13, Volume I). For example, in countries where English is spoken there are nursery schools and kindergartens, but it is only a small percentage of an age group which attends. In the countries where French is spoken it is estimated that approximately 50 percent of an age group attend the *école maternelle* (or *jardin d'enfants*). Thus, the differences in amounts of preschooling must be borne in mind when comparing at a later stage the performance of pupils from countries with different mandatory ages of entry to school.

Previous Research

There are three cross-country studies which have examined, in part, the effect of differing amounts of formal schooling to which children in different countries have been exposed. I. H. Anderson (1964) has suggested that the superior performance of English and Scottish over American pupils at the age of 7 can be attributed to the extra year of schooling. But when differences occurred at ages 10 and 14, he preferred to explain these in terms of differences in instruction. Similarly, Pidgeon (1958), although finding English 11-year-old children superior to 11-year-old Californian children, states that the main reasons for the different levels in performance are probably that formal teaching is introduced at an earlier age in England and that there is a difference in the standards in the two systems. He points out that in the United States more limited objectives are formulated for children of primary school age and less emphasis is placed on progress in mechanical arithmetic than is customary in England.

A national study which has relevance to this problem is that carried out by Mogstad (1958) in Norway. In a rural region of Norway 1:

year-old students were taught in two parallel groups. One group received the full regular schooling for two years. The second group received only half as much formal instruction, although they undertook much more homework due to the fact that they were in sparsely populated areas and could attend school for only half the time. In specially constructed achievement tests, the second group was only slightly inferior in performance at the end of the two years to the first group.

Present Study

This IEA study is the first study undertaken where it has been possible to examine differences between the performance of fully representative samples (see Chapter 9, Volume I) from more than three countries in a particular school subject (in this case, mathematics). Here it has been possible to compare the performances of 13-year-olds in countries having mandatory ages of entry to school at 5 (two countries), 6 (six countries), or 7 (two countries).

There are two variables involved. The first is Total Mathematics Score (for details of this see Chapter 5, Volume I); the second is the mandatory age of entry to school, the source of this information being the National Case Study Questionnaire.

The two populations which it is relevant to examine in connection with this problem are the 13-year-olds in each system (that is, the l a population) and students in the grade where most 13-year-olds are to be found (that is, the l b population). The mean ages and standard deviations by country are shown in Table 3.1.

The l a populations are chronologically almost completely comparable and are directly related by age to the mandatory age of entry to school. The variances about the country means are uniform. If the various lengths of schooling up to the age of 13 years make a difference, then it should be apparent in this analysis. Germany and Israel did not test Population l a in their countries and therefore are omitted from the l a analysis.

The second population is the grade population in which most 13-year-olds were to be found. All countries were here represented. The actual grades tested are given in Volume I, Chapter 13. As can be seen from Table 3.1, there is a considerable range in age both between and within countries, and the differences between countries promoting by age and promoting by grade are brought into prominence in the column of standard deviations.

In Table 3.2, countries are grouped into three groups according to whether the mandatory age of entry is 5, 6, or 7 years of age. The median

TABLE 3.1. *Mean Ages and Standard Deviations of Age for Populations 1a and 1b.*

	Population 1 a		Population 1 b	
	Mean Age in Months	Standard Deviation	Mean age in Months	Standard Deviation
Australia	161	3.5	159	7.7
Belgium	162	3.3	168	8.8
England	162	3.3	172	4.2
Finland	163	3.3	167	7.3
France	162	3.5	163	7.8
Germany	—	—	164	6.6
Israel	—	—	167	5.6
Japan	161	3.4	161	3.4
The Netherlands	163	3.1	157	11.6
Scotland	160	3.5	168	5.4
Sweden	163	3.4	164	4.9
United States	163	3.5	164	6.8
Median	162	3.4	164	6.7
Range	3	0.4	15	8.2

age of actual entry for each country is given. The source for these figures is the National Case Study Questionnaire.

Results

The means, standard deviations, and number of students for Populations 1a and 1b are given in Table 3.2, in which the countries are grouped by mandatory age of entry to school. They are also shown in diagrammatic form in Figure 3.1.

The differences in means are listed in Tables 3.3 and 3.4. There is no simple test which can be applied to test the statistical significance of these differences. It is probably more profitable to study figure 3.1 bearing in mind that each of the means portrayed there has a standard error of about half a point. Three of the six countries with an entrance age of six produced significantly higher mean scores for Population 1 a than did either of the two countries using an entrance age of five, or one of the two countries with an entrance age of seven; one of the six produced a significantly lower mean score and in two cases the differences are not significant. Of the two countries using an entrance age of seven, one had as high a score as the median of the six year group and much higher than either of the countries in the five year group, while the other had the lowest score of all.

TABLE 3.2. *Mean Scores and Standard Deviations of Scores in Mathematics for Different Ages of Entry.*

Country	Mandatory Age of Entry in years	Median Age of Entry		Population 1 a			Population 1 b		
		years	months	M.	S.D.	N.	M.	S.D.	N.
England	5	5	2	19.2	17.0	3,012	23.7	18.5	3,148
Scotland	5	5	2	19.1	14.6	5,256	22.3	15.7	5,718
				19.2			23.0		
Australia	6	5	7	20.2	14.0	2,916	18.9	12.3	3,078
Belgium	6	6	2	27.7	15.0	1,686	30.4	13.7	2,644
France	6	6	0	18.3	12.4	2,410	21.0	13.2	3,549
Germany	6	6	5	—	—	—	25.5	11.6	4,476
Israel	6	6	0	—	—	—	32.3	14.7	3,232
Japan	6	6	6	31.2	16.9	2,049	31.2	16.9	2,049
The Netherlands	6	6	5	23.9	15.9	428	21.4	12.1	1,444
United States	6	6	5	16.9	12.7	6,231	17.9	13.3	6,544
				23.0			24.8		
Finland	7	6	8	24.1	9.8	748	26.4	9.6	840
Sweden	7	7	0	15.7	10.8	2,553	15.3	10.8	2,828
				19.9			20.9		

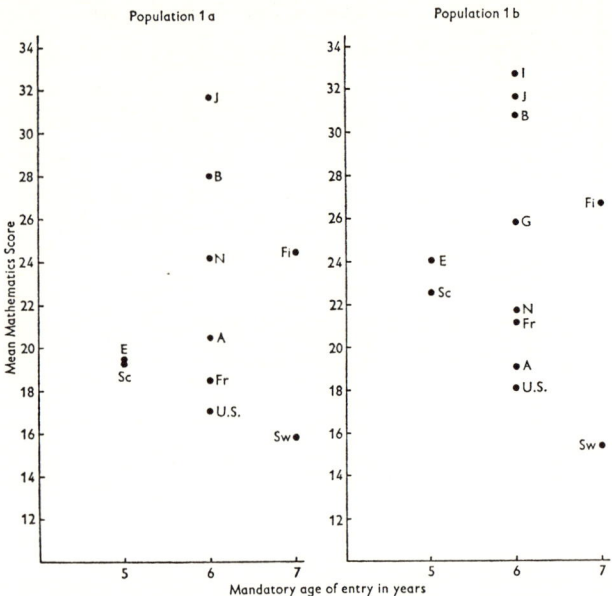

Figure 3.1. Mean mathematics score at the age of 13 for different ages of entry.

It was possible to break down the scores by social classes. Table 3.5 presents the scores for socio-economic class categories 1–6, and for categories 7, 8, and 9 separately. The definitions of the socio-economic class categories are given in Volume I, Chapter 8. Although it would appear that children from social categories 1 to 6 (parents being professional and white-collar workers) benefit more from early entry to school than do those for categories 7, 8, or 9 (farmers and blue-collar workers), it is difficult to draw firm conclusions in view of the heterogeneity of the scores within each of the age-of-entry groups. The table also facilitates comparison of scores of different social class groups within countries.

TABLE 3.3 *Differences Between Mean Scores — Population 1 a.*

(a) 6 years v. 5 years   Difference 3.8
(b) 6 years v. 7 years   Difference 3.1
(c) 7 years v. 5 years   Difference 0.7

TABLE 3.4 *Differences Between Mean Scores — Population 1 b.*

(a) 6 years v. 5 years   Difference 1.8
(b) 6 years v. 7 years   Difference 3.9
(c) 5 years v. 7 years   Difference 2.1

TABLE 3.5. *Mean Score in Mathematics by Socio-Economic Status.*

Country	Social Categories 1–6			Category 7			Category 8			Category 9		
	M.	S.D.	N.	M.	S.D.	N.	M.	S.D.	N.	M.	S.D.	N.
England	29.54	17.19	931	15.50	14.69	1,764	16.09	11.66	50	27.61	17.32	10
Scotland	26.33	14.88	1,456	17.13	13.57	3,180	17.04	13.77	122	13.27	12.54	171
	27.90	16.03	2,387	16.31	14.13	4,944	16.56	12.71	172	20.44	14.93	181
Australia	23.68	13.93	1,380	18.15	13.18	1,219	13.55	12.80	79	14.34	10.79	110
Belgium	31.62	14.17	863	24.83	14.72	662	24.49	21.99	9	21.19	13.92	107
France	21.88	13.31	895	16.85	11.09	1,249	15.27	10.59	39	13.82	12.05	80
Japan	33.30	16.61	1,406	28.05	16.27	485	23.07	14.87	45	21.68	17.52	24
The Netherlands	29.47	16.21	210	19.28	13.70	185	14.64	9.97	20	21.01	18.24	8
United States	20.17	13.62	2,916	13.89	12.06	2,645	12.23	10.45	102	12.89	11.78	28
	26.69	14.64	7,670	20.10	13.50	6,445	17.21	13.44	294	17.49	14.05	357
Finland	23.87	9.53	407	24.37	10.12	301	18.72	11.33	9	17.19	9.79	25
Sweden	17.62	11.13	1,226	14.45	10.15	1,075	11.42	7.92	99	12.21	8.33	49
	20.74	10.33	1,633	19.41	10.13	1,376	15.07	9.62	118	14.70	9.06	74

Differences in Mean Score

	1–6	7	8	9	Combined 8–9
5 years v. 6 years	1.21	−3.79	−0.65	2.95	1.19
5 years v. 7 years	7.16	−3.10	1.49	5.74	3.62
6 years v. 7 years	5.95	0.69	2.14	2.79	2.43

Conclusion

The results suggest, but do not establish beyond doubt, that school systems admitting pupils at the age of 6 produce mathematics scores at age 13 which are superior to those obtained by students in systems admitting children at 5 or delaying admission to 7. The extra year of schooling enjoyed by those entering at 5 would appear to be of no consequence as far as progress in mathematics by the age of 13 is concerned. The loss of a year of schooling between 6 and 7 appears to have a detrimental effect on mathematical attainment at 13.

To make the mandatory age of entry to school earlier (for example, from 6 to 5) will not in itself improve performance; it is what happens in that extra initial year that is important. It is the qualitative differences which should now be examined more systematically.

Age at Which School Education Ceases

This hypothesis was originally phrased in the form, "Student age is related to mathematics achievement across countries". As shown in Chapter 1 (Table 1.12), the correlations between the two variables are small, and the reasons for this are explored in Chapter 6. These correlations have been calculated on a "within country" basis or an "all countries" basis, while the discussion under the present hypothesis is aimed rather at a "between country" comparison and may be framed thus: *The mathematical achievement of each country is related to the mean age of the students forming the population within that country* (Hypothesis 02). The hypothesis will be tested for each of the Populations 1 b, 3 a, and 3 b.

The average ages of the various populations and their standard deviations have already been given in Volume I, Chapter 14, but are repeated in Table 3.6 for convenience.

The most striking feature of the table is the difference among countries both in mean age and scatter of ages for Populations 1 b, 3 a, and 3 b. For Population 1 a, differences are at a minimum, since that population was the age group of thirteen-year-olds. In Population 1 b (the grade containing most thirteen-year-olds) the means vary from 157 months for the Netherlands to 172 months for England, a difference of more than a year. In several countries the decision on the grade to be chosen to represent the thirteen-year-olds has been a difficult one; in one, the average age has turned out to be over 14. The standard deviations range from 3.4 months for Japan, where promotion from grade

TABLE 3.6. *Mean Ages and Dispersions of Students by Country and Population (in Months).*

Country	1 a		1 b		3 a		3 b	
Population...	Mean	S.D.	Mean	S.D.	Mean	S.D.	Mean	S.D.
...alia	161	3.5	159	7.7	206	9.2	—	—
...um	162	3.3	168	8.8	217	11.6	216	11.2
...nd	162	3.3	172	4.2	215	7.5	215	6.8
...nd	163	3.3	167	7.3	229	10.6	230	10.8
...ce	162	3.5	163	7.8	223	13.7	225	12.8
...any	—	—	164	6.6	238	8.4	237	8.8
...	—	—	167	5.6	218	8.5	—	—
...	161	3.4	161	3.4	212	3.6	212	3.7
Netherlands	163	3.1	157	11.6	218	11.7	223	11.3
...nd	160	3.5	168	5.4	210	8.0	205	6.2
...en	163	3.4	164	4.9	235	10.9	235	11.3
...d States	163	3.5	164	6.8	213	6.3	214	7.3
...ountries	162	3.6	166	7.7	217	12.9	214	10.4

to grade is by age, to 11.6 for the Netherlands, where a substantial number of students repeat grades.

In Populations 3 a and 3 b the differences in means are still greater, covering a range of over two years. Students are younger in Australia and Scotland (206 and 205 months) and older in Sweden and Germany (235 and 238 months).

The mean scores for these three populations, which were given in Tables 1.2–1.4, are plotted against their corresponding mean ages in Figures 3.2, 3.3, and 3.4. At first sight there would appear to be no relation of any educational significance between ages and attainments, but closer examination of the data reveals certain patterns.

For most of the countries in the inquiry, Population 1 b does not represent the last grade of compulsory schooling but was the highest grade common to all countries that could be chosen for this purpose. There is in most countries a very large overlap between the 1 a population (the thirteen-year-olds) and the 1 b population, and it is therefore not surprising that Table 3.7 shows a distinct association between the 1 a and 1 b mean scores for each country. It also shows a very strong association between the difference in the mean scores for the two populations and the difference between their mean ages. This suggests that the fairly low correlation of 0.3 between achievement and age for Population 1 b may conceal more important relationships to be discovered when differences in age and differences in initial score are treated separately.

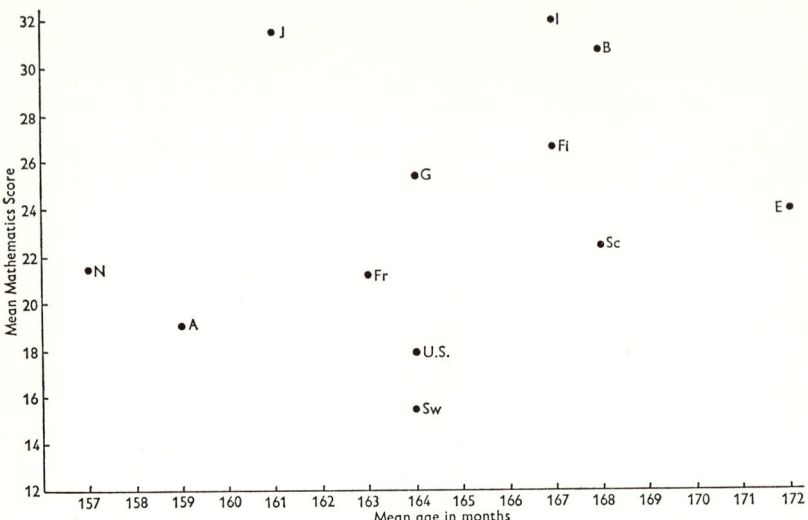

Figure 3.2. Relation of mathematics score to age by country (Population 1 b).

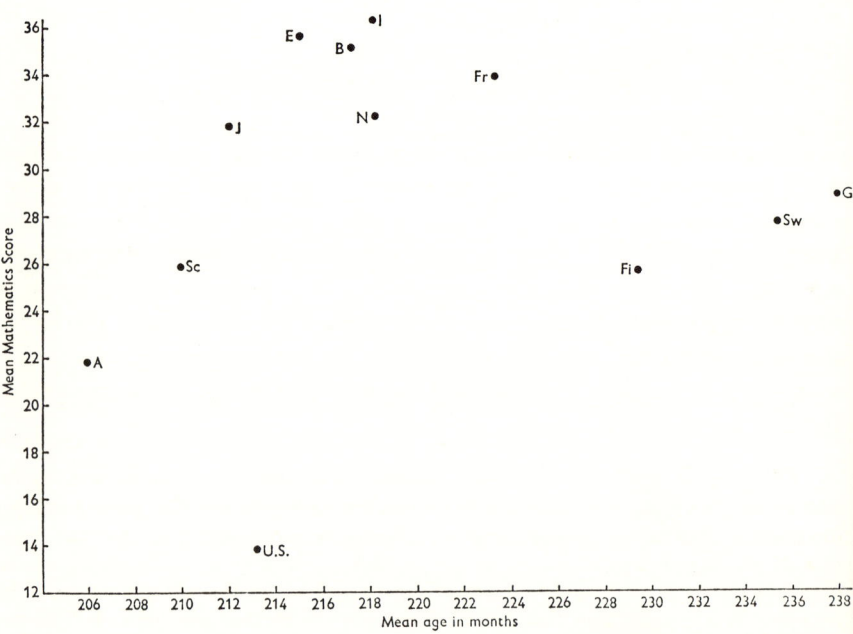

Figure 3.3. Relation of mathematics score to age by country (Population 3 a).

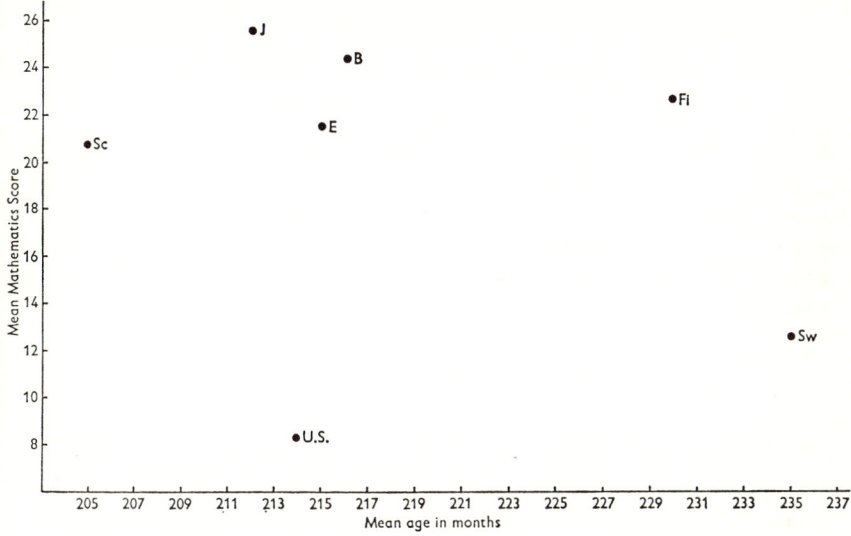

Figure 3.4. Relation of mathematics score to age by country (Population 3 b).

The regression equation connecting mean score for Population 1 b with the mean score for Population 1 a and the difference in ages of the two populations is, in standard units,

1 b score = 0.94 × 1 a score + 0.39 × age difference,

with a multiple correlation of 0.98.

The ratio of the two coefficients shows that the 1 a score is much more important than the difference in ages in estimating the 1 b score. It also shows that when the factor of inequalities in the 1 a score has been removed, the 1 b score is positively associated with differences in ages and, therefore, with the age of the 1 b population since the ages of the 1 a populations are relatively uniform. To that extent, the hypothesis is supported and the differences between countries in the 1 b scores can be accounted for partly by the differences in mean ages of the 1 b populations.[1]

[1] At this stage it may be as well to comment on the use of correlation and regression coefficients with as few as ten observations for each variable. The point has been discussed in the preface, and in Chapter 2, but it is worth repeating that the function of the application of the technique in this instance is to fit constants to the data and *not* to predict values for other countries of which the ten quoted could be regarded as a sample. The fact that the multiple correlations are high (for the regression quoted, $R = 0.98$) is partly a consequence of the small number of observations, since the maximum number of constants that can be fitted to ten observations is ten.

TABLE 3.7. *Relationship Between 1b Score, 1a Score, and Age.*

Country	1 b Score	1 a Score	Differences 1 b–1 a		1 b Score from Regression
			Score	Age	
Australia	18.9	20.2	−1.3	−2	19.7
Belgium	30.4	27.7	2.7	6	30.4
England	23.8	19.3	4.5	10	23.9
Finland	26.4	24.1	2.3	4	26.0
France	21.0	18.3	2.7	1	19.1
Japan	31.2	31.2	0	0	31.2
The Netherlands	21.4	23.9	−2.5	−6	21.6
Scotland	22.3	19.1	3.2	8	22.8
Sweden	15.3	15.7	−0.4	1	16.5
United States	17.8	16.2	1.6	1	17.0
Average	22.85	21.57	1.28	2.3	22.82

When the scores of the 3 a populations are being considered, it must be borne in mind that this population is very differently structured in each country. In about half of the countries, the 3 a population represents 5 percent or less of the age group; in the remainder, the percentage is larger, increasing to 18 for the United States. This factor (termed retentivity) is considered in greater detail in later hypotheses (09, 10, 11) but it has to be introduced here on account of its effect on the variables under consideration. In Table 3.8 the percentages are given to the nearest whole number and the differences between the ages of the 1 a and 3 a populations are given in months. These differences are derived from the average ages for 3 a and 1 a students given in Table 3.6.

The first order correlation between the 3 a score and the difference between 1 a and 3 a ages is 0.22. The regression equation relating 3 a score to the remaining three variables is

$$3 \text{ a score} = 0.14 \times 1 \text{ a score} + 0.27 \times \text{age difference} - 0.73 \times \text{retentivity}$$

and the corresponding multiple correlation is 0.84. The 3 a scores calculated from this regression equation are shown in the last column of Table 3.8. The 1 a score and the difference in age each account for about 6 percent of the variance and retentivity for 58 percent. The differences in age, and hence the age of the 3 a population, are seen to have about twice the weight of the 1 a score but only about one third

TABLE 3.8. *Relationship Between 3a Score, 1a Score, Age, and Retentivity.*

Country	3 a Score	1 a Score	Age Difference	Retentivity	3 a Score from regression
Australia	21.6	20.2	45	14	20.4
Belgium	34.6	27.7	55	4	33.3
England	35.2	19.3	53	5	30.4
Finland	25.3	24.1	66	7	32.3
France	33.4	18.3	61	5	32.0
Japan	31.4	31.2	51	8	29.3
The Netherlands	31.9	23.9	55	5	31.7
Scotland	25.5	19.1	50	5	29.7
Sweden	27.3	15.7	72	16	23.6
United States	13.8	16.2	50	18	17.0
Average	28.0	21.57	55.8	8.7	28.0

TABLE 3.9. *Relationship Between 3b Score, 1a Score, Age, and Retentivity.*

Country	3 b Score	1 a Score	Age Difference	Rentenivity	3 b Score from Regression
Belgium	24.2	27.7	54	9	27.1
England	21.4	19.3	53	7	19.8
Finland	22.5	24.1	67	7	21.6
Japan	25.3	31.2	51	49	24.0
Scotland	20.7	19.1	45	12.6	20.2
Sweden	12.6	15.7	72	7	12.9
United States	8.3	16.2	51	52	9.5
Average	19.3	21.9	56.1	20.5	19.3

of the weight of the retentivity factor. The hypothesis thus receives rather weak support. While increased age is associated with increased score, the effect is far outweighed by the negative association between score and retentivity.

The analysis for the 3b population follows similar lines, although only seven countries are available at this level (Table 3.9).

The first order correlation between 3b score and age difference is -0.29. The negative sign is not so surprising as it might at first appear, for in some countries students give up the study of mathematics soon after the 1 a stage and the age difference is not so much a measure of increased study as of time to forget. The regression equation is

3 b score = 0.87×1 a score $- 0.29$ age difference $- 0.57$ retentivity and the multiple correlation is 0.97.

For this population the hypothesis is not supported. On the contrary, the indications are that increased age is associated with lower scores and that it is proficiency in mathematics at the age of 13 which is most closely related to the proficiency of the nonmathematics students at the terminal stage.

Conclusion

The differences in mean score between countries can be partly accounted for by the differences in the mean ages within populations. For Populations 1 b and 3 a, increased age is associated with increased score, but the reverse is true for Population 3 b.

Size of School

The hypothesis to be tested here is: *The mean level of achievement in a school will be related to the total enrolment at the school* (Hypothesis 03).

Introduction

Research on the relation between level of academic achievement and size of school has been reported in the *Review of Educational Research*.[1]

The general finding from these studies, all carried out in the United States, is that very small schools have lower academic achievements than larger schools. Gray found an increase in quality as school size increased, with a plateau reached at about 400 pupils. Smith found that a size range of 800–1,200 students was the one at which favorable factors approached a maximum and unfavorable factors approached a minimum.

These investigators employed a number of measures of achievement whereas the criterion dealt with in our hypothesis is concerned only with levels of achievement in one subject of the secondary curriculum.

---

[1] Volume XXXI, No. 4 (October, 1961), p. 385, reports work by Hieronymus on achievement in basic skills as related to size of school and type of organization, and by Feldt on the relationship between pupil achievement and high school size. Volume XXXIV, No. 4 (October, 1964), pp. 475–476, refers to work by Gray who studied the relationship between secondary-school size and five qualitative and quantitative factors of education in schools in Iowa, and by Smith who investigated the relationship of secondary-school size to 21 selected factors.

It is clear from previous investigations that any relationship found is not likely to be a linear one. There is a minimum size below which a school cannot provide a reasonably well-qualified staff for the different subjects of the curriculum. There may in fact be an optimum size of school beyond which the general level of attainment falls.

A complicating factor is the type of school organization. A small comprehensive school is obviously a different educational unit from a selective school of the same size. Also, it may be that effects differ at different ages; what is an optimum size from the point of view of the older classes may not be so for the younger classes.

The main factors involved therefore are population, country, type of school,[1] and size of school. The first three of these provide definite categories, but there is no strong theoretical basis for classifying schools by size. All of the schools in the investigation were therefore divided into four approximately equal groups to form categories ranging from "very large" to "very small".

It was found that the four categories of size best fitting the data for Populations 1 a and 1 b were 1–200, 201–450, 451–800, and 801 upward. For Populations 3 a and 3 b the categories chosen were 1–400, 401–700, 701–1,100, and 1,101 upward.

Results

Mention has already been made in Chapter 1 that large size and high performance accompany each other quite consistently in the younger age groups in almost all of the countries, while in the older populations the relationship becomes quite inconsistent from country to country and disappears on the average. The actual correlation coefficients are given in Tables 6.14–6.17. The average numbers of pupils on the enrollment of each school in each country were given in Tables 14.1 A–D of Chapter 14 of Volume I but are repeated here for ease of reference (Table 3.10).

The tabulation of the mean scores for four populations, ten countries, four types of schools, and four categories of size would occupy an unwarranted amount of space. Many of the differences found were not statistically significant. The procedure has therefore been adopted of tabulating the mean scores for Populations 1 a and 3 a for all countries combined and for separate types of school where the number of students was sufficiently large (Table 3.11). Particular cases within countries which seemed to merit attention are shown separately in Tables 3.12–13, pages 77–78.

[1] For definitions of type of school, see pages 91–92.

TABLE 3.10. *Mean Enrollments per School by Country and Population.*

Population...	1 a		1 b		3 a		3 b
Country	Mean	S.D.	Mean	S.D.	Mean	S.D.	Mean
Australia	540	312	542	312	786	376	—
Belgium	408	378	412	380	446	216	509
England	501	311	499	311	644	266	619
Finland	340	277	340	277	672	244	688
France	384	300	381	298	928	294	—
Germany	—	—	295	235	590	173	584
Israel	—	—	532	254	563	146	—
Japan	985	535	985	535	1,146	494	1,185
The Netherlands	330	240	323	224	462	268	—
Scotland	548	415	552	414	768	318	769
Sweden	445	300	445	300	657	363	644
United States	662	492	658	488	1,434	1,166	1,346
All Countries	558	437	553	435	920	726	1,012
Range	655	—	690	—	972	—	837

For Population 1 a the general conclusion to be drawn is that in comprehensive, selective academic and selective vocational schools, the schools with enrollments above 800 have the highest scores. In the remainder type of school, on the other hand, schools with enrollments between 200 and 450 are more efficient than the largest schools with enrollments over 800.

In Population 3 a differences exist between the comprehensive and selective academic schools. In the comprehensive schools, mathematical achievement is significantly related to size of school, the larger the school the higher being the level reached. In the selective academic schools, however, schools with enrollments between 700 and 1,100 achieve significantly higher scores than any other size of school, including the very large ones with an enrollment exceeding 1,100.

These international averages conceal national differences which are occasionally marked. Those showing statistical significance for Population 1 a are shown in Table 3.12 and those for Population 3 a, in Table 3.13.

The general trend in Table 3.11 is for the larger schools to have the higher scores. One exception is Scotland, where the smallest schools show up best in the comprehensive group. A possible reason for this is that these schools are in isolated areas where the whole range of ability attends the local school and the risk of "creaming off" is very low.

TABLE 3.11. *Weighted Mean Scores by Population, Type of School and Size of School.*

Population	Type of School	Size of School				Significance[a]
		(1) 1–200	(2) 201–450	(3) 451–800	(4) 801+	
1 a	Comprehensive	16.26	16.76	17.84	22.31	4 > 1, 2, 3
	Selective Academic	23.71	25.23	30.13	32.96	{4 > 1, 2, 3; 3 > 1, 2}
	Selective Vocational	20.31	22.00	16.75	32.62	{4 > 1, 2, 3; 2 > 3}
	Remainder	12.54	13.32	11.70	10.72	2 > 4
		Size of School				
		(1) 1–400	(2) 401–700	(3) 701–1,100	(4) 1,101+	
3 a	Comprehensive	10.96	17.66	21.63	24.45	4 > 3 > 2 > 1
	Selective Academic	29.54	29.38	31.27	28.21	3 > 1, 2, 4

TABLE 3.12. *Mean Scores Showing Significant Differences Within Countries by Type of School and Size (Population 1a).*

Type of School	Country	Size of School				Significance[a]
		1–200	201–450	451–800	801+	
Comprehensive	Australia	16.73	14.73	17.54	22.64	4 > 1, 2, 3
	Scotland	33.68	18.27	16.26	18.08	1 > 2, 3, 4
	United States	14.92	15.52	15.31	18.87	4 > 1, 2, 3
Selective Academic	Belgium	27.49	32.30	33.24	39.23	{4 > 1, 2, 3; 3 > 1}
	England	27.99	40.80	40.85	42.56	2, 3, 4 > 1
	Finland	22.80	21.97	26.70	29.72	{4 > 1, 2; 3 > 1}
	France	23.14	16.22	23.86	28.15	1, 3, 4 > 2
	United States	18.40	38.19	—	28.71	2 > 1
Selective Vocational	Belgium	17.65	20.44	12.24	32.36	{4 > 1, 2, 3; 2 > 3}
	The Netherlands	27.89	23.23	21.05	40.64	4 > 2, 3
Remainder	Australia	18.24	28.79	23.08	—	2 > 1
	England	7.85	12.27	13.65	12.71	2, 3, 4 > 1

[a] See footnote to Table 3.13.

TABLE 3.13. *Mean Scores Showing Significant Differences Within Countries by Type of School and Size (Population 3a).*

Type of School	Country	Size of School				Significance
		1–400	401–700	701–1,100	1,101+	
Comprehensive	Australia	21.84	15.84	22.66	16.14	3 > 4
	Japan	36.00	32.12	27.55	34.98	4 > 3
	United States	7.34	3.70	13.19	18.26	4 > 3 >
Selective Academic	Belgium	28.90	34.48	38.73	—	3 > 1
	Finland	35.39	24.19	26.19	21.06	1 > 2, 3
	Germany	32.96	26.19	29.43	—	1, 3 > 2
	Israel	27.20	37.11	41.87	—	3 > 1
	The Netherlands	29.49	31.06	34.08	35.71	3,4 > 1
	Sweden	27.47	22.15	29.04	32.52	1, 3, 4 >
	United States	16.30	—	19.90	47.20	4 > 1, 3
Selective Vocational	Japan	48.88	—	22.06	32.48	1 > 3, 4
Remainder	United States	3.41	—	8.32	37.30	4 > 1, 3

a In the final columns of these tables the position regarding significance is summarized thus: 1 > 2, 3, 4 indicates that the mean score for students in column 1 is significantly higher than those of students in other sizes of school, but that the differences between the mean scores in columns 2, 3, and 4 are not statistically significant.

Some countries have few or none of the very large schools, with enrollments exceeding 1,100. A striking feature is the relatively poor performance of the schools with enrollments between 400 and 700 whether comprehensive or selective academic. This contradicts the finding of Gray, referred to in the introduction.

There are curious reversals; for example, in Finland the schools with average rolls above 800 make higher scores in Population 1 a, while in Population 3 a the best scores are made by schools with enrollments below 400. In Belgium and the United States the large schools are consistently superior at both levels. Further analyses of the relationship between size of school and performance have been made in some of the IEA *National* Reports.

Conclusion

The general conclusion is that within the range of sizes considered in this investigation, the best performances in mathematics by the younger students were given in schools with enrollments exceeding 800. Where the older students were concerned, the evidence was more conflicting and national differences appeared to play an important part.

Size of Class
Introduction and Results

The hypothesis states: *The level of mathematics achievement is not related to size of class* (Hypothesis 04).

Although there has been considerable research in the past on the relationship between size of class and level of achievement, it was nevertheless decided to explore this problem again in the IEA study because of the number of school systems involved. Results of past research can be summarized as indicating that class size has not been an important factor. Marklund (1962) has an up-to-date summary of research in this field. In the IEA study, each student was asked to report the size of his mathematics class. The mean sizes of class and standard deviations for each population in each country are given in Tables 14.7 A, B, C, and D in Chapter 14 of Volume I. The range of average size is from 24 in Belgium to 41 in Japan for Populations 1 a and 1 b, from 12 in England to 41 in Japan for Population 3 a, and from 15 in Germany to 41 in Japan for Population 3 b. The rank order correlations by country between average mathematics score and average size of class are: 1 a, +.29; 1 b, +.43; 3 a, −.41 and 3 b, −.23. In other words, students in countries with larger classes score higher at the 13-year-old level but the reverse is true at the preuniversity level.

For the following analysis, three sizes of class were used and designated "small", "medium", and "large". For Populations 1 a and 1 b, small meant less than 25; medium, 25–34; and large, 35 and over. For Populations 3 a and 3 b, the categories were, respectively, 19 or fewer, 20–34, and 35 or over. The limitations of this sort of trichotomy have been stated in Chapter 9 of Volume I.

Since the relationship between class size and achievement may differ according to school type, the analysis was carried out separately for comprehensive, selective academic, selective vocational, and other schools. Tables 3.14–3.17, pages 80–83, present the means, standard deviations, and number of students for the various groups. (Where the numbers in all three class sizes within a school type in a country were very small, these groups have been omitted from the tables.) For most countries, there were no statistically significant differences in score among pupils from the small, medium, or large groups when in fact there were sufficient pupils in the various groups to test the between-group differ-

TABLE 3.14. Mean Mathematics Scores for Different Sizes of Class (Population 1a).

Type of School	Comprehensive									Selective Academic								
Size of Class	Small			Medium			Large			Small			Medium			Large		
	M.	S.D.	N.	M.	S.D.	N.	M.	S.D.	N.	M.	S.D.	N.	M.	S.D.	N.	M.	S.D.	N.
Australia	—	—	—	10.18	11.88	213	21.20	14.04	1981	—	—	—	—	—	—	21.62	12.60	382
Belgium	—	—	—	—	—	—	—	—	—	29.87	15.27	150	32.18	12.48	575	32.75	9.70	271
England	13.02	12.85	10	12.87	13.09	97	19.03	13.72	140	33.82	12.70	54	40.77	12.44	264	41.41	10.71	360
Finland	—	—	—	—	—	—	—	—	—	—	—	—	22.68	9.64	36	24.87	9.49	677
France	—	—	—	—	—	—	—	—	—	18.71	10.47	120	17.84	12.27	646	18.92	13.32	692
Japan	—	—	—	—	—	—	31.34	16.81	2025	—	—	—	—	—	—	—	—	—
The Netherlands*	17.00	8.14	26	18.03	9.06	33	15.66	8.94	60	42.51	4.71	27	44.35	11.29	40	—	—	—
Scotland	17.40	14.84	197	13.53	11.68	537	20.23	11.08	1375	27.90	6.70	15	36.79	11.47	145	33.87	11.08	902
Sweden	11.32	8.50	423	13.98	9.75	968	16.65	11.03	295	—	—	—	14.14	9.86	256	21.94	9.52	416
United States	15.18	14.05	595	16.96	13.33	2075	16.76	12.89	2926	25.85	10.30	43	—	—	—	16.42	17.98	19

Type of School	Selective Vocational									Remainder								
Size of Class	Small			Medium			Large			Small			Medium			Large		
	M.	S.D.	N.	M.	S.D.	N.	M.	S.D.	N.	M.	S.D.	N.	M.	S.D.	N.	M.	S.D.	N.
Australia	—	—	—	—	—	—	—	—	—	14.93	12.34	30	16.72	10.91	34	21.88	12.66	206
Belgium	18.92	11.65	131	22.56	12.01	273	22.17	10.54	115	—	—	—	—	—	—	—	—	—
England	—	—	—	31.33	11.28	31	32.43	9.31	41	4.06	7.57	145	7.62	10.84	529	16.92	11.55	1146
Finland	—	—	—	—	—	—	—	—	—	—	—	—	—	—	—	—	—	—
France	—	—	—	—	—	—	—	—	—	—	—	—	17.05	10.36	217	17.53	9.93	244
Japan	—	—	—	—	—	—	—	—	—	—	—	—	—	—	—	—	—	—
The Netherlands*	17.85	8.58	16	25.00	12.18	110	—	—	—	—	—	—	—	—	—	—	—	—
Scotland	—	—	—	—	—	—	—	—	—	10.37	9.95	452	9.11	8.57	557	11.30	9.00	872
Sweden	—	—	—	—	—	—	—	—	—	—	—	—	—	—	—	—	—	—
United States	—	—	—	—	—	—	—	—	—	—	—	—	11.37	7.90	110	14.74	9.79	99

Type of School...	Comprehensive												Selective Academic									
	Small			Medium			Large			Small			Medium			Large		
Size of Class...	M.	S.D.	N.	M.	S.D.	N.	M.	S.D.	N.	M.	S.D.	N.	M.	S.D.	N.	M.	S.D.	N.
Australia	—	—	—	8.78	10.91	219	19.30	11.92	2,100	—	—	—	—	—	—	22.82	11.80	363
Belgium	15.22	17.48	16	17.95	13.65	116	22.38	14.85	137	34.24	13.17	217	32.96	11.59	869	37.58	11.42	431
England	—	—	—	—	—	—	—	—	—	38.19	14.40	76	44.49	11.30	347	46.66	10.63	342
Finland	—	—	—	—	—	—	—	—	—	—	—	—	27.98	4.75	19	27.02	8.98	785
France	—	—	—	—	—	—	—	—	—	20.05	9.28	147	20.48	12.25	861	21.37	12.83	1,056
Germany	24.03	10.88	180	26.53	12.27	644	36.44	11.97	1,244	—	—	—	36.44	9.12	301	31.35	7.68	363
Israel	—	—	—	—	—	—	31.34	16.81	2,025	—	—	—	—	—	—	—	—	—
Japan	17.38	7.47	56	17.55	8.29	209	19.46	9.64	326	—	—	—	39.26	9.76	132	40.52	3.99	16
The Netherlands	20.22	16.05	305	19.39	13.47	705	22.33	12.08	1,412	32.49	9.91	60	40.71	10.38	246	37.25	10.84	923
Scotland	11.43	8.67	463	13.66	9.87	1,075	16.60	10.74	321	—	—	—	14.27	9.80	273	20.43	10.17	462
Sweden	—	—	—	—	—	—	—	—	—	—	—	—	—	—	—	—	—	—
United States	16.93	13.34	699	18.45	13.68	2,334	18.27	12.62	2,901	—	—	—	—	—	—	—	—	—

Type of School...	Selective Vocational									Remainder								
	Small			Medium			Large			Small			Medium			Large		
Size of Class...	M.	S.D.	N.	M.	S.D.	N.	M.	S.D.	N.	M.	S.D.	N.	M.	S.D.	N.	M.	S.D.	N.
Australia	—	—	—	—	—	—	—	—	—	14.36	12.33	33	11.67	8.91	20	20.98	10.99	267
Belgium	23.46	12.09	202	28.11	11.10	438	26.44	9.00	167	14.80	8.19	134	10.53	5.20	7	—	—	—
England	—	—	—	33.49	8.92	27	41.56	9.28	57	5.65	6.97	146	14.60	14.19	539	18.54	13.50	1,142
Finland	—	—	—	—	—	—	—	—	—	—	—	—	—	—	—	—	—	—
France	—	—	—	—	—	—	—	—	—	17.89	5.57	151	17.08	9.68	267	19.49	9.69	353
Germany	—	—	—	—	—	—	—	—	—	23.57	10.8	335	26.33	11.60	1,447	23.11	11.04	1,989
Israel	—	—	—	—	—	—	—	—	—	—	—	—	30.32	9.64	46	28.67	9.38	39
Japan	—	—	—	—	—	—	—	—	—	—	—	—	—	—	—	—	—	—
The Netherlands	18.43	8.96	41	17.80	10.21	200	27.40	7.54	52	10.24	—	—	10.24	5.65	50	22.49	7.18	33
Scotland	—	—	—	—	—	—	—	—	—	12.55	10.24	496	11.22	9.55	495	14.31	9.64	867
Sweden	—	—	—	—	—	—	—	—	—	—	—	—	—	—	—	—	—	—
United States	—	—	—	—	—	—	—	—	—	—	—	—	12.11	8.70	169	17.11	10.1	76

TABLE 3.16. Mean Mathematics Scores for Different Sizes of Class (Population 3a).

Type of School...	Comprehensive											Selective Academic									
Size of Class...	Small			Medium			Large			Small			Medium			Large					
	M.	S.D.	N.	M.	S.D.	N.	M.	S.D.	N.	M.	S.D.	N.	M.	S.D.	N.	M.	S.D.	N.			
Australia	17.35	9.76	324	22.64	9.56	342	19.69	9.52	97	27.03	9.96	73	24.29	9.63	122	25.53	9.39	90			
Belgium	—	—	—	—	—	—	—	—	—	34.23	12.14	275	36.08	10.64	123	33.04	9.89	35			
England	34.15	15.08	26	—	—	—	—	—	—	36.21	11.74	816	33.14	11.04	53	—	—	—			
Finland	—	—	—	—	—	—	—	—	—	26.88	10.18	82	25.53	8.71	216	27.24	5.27	72			
France	—	—	—	—	—	—	—	—	—	27.54	9.94	31	28.80	8.22	38	35.40	11.40	57			
Germany	—	—	—	—	—	—	—	—	—	28.46	9.84	593	33.37	7.97	55	—	—	—			
Japan	34.09	13.44	11	31.97	9.14	23	31.37	14.18	721	—	—	—	—	—	—	—	—	—			
The Netherlands	—	—	—	—	—	—	—	—	—	33.62	7.04	195	29.20	7.82	243	—	—	—			
Scotland	27.05	11.58	310	21.35	7.07	246	20.75	7.51	71	28.49	11.80	285	24.12	9.19	383	25.59	8.35	127			
Sweden	—	—	—	—	—	—	—	—	—	28.96	11.70	293	27.20	12.12	378	22.98	10.12	78			
United States	14.22	11.39	610	12.65	10.87	485	13.08	11.97	253	—	—	—	—	—	—	—	—	—			

Type of School...	Selective Vocational									Remainder											
Size of Class...	Small			Medium			Large			Small			Medium			Large					
	M.	S.D.	N.	M.	S.D.	N.	M.	S.D.	N.	M.	S.D.	N.	M.	S.D.	N.	M.	S.D.	N.			
Australia	—	—	—	—	—	—	—	—	—	22.16	11.69	43	—	—	—	—	—	—			
Belgium	18.84	4.56	6	21.26	7.87	38	21.47	9.42	21	—	—	—	—	—	—	—	—	—			
England	24.05	6.57	46	—	—	—	—	—	—	—	—	—	—	—	—	—	—	—			
France	—	—	—	30.84	5.80	19	—	—	—	—	—	—	—	—	—	—	—	—			
Japan	—	—	—	—	—	—	30.99	11.51	60	—	—	—	—	—	—	—	—	—			
United States	—	—	—	—	—	—	—	—	—	11.33	4.55	13	30.15	9.56	54	21.82	11.72	32			

TABLE 3.17. Mean Mathematics Scores for Different Sizes of Class (Population 3b).

Type of School	Comprehensive									Selective Academic								
Size of Class	Small			Medium			Large			Small			Medium			Large		
	M.	S.D.	N.	M.	S.D.	N.	M.	S.D.	N.	M.	S.D.	N.	M.	S.D.	N.	M.	S.D.	N.
Belgium	—	—	—	—	—	—	—	—	—	23.92	9.51	293	25.66	8.23	380	25.69	9.48	260
England	21.08	11.12	8	21.48	8.22	21	20.08	9.01	13	23.09	9.89	577	22.17	8.94	657	21.21	8.51	340
Finland	—	—	—	—	—	—	—	—	—	23.23	6.51	60	22.45	8.34	257	20.61	6.51	81
Japan	31.65	10.78	7	34.02	8.10	27	29.07	14.20	2,136	—	—	—	—	—	—	—	—	—
Scotland	20.80	8.69	450	20.87	9.76	386	15.98	8.52	109	21.21	10.15	376	21.84	8.67	519	20.13	9.22	253
Sweden	—	—	—	—	—	—	—	—	—	12.05	7.99	15	12.66	6.04	127	12.33	5.69	69
United States	7.47	8.06	240	8.29	8.53	900	8.85	8.96	562	—	—	—	—	—	—	—	—	—

TABLE 3.18. *Significant Differences in Score by Different Sizes of Class.*

Population	Type of School	Comparison	Country	Difference[a]
1a	Comprehensive	Large with medium	Australia	11.02
	Comprehensive	Large with medium	England	6.16
	Selective Academic	Large with medium	Sweden	7.80
	Remainder	Large with medium	England	9.30
	Comprehensive	Large with small	Sweden	5.33
	Selective Academic	Large with small	England	7.59
	Remainder	Large with small	England	12.86
	Selective Academic	Medium with small	England	6.95
	Selective Academic	Medium with small	Scotland	8.89
1b	Comprehensive	Large with medium	Australia	10.52
	Comprehensive	Large with medium	Israel	9.91
	Selective Academic	Large with medium	Germany	−5.09
	Remainder	Large with medium	England	3.94
	Selective Academic	Large with small	England	8.47
	Remainder	Large with small	England	12.89
	Selective Academic	Medium with small	England	6.30
	Selective Academic	Medium with small	Scotland	8.22
3a	Comprehensive	Small with medium	Australia	−5.29
	Comprehensive	Small with medium	Scotland	5.70
	Comprehensive	Small with large	Scotland	6.30
	Selective Academic	Small with large	Sweden	5.98

[a] A difference is positive when it favours the first-named size of class.

ences. In some cases all pupils were in one group and therefore no comparisons could be made. For example, in Populations 1 a and 1 b Japanese pupils were in classes of 35 or over. However, within most countries there is a general tendency in Populations 1 a and 1 b for pupils with higher scores to be in larger classes, whereas the reverse is true in Populations 3 a and 3 b. This confirms the between-countries trends. Where differences are statistically significant, they have been reported in Table 3.18.

One or two additional features of the data presented are worthy of note. In Populations 1 a and 1 b in England, students in larger classes perform better. This may be explained by the fact that streaming is prevalent in these schools, and in general the duller children are put into the "C" streams in which classes are usually smaller than in the "A" or "B" streams. The fact that Australian 3 a students in comprehensive schools from medium size classes perform better than those in small classes may be because larger classes in comprehensive schools are usually to be found in urban areas. The reversal of the trend for medium and large classes in Germany's Population 1 b is interesting, but it should be noted that there was no statistically significant difference in mean score between students from medium and small classes, nor between those in large and small classes.

Conclusion

In the majority of the cases listed in Tables 3.15–3.18, size of class is not related to mathematics achievement; in some there are significant differences in score between students from larger and smaller classes in the 1 a and 1 b populations in favor of the students in the larger classes. In some cases in 3 a and 3 b populations, students from smaller classes tended to score higher marks than students from larger classes; this is in keeping with expectation since in the preuniversity year there is a tendency to have smaller classes for the more advanced students.

It is easy to account for the positive correlation between attainment and class size within an administrative unit. There is a general pressure to keep classes to a relatively uniform size, the exceptions being where a backward class is reduced in size to assist progress and where good teachers are given larger classes because they are more able to cope with them. These two factors seem to operate to produce the positive correlation evident in Populations 1 a and 1 b. But when data for administrative units are pooled, as they are within a country and even more so in this group of countries, administrative pressures vary and it might have been expected that the effect of class size might have shown through more clearly.

The fact that the correlations are negative where the older and smaller classes are concerned suggests that a reduction in class size may operate to advantage only when the size is reduced to twenty or fewer. To the average teacher, a class of 25 may mean much the same as a class of 35 or 45, the same teaching methods being used. It appears that this problem can be solved only by experiments on the classroom floor, in which the situation can be more rigidly controlled.

Specialization

This hypothesis states: *In Population 3 a the level of mathematics achievement will be higher where the number of subjects studied is smaller* (Hypothesis 05).

TABLE 3.19. *Variation of Score with Number of Subjects Studied (Population 3a)*

Country	Number Subjects Studied	Mean Score	Standard Deviation	Number of Students
Australia	6	21.6	10.5	1,089
Belgium	9+	34.6	12.6	519
England	3	35.2	12.6	967
Finland	9	25.3	9.6	369
France	9+	33.4	10.8	222
Germany	9	28.8	9.8	649
Israel	8	36.4	8.6	146
Japan	9+	31.4	14.8	818
The Netherlands	9+	31.9	8.1	462
Scotland	4	25.5	10.4	1,422
Sweden	9	27.3	11.9	776
United States	4	13.8	12.6	1,568

Introduction and Results

The data from the school questionnaire on the average number of subjects studied are limited in application because in some countries different interpretations have been put on the word "subject" by different head teachers. Some have included all "subjects", including sport and drama, whereas others have included academic "subjects" only. However, the data given in the case study questionnaire on the "number of subjects studied" would appear to be in order. Table 3.19 indicates the average number of subjects studied per country (according to the case study questionnaire), the mean corrected mathematics score in each country, the standard deviation, and the number of students.

If the eight countries showing eight or more subjects studied are combined to form one group and the three countries showing four or fewer subjects are combined to form a second group, then the mean scores of the two groups are found to be 31.1 and 24.8, respectively, giving a difference of 6.3 which is highly significant. Students from countries where eight or more subjects (including mathematics) are studied at the preuniversity level perform better in mathematics than students from countries where only four or less subjects (including mathematics) are studied. This is contrary to expectations.

However, there are complications. The United States system is not as specialized as it would appear from the entry in the table. The student with four subjects may well have studied a different four in the preceding session. If that country is omitted from the specialist group, the average score of that group becomes 30.4, which is not significantly different from the average of 31.1 of the first group. The question of age has been already considered in Hypothesis 02. It is noteworthy that the average ages of the students in the eight countries are with one exception (Japan) over 18, while the average ages of the students in the four remaining countries are all under 18. Carrying more subjects is thus associated with a higher age of presentation for the tests. The presumption is that students must prolong their school education to be able to carry the extra load. The question of differing proportions of the age groups in the different countries is also relevant, but this is dealt with in a later series of hypotheses (09, 10, 11).

Conclusion

The conclusion is that the hypothesis now under consideration is not confirmed. Specialization, in the sense of restricting the number of subjects studied in the preuniversity year, is not necessarily associated with higher scores in mathematics.

Achievement in Selective and Comprehensive Systems

The hypothesis to be examined is: *The level of mathematics achievement for students will be higher and the variability lower in specialized schools than in comprehensive schools* (Hypothesis 06).

Introduction

This hypothesis, in effect, asserts that students pursuing academic, vocational, or general courses of study will achieve a higher mean level of performance in mathematics (and display less variability about that mean) if they are attending schools specifically designed to provide for one of these courses only, than if they are receiving their education in a school which provides for all three types of course. In the English educational system, for example, the hypothesis states that students in grammar, technical, and modern schools will consistently achieve higher performances in mathematics than will comparable students in comprehensive schools.

This is, of course, related to the important and controversial issue of comprehensive versus selective education—an issue in which many countries have an interest at the present time. A number of countries, particularly in Europe, over the past century or so, have developed systems of education which provide specialized courses in separate schools, particularly for an academic "elite". Such systems have demonstrably served their countries well. In recent years studies stressing the close relationship which exists between the kind of education system employed and the needs of a changing society in the modern technological age (Trow, 1961) have shown that a selective system providing a high level of education for only a small proportion of an age group will fail to satisfy the ever-increasing need for trained manpower (Husén, 1965).

It should be made clear at this point that there are two main senses in which a school can be said to be "comprehensive". In the first place, for countries in which the basic philosophy of education has been to provide an advanced academic education for a selected elite, and a more general or vocational education for the remainder, the comprehensive school is regarded essentially as one in which pupils of all ranges of ability and from all sections of the community are represented. The type of education offered in such schools, however, may differ little from that previously given in separate specialized schools. That is to say, selection still takes place on entry and pupils are allocated, according to their abilities and aptitudes, to quite distinct academic, general, or vocational courses. The philosophy governing the beliefs and attitudes of the teachers may remain largely the same as in the differentiated system.

There is, however, another sense in which the comprehensive school can be regarded as providing "comprehensive education" to all pupils. Essentially, this is a question of a different philosophical outlook on the part of teachers and administrators. The practical effects of this different outlook are seen in the internal organization of the school, where no differentiation into "tracks" or "streams" pursuing different kinds of educational courses takes place until toward the end of compulsory schooling.

Even within this latter type of comprehensive school, different approaches to teaching and learning are possible. It is possible for all children to be given virtually the same instruction at any given age. It is possible also for children to be regarded as different and an individual approach taken, children being grouped together only for administrative purposes and not for teaching. Education in both these approaches can be called "comprehensive" in that all children are exposed to the same "treatment", the difference being that in the former case all will be learning the same thing at the same time and an authoritarian attitude to teaching would be necessary, while in the latter, each will learn at his own rate in a permissive learning situation and some may take considerably longer than others to complete a year's course. Those views represent extremes, and most countries today offering "comprehensive education" probably fall somewhere between them.

It is clear from the foregoing that the term "comprehensive" does not have precisely the same meaning when applied to schools in different countries. Nevertheless, with certain reservations which will be noted later, there is sufficient agreement to make a limited comparison across countries a worthwhile exercise.

In many of the countries with existing selective systems the opposition to the introduction of a comprehensive system is still very strong. While much of this opposition stems from an unwillingness to change from a system which has adequately served a society itself differentiated in terms of social class, arguments have been put forward that any attempt to change to a comprehensive system of secondary education will inevitably result in a "lowering of academic standards" and hence of the number of students able to go forward to universities and other institutions of higher learning.

Previous research

Only three studies have been found which attempt to compare the two types of secondary organization under discussion. Two of these occurred in England, and both were carried out by overseas students. The third, a far more ambitious investigation, was carried out in Sweden, a country dedicated to educational reform and to the research which should precede it. No relevant research has been found in the United States or any other country.

Koshe (1957), in a small study, attempted to compare the attainments and intelligence of thirteen-year-old students attending comprehensive, modern, and grammar schools. Using five selected schools and matching students on the basis of their 11 + intelligence (verbal reasoning) test results, Koshe concluded that students in the comprehensive schools were superior in his tests at age 13 to their counterparts in modern schools. The comparisons between the comprehensive and grammar school students were largely inconclusive but suggested that "girls seemed to have been more susceptible to the influence of the type of school than boys".

Miller (1958) also compared students in the two types of school organization, but was primarily concerned with student attitudes. He found, for example, that while the comprehensive school may well contribute to the development of greater cultural unity (as measured by students' attitudes to academic and practical subjects) this "is not achieved by bringing children of all abilities together in one type of school, i.e., it is not an inevitable outcome of comprehensive secondary education".

The Swedish investigation by Svensson (1962) was designed to follow up over a five-year period both students selected for academic classes and those not so selected. Three types of schools were studied. In the first, selection took place after grade 4 (11+), selected students transferring to the 5-year *realskola*; in the second, selection took place after grade 6 (13+), and some selected students only were transferred to the 3-year *realskola*; in the third, the comprehensive school, no formal selection took place, although, in practice, selective and nonselective classes were formed after grade 8 (15+), all students remaining in the school. Before selection took place, students were all in unstreamed elementary school classes, that is, classes heterogeneous for ability. All students were initially tested when in grade 4, and subsequently random samples were retested up to the end of grade 9. Svensson concluded from his study that the attainments of students in selective academic classes did not correlate in any way with the kind of previous schooling they had received. A slight tendency toward the superiority of students in early streamed classes, observed early in the study, was erased in grades 8 and 9. With regard to students in nonselected classes, the time of streaming appeared to have no demonstrable effect on their attainments. Furthermore, whether classes were in the 5-year *realskola*, the 3-year *realskola*, or the comprehensive school had no bearing on the achievement of students in the long run.

The results of this investigation throw some light on the question of whether early or late differentiation is important. They also offer some support for the hypothesis that, within a country changing from a differentiated to a comprehensive system, little difference will be found in the performance of academic students in different school types.

Results

The two variables of importance in the analysis of the data pertaining to this hypothesis are *Type of School* and *School Program*. Information on the former was obtained from the school questionnaire, in which each participating school was asked to classify itself in one of four categories: Type 1: Comprehensive; Type 2: Selective-Academic; Type 3 Selective-Vocational; and Type 4: Takes Remainder After Selection. Unfortunately, there was not complete agreement across all countries in the classification of a comprehensive school. For the purposes of this hypothesis a comprehensive school must be regarded basically as one containing all ranges of ability and from which no selection has occurred. Difficulties arise since many of the countries involved in this study do not have schools which comply with this definition. In France, and the Netherlands, for example, the students not selected for a specialized academic or vocational course in a separate school remain in their elementary or primary school and, although in many cases this latter school was labeled Type 1, it cannot be considered comprehensive at age 13 and above in the sense demanded here. Belgium and Germany do not have any comprehensive schools. Japan, on the other hand, is 100 percent comprehensive, and all the students in this country in Populations 1 a and 1 b are following a "general" course.

Information concerning school programs was supplied by the students themselves in the student questionnaire. They were asked to state whether the course they were now following in school was essentially one which could prepare them for a university (academic), one which could prepare them for a specific trade occupation (vocational), or was a general course with no specific occupation or course of study in view (general). Unfortunately, this again led to complications since in many instances the decision of a student to indicate a particular course was based more on aspiration than on actual fact. Thus, a student in one country following a particular course might have indicated that he was taking an "academic" program, whereas a student in another country following a similar course would have stated that it was a "general" one.

Because of variations within different countries with regard to the organization of schools and programs within schools, the data obtained in order to test this hypothesis had to be treated with some care. The attempt was made in the first place to collate the data in such a way as to compare (over all relevant countries) students taking the three separate types of courses within comprehensive schools with students taking these courses in specialized schools. In order to do this certain regroupings of the data had to be made. What these were and the results obtained are given below. This part of the analysis was confined to Populations 1 a and 3 a, and for clarity of presentation they are dealt with separately.

Population 1 a

COMPREHENSIVE SCHOOLS

Only five countries possess schools which, according to liberal criteria, fulfill the adopted definition of comprehensive: England, Scotland, Sweden, the United States and Australia. Japan had to be omitted for the purposes of this hypothesis because all students take the same "general" course. Some reservation should be made, however, with regard to the comprehensive schools in England in that the majority have suffered some degree of selection, either to maintained grammar schools or to independent schools. In these comprehensive schools some slight doubt might be voiced about the authenticity of students' claims concerning the course or program they were following, but no changes were made.

SELECTIVE ACADEMIC SCHOOLS

Nine countries have selective academic schools in this population. In most of the countries concerned some pupils indicated they were following "general" or even "vocational" courses, but for the purposes of this analysis their results were grouped together with those indicating an "academic" course. This procedure of amending the students' responses seemed justifiable in view of the nature of the school, and the fact that the numbers were in any case fairly small and in most cases their performance did not differ greatly from that of "academic" pupils.

SELECTIVE VOCATIONAL SCHOOLS

Only three countries have vocational schools in this population: Belgium, England, and the Netherlands. The "technical" schools in England give courses that are often more "academic" than vocational, and some pupils in these schools indicated this in their choice of program; however, as with choices other than vocational in the other two countries, all pupils were grouped as taking a "vocational" course.

SCHOOLS TAKING REMAINDER AFTER SELECTION

As mentioned earlier, the schools categorized as Type 1 in France and the Netherlands have been regrouped as Type 4 as were those in Finland. In these schools, as in some of the Type 4 schools in the 6 other countries concerned, a small proportion of students indicated that they were following courses other than "general". (In England, for example, pupils in secondary modern schools (clearly Type 4) may be in an "A" stream pursuing a course leading to the GCE examination in some academic subjects or may be taking a course including such subjects as metal work or domestic science. Some students taking such courses said they were taking either an academic or a vocational program.) In these and in similar cases in other countries such students were all regarded as following a "general" course. Again, this procedure was felt to be justified since the numbers of students involved was small

TABLE 3.20. *Mean Corrected Mathematics Test Scores for Students Following Academic, Vocational, and General Programs in both Comprehensive and Specialized Schools in Population 1a.*

	Program								
	Academic			Vocational			General		
ol Type	M.	S.D.	N.	M.	S.D.	N.	M.	S.D.	N.
prehensive	20.31	13.56	5,680	17.88	12.63	639	13.98	11.51	5,556
tive Academic	27.11	11.65	6,451	—	—	—	—	—	—
tive Vocational	—	—	—	23.44	12.46	799	—	—	—
inder	—	—	—	—	—	—	13.07	10.41	5,255

and their mean performance did not differ greatly from those students in the same country who stated they were following a general course.

With these reservations in mind the results given in Table 3.20 now can be examined.

All differences of means between comprehensive and selective schools *within* programs are significant. Thus, the hypothesis as originally stated is supported, so far as mean scores are concerned, for pupils taking academic and vocational courses; for students taking "general" courses, however, the evidence from Table 3.20 is in the opposite direction—students in the comprehensive school do better than their counterparts in specialized schools. On the other hand, in accordance with the hypothesis, students in all three programs within the comprehensive school show greater variability than do students following similar courses in specialized schools; however, the difference for students in vocational courses is not significant.

Population 3 a

At this level, comparisons of the kind specified in the hypothesis can only be made within academic courses, because for the most part students taking either vocational or general courses are no longer in school and are not represented in the sample tested.

COMPREHENSIVE SCHOOLS

Sweden's data cannot be used at this level as all "academic" students are to be found in gymnasia. Japan, however, comes in, although the majority of students still at school are in fact still taking a "general" course, and only a small proportion are taking an "academic" one. In

Japan, as in the other four countries, only those students taking an "academic" course have been included.

SELECTIVE ACADEMIC SCHOOLS

There are ten countries with selective academic schools at this level. In all countries except England and Australia nearly all students indicated that they were following an academic course, and the few exceptions have been ignored. In England and Australia, a number of students, who said they were taking a vocational or general course, although the course they were following was obviously academic, have been grouped as "academic".

The results in Table 3.21 show a highly significant difference in favor of the selective schools, as the original hypothesis specified. The difference in variability, however, is small and nonsignificant.

While the differences in Tables 3.20 and 3.21 are statistically significant—highly so in some cases—they must be viewed with considerable reservation. The means compared in Tables 3.20 and 3.21 have been calculated from a weighted average of the country means given in Tables 3.22–3.24, and there are large between-country differences, particularly in Population 3 a. Students in comprehensive schools following an academic program in Japan, for example, obtain a mean score of 47.8, whereas for the similar group of students in the United States, the mean is only 14.0. In comparing these figures it must be remembered that the Japanese group is highly selected, comprising somewhat less than 1 percent of the population at this level, as contrasted with 16 percent in the United States.

Results Within Countries

Because the between-country differences are large and the groups represented by the means in each country vary with respect to the degree of selectivity, it was decided to carry out a further analysis within countries. This could be done only in countries which had both com-

TABLE 3.21. *Mean Corrected Mathematics Test Scores for Students Following Academic Programs in Comprehensive and Specialized Schools in Population 3a.*

School Type	Academic Program		
	M.	S.D.	N.
Comprehensive	19.34	10.94	2,695
Selective Academic	29.89	10.69	4,453

TABLE 3.22. *Mean Corrected Mathematics Test Scores for Students Following Academic, Vocational, and General Programs in Comprehensive Schools in Various Countries in Population 1a.*

	Program								
	Academic			Vocational			General		
ry	M.	S.D.	N.	M.	S.D.	N.	M.	S.D.	N.
lia	24.5	14.7	851	18.6	12.6	564	16.7	13.6	792
d	25.8	15.2	58	12.9	13.0	75	13.5	10.8	116
d	27.5	11.0	918	—	—	—	11.9	9.0	1,212
n	17.1	10.8	720	—	—	—	11.0	8.1	995
States	17.7	13.8	3,131	—	—	—	15.4	12.3	2,439

TABLE 3.23. *Mean Corrected Mathematics Test Scores for Students Following Academic, Vocational, and General Programs in Specialized Schools in Various Countries in Population 1a.*

	Program and School								
	Academic-Selective			Vocational-Selective			General-Remainder		
y	M.	S.D.	N.	M.	S.D.	N.	M.	S.D.	N.
ia	21.2	12.7	386	—	—	—	19.4	12.0	270
1	32.7	12.3	1,001	22.3	12.2	620	11.2	8.7	49
d	40.0	11.6	679	30.6	10.5	75	12.6	11.5	1,869
	24.7	9.5	716	—	—	—	10.7	8.2	32
	18.5	12.7	1,787	—	—	—	17.8	10.2	735
therlands	42.8	10.4	67	24.8	12.8	124	16.0	10.8	194
d	34.2	11.5	1,068	—	—	—	10.4	9.3	1,894
	21.4	12.2	679	—	—	—	14.7	11.9	47
States	30.8	11.4	68	—	—	—	14.0	9.9	214

prehensive and selective schools, and this restricts the analysis to England, Scotland and Australia. Although the United States and Sweden have both types of school, there are too few selective schools in the amples for these countries to produce meaningful results. Again, only n England could comparisons be made in vocational programs.

ENGLAND

The relevant data for Populations 1 a, 1 b, 3 a, and 3 b are given in Table 3.25.

The differences between the comprehensive and specialized groups

TABLE 3.24. *Mean Corrected Mathematics Test Scores for Students Following Academic Programs in Comprehensive and Specialized Schools in Various Countries in Population 3a.*

Country	Comprehensive Schools			Selective Academic Schools		
	M.	S.D.	N.	M.	S.D.	N.
Australia	21.1	10.6	666	24.4	10.0	284
Belgium	—	—	—	35.7	12.0	433
England	34.8	15.0	27	35.9	11.7	907
Finland	—	—	—	25.3	9.6	368
France	—	—	—	34.3	11.1	125
Germany	—	—	—	28.8	9.8	648
Israel	—	—	—	35.9	9.0	110
Japan	47.7	8.9	76	—	—	—
Scotland	24.5	10.0	624	25.9	10.9	793
Sweden	—	—	—	27.4	11.9	752
United States	14.0	11.6	1,302	22.3	9.5	36

taking academic and vocational courses in Populations 1a and 1b are significant. The differences between groups of students from the two types of schools taking general courses in Populations 1a and 1b or taking academic courses in Populations 3a and 3b are not significant. In considering these results it must be remembered that in England some selection will have been made from the comprehensive schools in most areas of the country both to maintained grammar schools and to independent schools. This may account in part at least for the large differences in Populations 1a and 1b for students taking the academic program, although a further reason may well be that the "academic" students in the comprehensive schools differ somewhat as a group from the "academic" students in the selective schools. That is, the greater flexibility within the comprehensive schools has permitted students with a wider range of ability to take academic courses, and some of these students, had they been educated in separate specialized schools would have been classified "vocational" or even "general". The rather larger standard deviations for the comprehensive schools students offer support for this contention. It is noticeable, however, that the "academic" students from the two types of school do equally well at the preuniversity level.

The comparison within the vocational program is probably not a legitimate one. Whereas the students in the technical schools have been

TABLE 3.25. *Mean Corrected Mathematics Test Scores for Students Following Academic, Vocational, and General Programs in Either Comprehensive or Specialized Schools in England.*

		Program								
		Academic			Vocational			General		
Level	School Type	M.	S.D.	N.	M.	S.D.	N.	M.	S.D.	N.
1a	Comprehensive	25.8	15.2	58	12.9	13.0	75	13.5	10.8	116
	Specialized	40.0	11.6	679	30.6	10.5	75	12.6	11.5	1,869
1b	Comprehensive	29.8	14.5	78	16.7	13.1	68	16.2	12.8	124
	Specialized	44.7	11.5	759	38.7	9.5	86	16.3	13.1	1,856
3a	Comprehensive	34.8	15.0	27						
	Specialized	35.9	11.7	907						
3b	Comprehensive	21.0	8.4	62						
	Specialized	22.2	9.3	1,634						

selected and are, in reality, mostly following an academic course, those within the comprehensive schools may merely be taking additional subjects with a vocational bias.

If, within the comprehensive schools, students with a wider ability range are taking "academic" courses, it might have been expected that the remainder, taking a "general" course, would not do as well as the corresponding "general" students in secondary modern schools. In fact, there is no significant difference between the mean scores of the two groups.

SCOTLAND

The Scottish data are presented in Table 3.26 below. In general the Scottish data parallel those for England—superior results from the specialized school students taking academic courses at the 13–14-year-old level, and no difference between the school groups at the preuniversity level. There would appear to be no definitive evidence as to why the selective school students should have done better than their counterparts in comprehensive schools at the younger age. In Scotland, a fairly rigid distinction is maintained between the academic and nonacademic students in the comprehensive school,[1] in much the same way as separate buildings effect this for specialized schools. This fact may account for the similar standard deviations in the two types of school.

The main difference between the Scottish and English results is the

---

[1] This was true in 1964, but the position is now greatly changed.

TABLE 3.26. *Mean Corrected Mathematics Test Scores for Students Following Academic and General Programs in Either Comprehensive or Specialized Schools in Scotland.*

		Program					
		Academic			General		
Level	School Type	M.	S.D.	N.	M.	S.D.	N.
1a	Comprehensive	27.5	11.0	918	11.9	9.0	1,212
	Specialized	34.2	11.5	1,068	10.4	9.3	1,894
1b	Comprehensive	33.2	11.4	919	14.9	10.1	1,527
	Specialized	38.2	10.7	1,239	13.2	10.2	1,870
3a	Comprehensive	24.5	10.0	624			
	Specialized	25.9	10.9	793			
3b	Comprehensive	20.1	9.5	955			
	Specialized	21.4	9.4	1,158			

significant superiority of the comprehensive school students taking a "general" program over those taking the same program in a junior secondary school in Scotland. In England, although these students did better than their modern school counterparts, the difference in the means did not reach significance.

AUSTRALIA

Table 3.27 gives the results for the three populations for which this country was represented.

The results from Australia show some inconsistency and vary from

TABLE 3.27. *Mean Corrected Mathematics Test Scores for Students Following Academic and General Programs in Either Comprehensive or Specialized Schools in Australia.*

		Program					
		Academic			General		
Level	School Type	M.	S.D.	N.	M.	S.D.	N.
1a	Comprehensive	24.5	14.7	851	16.7	15.6	792
	Specialized	21.2	12.7	386	19.4	12.0	270
1b	Comprehensive	20.7	12.4	940	17.4	12.5	853
	Specialized	22.7	12.1	372	20.8	11.1	322
3a	Comprehensive	21.1	10.6	666			
	Specialized	24.4	10.0	284			

those in England and Scotland. The performance of "academic" students in comprehensive schools is significantly superior to those in specialized schools in Population 1 a. In Population 1 b this is reversed, although the difference is not statistically significant. In both populations, however, the students taking a "general" program in the specialized schools do better than their counterparts in comprehensive schools—significantly so in Population 1 b. In Population 3 a the "academic" students in specialized schools are again superior to those in comprehensive schools. These results appear to reverse the trend shown in England and Scotland, but reservations must be made about the figures from the comprehensive schools as there is some doubt concerning the judgments of the students about the actual courses they are following.

Conclusion

It is clear that the results presented here must be viewed with extreme caution. The main difficulty in interpretation is the probable lack of comparability between the groups of students classified as "academic", "vocational", and "general" attending comprehensive schools on the one hand and specialized schools on the other. It must be recognized that a fairly high correlation exists between "general ability" and mathematics performance and, hence, if comparisons are to be made on the latter, any differences in the former must necessarily be taken into account.

Consider first those students taking "academic" courses. It seems highly probable that such students in specialized schools represent a different "population" from those taking similar courses in comprehensive schools. If it is true that the greater flexibility within the comprehensive school allows students of lower general ability to follow "academic" courses, then it might be expected that the mean mathematics score of all "academic" students would indeed be lower in this type of school than in schools taking selected students only. Comparisons, therefore, would depend upon both the "flexibility" of the comprehensive schools and the degree of selection that had taken place in the selective system. Clearly, these vary from one country to another and even from one school to another within a country. Any interpretation of the results obtained can only be made in the light of a fairly intimate knowledge of the circumstances prevailing in each educational system represented. No general conclusions can therefore be drawn from the data concerning "academic" students.

For England alone, there would appear to be two factors operating to influence the score of "academic" students—both operating against the comprehensive schools. In the first place, most of the comprehensive schools have lost the "cream" of high ability students, either to independent schools or to selective grammar schools; such students would be represented in the specialized school sample only. Second, it is highly probable that students of somewhat lower general ability are following "academic" courses in the comprehensive schools as compared with selective schools. Differences in favour of the specialized schools are therefore only to be expected. What is perhaps remarkable is that the differences obtained in Populations 1 a and 1 b have largely disappeared in Populations 3 a and 3 b.

Because of differences in the organization of schools in Scotland it is unlikely that the two factors mentioned above operate to the same extent, if at all. The superiority of the students in the specialized schools is still present in Populations 1 a and 1 b, but the sizes of the differences on mean scores are greatly reduced. Whatever the reason for this superiority it has disappeared, as in England, with the preuniversity students.

The situation in Australia is again different since the organization of schools varies as between states and a selective factor associated with the area in which the schools are situated may be operating. An added factor perhaps is the lack of clear distinction made by the younger students in the comprehensive schools concerning the precise courses they are following. Because of this, less reliance can be placed upon the results obtained.

So far as the analysis of the data within countries is concerned, the only conclusions that might be drawn are that the variability of performance of the "academic" student in comprehensive schools appears to be greater than that for students in selective schools, and that this factor, as well as others that might be operating, results in the lower mean performances of 13 to 14-year-olds. In England and Scotland, the preuniversity students in comprehensive schools appear to do as well as those in selective schools.

Little can be said concerning students taking "vocational" courses. Clearly, these courses vary considerably from one school to another both within and between countries, and they cater to pupils of varying levels of ability. No conclusions can be drawn from the data.

There are clearly differences among students following "general" courses in different schools in different countries. The case of Japan, where all students in Population 1 a are said to be taking a "general" course, illustrates this point well. If there is anything common, however, about the "general" courses in most countries, it is that they are often taken, for reasons that must be largely associated with ability, by students who have not been selected for either "academic" or "vocational" courses. It would follow, therefore, that the general caliber of students following "general" courses is dependent upon both the extent and effectiveness of the selection process, whether this takes place in a selective system or within a comprehensive school. The greater the selection ratio and the more effective the procedure, the lower will be the performance of the "remainder after selection". It might be argued from this that if there are relatively more students taking academic and vocational courses in comprehensive schools than in a selective system with specialized schools, a lower mean test score might be expected from the "general" students in the comprehensive schools.

It was pointed out above that the variability in performance of "academic" students in England in comprehensive schools was greater than that of their counterparts in selective schools. It may also be noted that the percentages of students in the sample tested taking "academic" or "vocational" courses in comprehensive schools in this country was 54 percent, compared with only 29 percent taking these courses in separate schools. Thus, there would appear to be a clear expectation for a lower mean test score by the "general" students in the comprehensive schools. In fact, the results show a tendency for these students to perform somewhat better than the "general" students attending separate schools. It might be concluded, therefore, that in England, at least, there is an advantage, so far as mathematics is concerned, for students not taking either "academic" or "vocational" courses to be in comprehensive schools rather than in separate specialized schools.

The evidence for this study has not provided any definitive answer to the problem of whether comprehensive schools are more efficient institutes of learning than separate specialized schools or whether a comprehensive system of education provides better "results" for all students than a highly selective one. Because of the variety of factors operating in the selection process in different countries and because of the varying ways in which both comprehensive schools and the courses followed within them are perceived, it is not possible to generalize from the evidence produced. Where comparisons have been legitimately made within the countries there is some evidence to suggest that any initial superiority gained by "academic" students in specialized schools over their counterparts in comprehensive schools is overcome by the preuniversity level, in spite of the possible inclusion in the latter of students of lower general ability. Furthermore, students following "general" courses in comprehensive schools appear to do rather better than might be expected when compared with students following similar courses in separate schools.

Interest in Mathematics in Selective and Comprehensive Systems

Introduction

Although a fair amount has been written in some countries on the good and bad effects on academic attainment resulting from selective education and differentiation in schools, little seems to have been recorded about the effects of these policies on the interests of the pupils in the various school subjects (but see G. Boalt and T. Husén, 1964). The data of the study were therefore used to test the following hypothesis:

Students at the 13-year-old level will have more favorable interests in mathematics in schools and countries which emphasize comprehensive and nonselective secondary education (Hypothesis 07).

For the purpose of this hypothesis, "interest in mathematics" was measured by a composite score derived from responses in the students' questionnaires and the student opinion booklet. Credit points were awarded for including mathematics as a best-liked subject, as a subject in which highest grades were awarded, and one in which additional courses were desired; they were deducted where the contrary was the case, and a constant of five was added to eliminate negative scores. (See Chapter 12, Volume I.) The national means were found to vary from 4.4 to 7.3 with standard deviations ranging from just under 1 to just over 2.

Emphasis on comprehensive and nonselective secondary education was assessed on a twofold basis from the responses by head teachers to two items:

1. Item 7 of the school questionnaire (SCH I) asked for a classification of the particular school into one of four categories which were afterwards telescoped to three, i.e., comprehensive; selective; dealing with remainder after selection;
2. Item 8 of the same questionnaire asked principals to rate the extent to which educational differentiation took place within their school. Educational differentiation was defined as "setting, streaming, ability grouping", and a four-point scale, afterward telescoped to three, asked the principal to give a rating on the extent of use of systems of this type.

It appeared likely that the two assessments would be linked. A school that is already selective has less scope for differentiation; one that is comprehensive would apparently require more differentiation unless the philosophy of equality were applied so that all students of the same age were given the same educational fare. A study of the responses to the second and thirteenth items of the national case study questionnaire did, in fact, show that there was a certain amount of linkage between the two ratings.

On the other hand, the data provided by the schools from the items described above did not show so clear a pattern, and there was unfortunately a substantial amount of misunderstanding among teachers on the terms used. In some countries known to have very few, if any, comprehensive schools, there was a considerable number of teachers describing their schools as comprehensive. It is very probable that these schools should properly have been included with the schools for that part of the age-group remaining after the selective schools have withdrawn their contributions (that is, the "remainder" schools).

As it was impossible at that stage of the inquiry to go back to the schools to amend these entries, it was decided, after consultation with the representatives of the countries concerned, to discard the information from some countries. The following analysis is therefore confined to six countries (Australia, England, the Netherlands,[1] Scotland, Sweden, and United States) in the case of Population 1 a, and to seven (above, plus Israel) in the case of Population 1 b.

Results

Table 3.28 shows the number of pupils classified into the nine categories formed by a three-way classification of school types and a three-way classification of degrees of differentiation.

Inspection of this table shows that there is a slightly greater degree of differentiation within comprehensive schools than within selective schools. There is what appears at first sight to be a surprisingly large degree of differentiation within the "remainder" schools, but this is not so surprising when one bears in mind the large proportion of an age group which is generally accommodated in schools of this type in countries favoring a selective plus remainder organization of secondary education.

The mean scores in the variable defined as "interest in mathematics" are shown in Table 3.29.

In neither population does the extent of differentiation within schools

---

[1] From information received after this table was printed it is clear that the Netherlands should not have been included.

TABLE 3.28. *Number of Students Classified by Type of School and Degree of Differentiation of Courses.*

Population 1 a.

Type of School	Degree of Differentiation			Totals
	Much	Some	None	
Selective	458	366	140	964
Comprehensive	2,085	792	351	3,228
Remainder	948	163	77	1,188
Totals	3,491	1,321	568	5,380

Population 1 b.

Type of School	Degree of Differentiation			Totals
	Much	Some	None	
Selective	518	411	208	1,137
Comprehensive	2,485	1,281	686	4,452
Remainder	957	219	109	1,285
Totals	3,960	1,911	1,003	6,874

make a statistically significant difference in the mean scores. Therefore we can concentrate our attention on the variable "type of school".

In both populations the mean score made by students in selective schools is significantly higher than that made by students in comprehensive schools or by students in the "remainder" schools. In Population 1 b the mean score made by students in comprehensive schools is significantly higher than that made by students in "remainder" schools, but in Population 1 a the difference is not statistically significant.

Since the presence of selective schools necessitates the provision of "remainder" schools, a fairer comparison, where the country as a whole is concerned, is to combine the scores of the students in the selective schools in that country with those of students in "remainder" schools and compare the combined score with that made by students in comprehensive schools. In Population 1 a the respective scores are 5.69 and 5.66, and in Population 1 b, 5.80 and 5.84, neither difference being significant.

Therefore, we conclude that interest in mathematics is a little stronger in the selective type of school and a little weaker in the type of school providing for the remainder of the students after selection has operated,

TABLE 3.29. *Mean Scores on "Interest in Mathematics" for Different Types of School and Degrees of Differentiation.*

Population 1 a.

Type of School	Degree of Differentiation			Mean Score
	Much	Some	None	
Selective	5.93	5.89	5.94	5.92
Comprehensive	5.63	5.69	5.73	5.66
Remainder	5.52	5.15	5.99	5.50
Mean Score	5.64	5.68	5.82	5.67

Population 1 b.

Type of School	Degree of Differentiation			Mean Score
	Much	Some	None	
Selective	5.96	6.07	6.53	6.10
Comprehensive	5.88	5.82	5.73	5.84
Remainder	5.53	5.15	6.26	5.53
Mean Score	5.81	5.80	5.95	5.83

with comprehensive schools in an intermediate position. When the comprehensive system is compared with the alternative system comprising selective schools and remainder schools, there is no measurable difference in interest in mathematics between the two systems.

The hypothesis deals not only with schools but with countries. When the data were analysed from this point of view a different result was obtained. The emphasis within each country on comprehensive and nonselective secondary education was assessed by calculating the percentage of students in each of the populations who were in comprehensive schools. The mean score in "interest in mathematics" was then calculated for each country (Table 3.30). The table also shows the percentage of students in schools in which no differentiation was practiced.

At first sight there would seem to be almost an exact correspondence between the rank orders in columns 2 and 4, only England being out of place, but it must be borne in mind that some of the differences between the interest scores are not statistically significant. The six countries in Population 1 a fall into three groups, the mean score in the United States being significantly higher than that in any other country, and those in Scotland and the Netherlands being significantly below

TABLE 3.30. *Relation Between Interest Score and Emphasis Within a Country on Comprehensive and Undifferentiated Education.*

Population 1 a.

Country	Percentage of Students in		Mean Interest Score
	Comprehensive Schools	Schools with No Differentiation	
U.S.	92	14	6.17
Australia	70	9	5.93
Sweden	64	31	5.84
The Netherlands	—*	24	5.39
Scotland	44	1	5.27
England	9	3	5.73

Population 1 b.

Country	Percentage of Students in		Mean Interest Score
	Comprehensive Schools	Schools with No Differentiation	
Israel	91	39	6.79
U.S.	91	14	6.10
Australia	72	9	6.07
Sweden	65	31	5.83
The Netherlands	—*	22	5.55
Scotland	46	1	5.17
England	6	2	5.75

* See footnote on page 103.

those in the other countries. In Population 1 b the groups were (1) Israel; (2) United States, Australia, Sweden, and England; (3) the Netherlands; and (4) Scotland. Even with this caution on the interpretation of the rank orders, the evidence strongly suggests that, in countries which emphasized comprehensive and nonselective secondary education, interest in mathematics was greater. On the other hand, there is no relation between interest in mathematics and proportion of students in schools showing no differentiation.

Conclusion

Interest in mathematics is strongest in the selective schools, less strong in the comprehensive schools, and weakest in the schools for the remainder of the students after selection has operated. There is no measurable difference in interest in mathematics between the comprehensive system and the combined system of selective schools and remainder schools.

The countries with the greatest emphasis on comprehensive education are those with the greatest interest in mathematics.

Socio-Economic Status in Selective Systems

The hypothesis to be tested here is: *There will be a systematic difference in socio-economic status between students in Populations 1 and those in Populations 3* (Hypothesis 08).

Introduction

In what is now sometimes called "the educative society" there is an increasingly marked correlation between the amount of formal education an individual has received and the level of occupational qualification he reaches. If occupations are put in a hierarchical order, which is done when we order them according to some kind of socio-economic scale, we find that the occupational structure in highly developed countries is shaped more like an egg than a pyramid. At the bottom there is a decreasing number of occupations that require little formal schooling and practically no vocational training. In the middle there is an increasing proportion of occupations requiring a formal education extended to the ages of 16 to 18 and a specialized vocational training. At the top, finally, the number of persons with professional and technological education increases rapidly.

This change in occupational structure is discernible in all highly industrialized countries and is reflected in the structure of their school systems and the enrollment at various levels. A modern economy demands mass education at what were earlier considered to be advanced levels. The ensuing educational explosion reflects this rapidly growing need for highly trained manpower and an increasing propensity of prosperous peoples to "consume" education. These new forces conflict, however, with a school organization and a curriculum designed for another type of economy, one which was static and associated with a rather rigid social structure.

Until recently, both the occupational status structure and the school systems in many countries, especially those in Europe, could be described as pyramids. At the bottom there was a majority of manual workers, unskilled or semiskilled. Most of these had a very modest formal education, as a rule in the compulsory elementary school, lasting from six to eight or nine years. The next level consisted mainly of white-collar workers, such as clerical and sales workers, supervisors in industry, and nurses. Their formal education as a rule lasted a few years more than the compulsory elementary schooling, and was given in most cases in some kind of middle school with graduation at the age of 16 or 17, which did not qualify for university entrance without additional secondary schooling. These middle schools were either separate establishments or consisted of the lower section of the preuniversity academic school and entrance was based upon competitive selection, such as the 11 + examinations in England. At the top level, a small percentage of the age group graduated from the preuniversity school and after university education was completed comprised the professional and technological specialists. Some executives and working proprietors without a university education are to be found in this stratum by virtue of the status of their occupations.

The school structure in several European countries, to some extent, mirrors the occupational and/or the class structure of the society (Husén 1966). Until the nineteenth century, formal schooling at the secondary level was provided on the whole only for the professionals. The program was academic; it consisted of liberal arts with an emphasis on the classical languages. In some countries the enrollment was limited mainly but not entirely to the upper strata of the society, nobility, leading merchants, and wealthy farmers. Even if the school in most cases was supposed to prepare directly for entrance to university, it also gave a certain preparation for entry into commerce and administration.

When elementary education, consisting mainly of the three R's and some orientation in scripture, science, history, and civics was made compulsory in the nineteenth century, it was designed for the populace and not for the classes which had previously entered academic school. As a rule, several grades in the compulsory school ran parallel with the preuniversity school; in some cases the parallelism was complete, and children from more privileged homes attended private preparatory schools. During the present century the parallelism has been confined mainly to the age range from 10 or 11 and upward. The pattern in Germany has been that the favored children transferred after the fourth grade of the elementary school (*Volksschule*) to the academic secondary school (the nine-year *Gymnasium*). Those in England or France transferred at the age of 11, that is, after five or six years in the elementary school, to the grammar school and *lycée* respectively. These secondary schools during recent decades have been characterized by competitive entrance and selectivity on the basis of examinations. Thus, for example, in England and Germany roughly only one fifth of an age cohort has until recently been admitted to the secondary academic schools.

In theory, the selective schools are supposed to admit and promote their students solely on the basis of ability and not on the basis of social background. It has been assumed until recently that, if tuition is free, equality of opportunity is operating adequately. A comprehensive body of research, however, has shown that criteria of selection are loaded in varying degrees with social factors, such as parental education and income and geographical accessibility of the school (Husén, 1965). There is evidence to show beyond any doubt that in the selection procedures for academic secondary education and in the screening-out of pupils during the course of study there are built-in mechanisms which handicap children from a less privileged social background. In a society for which an effective utilization of the pool of ability is of prime concern for the individual and for the economy, a school structure which does not promote the ability where it is to be found, irrespective of home background, will have to be revised. The demand for mass education, especially for an expansion of university enrollment, makes it imperative that built-in barriers which prevent ability from being utilized should be removed (compare Halsey, 1961).

The earlier a selection is made, and the earlier parents have to decide whether their children are to transfer to the academic secondary school, the more intensively social selection may be expected to operate. A child at the age of 10 or 11 cannot reasonably be involved in its future educational and vocational career to the same extent as a young person of 15 or 16. Therefore, it can be expected that in countries where transfer to secondary (selective) education takes place at an early age the social structure of the students in these schools will differ more from that of the general population than in countries where transfer takes place later or where education during the compulsory school period is comprehensive. On the other hand, home influence is cumulative, and it could be said that later transfer would only increase the differential effect of social factors.

The students in this study indicated in the student questionnaire their fathers' and mothers' occupations as well as their education. The occupations were coded in nine groups according to a categorization scheme developed by Professor C. Arnold Anderson (see Chapter 8 of Volume I). Group 1 consisted of higher professional and technical occupations. Group 9 comprised unskilled manual workers (excluding agriculture, forestry, and fishing, which were in Group 8). Group 0 includes those unclassifiable. This categorization is not a scale in the metric sense. As one moves from category 1 to 9, the amount of formal education and vocational training decreases. However, neither the prestige status nor the economic level of the various occupations display exactly the same order. Thus, for example, proprietors and managers in agriculture, forestry, and fishing were allocated to Group 5 below small working proprietors and technical personnel such as supervisors in industry. One might therefore have doubts about the meaningfulness of calculating the means or standard deviations of occupational distributions. Most of the results presented below are given as percentage distributions.

Results

The percentages of students in each occupational category for Populations 1 a, 3 a, and 3 b in each country are given in Table 3.31. Australia and Israel have been omitted since Population 3 b was not tested in these countries, and it would not be possible to judge all preuniversity students from those enrolled only in the mathematics-science program (that is, Population 3 a).

The distribution of parental occupations for Population 1a can be regarded as the best available national estimate of the occupational distribution of a given country, since all the students at the age of 13 are in full-time schooling. The percentage of occupations belonging to a given group or combination of groups can therefore be used as a base figure.

Points to note in Table 3.31 are the fairly close agreements between countries in Population 1 a, except in categories 5 and 7, in the latter of which the percentages range from 62 for England to 24 for Japan. Similarly, there is a fairly close agreement within countries between the 3 a and 3 b percentages. The most striking feature both within countries and across all countries is the differences in distribution between the 1 a population and the 3 a and 3 b populations. These differences have been highlighted by the use of a retentivity index which has been derived by dividing the average of the percentages for a given category in Populations 3 a and 3 b by the percentage for that category in Population 1 a, and multiplying the ratio by 100. Thus, one obtains an expression for the relative chances of students with a given occupational background reaching the last grade of the preuniversity school. The indices are given in Table 3.32.

There is a very high degree of association between this index of retentivity and the emphasis within a country on comprehensive educa-

3.31. *Percentages of Students by Country, Population, and Occupational Category of Parent.*

		Occupational Category									Total	
	Population	1	2	3	4	5	6	7	8	9	0	Percent	N.
	1a	4	4	9	12	5	18	40	0	7	1	100	1,656
	3a	10	7	16	15	5	19	25	0	3	0	100	516
	3b	17	8	11	14	6	23	17	0	3	1	100	993
	1a	4	1	8	9	2	8	62	2	1	3	100	2,899
	3a	15	12	21	13	2	15	20	1	1	0	100	960
	3b	20	15	19	14	2	10	17	1	1	1	100	1,763
	1a	5	4	14	5	25	2	40	1	4	0	100	743
	3a	12	14	22	3	26	1	17	0	4	1	100	367
	3b	6	17	15	7	23	1	27	0	3	1	100	395
	1a	2	2	11	9	7	7	53	1	4	4	100	2,292
	3a	17	6	19	17	11	7	18	1	1	3	100	213
y	1a	4	5	8	9	9	14	42	1	5	3	100	4,318
	3a	33	19	15	5	4	17	6	0	0	1	100	636
	3b	37	17	10	9	5	14	5	0	0	3	100	622
	1a	3	10	6	16	24	10	24	2	2	3	100	1,969
	3a	8	25	6	19	15	10	10	1	1	5	100	772
	3b	4	16	5	21	25	12	13	0	1	3	100	4,209
herlands	1a	4	7	8	6	11	13	43	5	2	1	100	423
	3a	16	20	17	9	7	17	12	2	0	0	100	116
	1a	5	2	8	4	3	6	61	2	4	5	100	4,972
	3a	21	8	20	10	3	10	24	1	2	1	100	1,401
	3b	13	9	13	9	3	10	33	3	3	4	100	2,017
	1a	3	5	13	7	16	4	43	4	2	3	100	2,458
	3a	17	12	22	12	6	5	21	1	1	3	100	748
	3b	15	11	25	10	4	7	19	1	2	6	100	207
tates	1a	8	7	10	6	5	12	43	2	1	6	100	5,806
	3a	15	4	17	8	12	14	27	0	3	0	100	1,525
	3b	10	6	13	7	6	8	43	0	2	5	100	1,920
tries	1a	4	5	10	8	11	9	45	2	3	3	100	
	3a	16	13	17	11	9	12	18	1	2	1	100	
	3b	15	12	14	11	9	11	22	1	2	3	100	
	1a	6	9	8	12	23	16	38	5	6			
	3a	25	21	24	16	24	18	21	2	4			
	3b	33	11	20	14	23	22	38	3	3			

(occupational categories) 1 = higher professional and technical, 2 = administrators, executives, and proprietors, large and medium scale, 3 = sub-professional, technical, 4 = small working proprietors an agriculture, forestry, and fishing), 5 = proprietors and managers in agriculture, forestry, and 5 = clerical and sales workers, 7 = manual workers, skilled and semi-skilled, 8 = laborers in agri- forestry and fishing, 9 = unskilled manual workers (excluding agriculture, forestry and fishing).

tion. For a measure of this emphasis we use the percentage of comprehensive schools given in Chapter 14 of Volume 1 and repeated in column 2 of Table 3.32.

The occupational groups in Table 3.31 have been combined in Table 3.32 in order to exemplify the usual division into upper, middle, and lower class. Thus, groups 1 and 2 make up the upper class; groups 3, 4, and 6, the middle; and groups 7 and 9, the lower. In order to account for differences between countries in the distribution of urban and rural inhabitants, groups 5 and 8 (farm proprietors and farm laborers) have been kept separate. Moreover, farm proprietors do not fit readily into a status order.

In the first column of Table 3.32 the eight countries testing students in the preuniversity year have been ranked according to the percentage of the total age group which is to be found in this grade. Columns 3 to 6 give the retentivity indices for the four composite groups mentioned above. Rank-order correlations have been computed between the first and each of the other columns. The rank correlations between the retention in terms of the proportion of the total age group being brought up to the preuniversity year on the one hand and the likelihoods of students coming from upper-class and lower-class homes, as measured by the indices used in Table 3.32, are high, being -0.70 and $+0.62$, respectively. Thus, one is justified in making the generalization that the more retentive a school system is, the less selective it is from a social point of view. On the other hand, the range of indices in column 3 for the percentages 10 to 15 in column 1 is very large, i.e., a low degree of retention can cover very different degrees of social selectivity in different countries. We must also bear in mind that, as retention approaches 100 per cent, social selectivity must disappear, since almost the whole age group is at school.

It should also be noticed that there is a striking difference between countries with a comprehensive system, where the students are kept within the same school type until the ages of 15 to 18, and the dualistic school systems, where the students at the age of about 10 or 11 are competitively selected for academic secondary education. In the latter type there is a strong predominance of upper-class students in the last grade of the preuniversity school. The assumption made above that an earlier selection would result in social bias is also confirmed by the findings reported here. Selectivity generally goes with a partition into different school types or programs at an early age.

Table 3.31 gives the percentage of students from each occupational category for Populations 1 a, 3 a, and 3 b in each country for which

TABLE 3.32. *Relationship Between Retentivity and Social Background of Enrollment at the Preuniversity Level (Populations 3a and 3b Combined).*

	(1)	(2)	(3)	(4)	(5)	(6)
				Retentivity Indices		
Country[a]	Percent of total age cohort in pre-university year	Percent of sample schools which are comprehensive	Professionals, high technical executives (groups 1 and 2)	Middle-class sub-professionals, clerks, working, proprietors, etc. (groups 3, 4 and 6)	Farm proprietors and farm laborers (groups 5 and 8)	Working class skilled, semi skilled, and unskilled (groups 7 and 9)
United States	70	94	117	120	129	85
Japan	57	100	204	114	79	48
Sweden	23	79	344	169	30	48
Finland	18	45	364	200	100	48
England	14	0	272	117	94	58
Belgium	13	0	263	126	110	51
Scotland	12	20	620	184	75	31
Germany	11	0	589	113	45	12
Rank Correlation With Column 1		0.86	−0.76	0.05	0.36	0.62

[a] Since adequate data were not available for Population 3b in Australia, France, Israel and the Netherlands, these countries have been omitted from this Table.

information was available. A second way of calculating the amount of social bias is to compare the proportions in low categories to those in high category groups at the 13-year-old level (the distribution of fathers' occupations for all 13-year-olds being regarded as typical of the school-going population), and then to compare this with the proportions in the same categories in the preuniversity year. We have chosen groups 1 (higher professional and technical) and 2 (administrators, executives, and working proprietors, large and medium scale) to represent the high categories, and groups 7 (manual workers; skilled and semiskilled) and 9 (unskilled manual workers, excluding agriculture, forestry, and fishing) to represent the low categories.

The following represents the initial calculation for each country, the letters $a, b, \ldots f$ representing the appropriate percentages.

	Groups 1 and 2	Groups 7 and 9
Population 1a	a	b
Population 3a	c	d
Population 3b	e	f

TABLE 3.33. *Indices of Social Bias.*

	3 a	3 b
Australia	4.7	—
Belgium	3.6	7.3
England	16.2	24.5
Finland	6.0	3.7
France	17.3	—
Germany	45.3	56.4
Israel	3.6	—
Japan	6.0	2.9
The Netherlands	12.3	—
Scotland	10.4	5.7
Sweden	2.1	7.0
U.S.	1.9	1.0

The index of bias for Population 3 a is, therefore, $(bc)/(ad)$ and for 3 b $(be)/(af)$.

The greater the size of the index, the more biased is that population in terms of the higher occupational categories (Table 3.33).

Tables 3.32 and 3.33 suggest a strong correlation between retentivity in general on the one hand (in terms of the proportion of an age group enrolled in the last grade of the preuniversity school) and social class composition of enrollment on the other hand. Thus, a social class bias seems to be inherent in the less retentive selective systems. From Table 3.33, however, it is interesting to note the relative biases for Populations 3 a and 3 b. In Belgium, England, Germany, and Sweden there are proportionately more lower category pupils in the 3 a (mathematics) population than in 3 b (nonmathematics). The opposite situation prevails in Finland, Japan, Scotland, and the United States.

There are obviously very few working-class students getting through to the preuniversity year in Germany, where the index is at least twice as large as in any other country. In England and France the bias is also quite high. It is interesting to note that there is little change in the proportion of pupils in low and high categories between Populations 1 a and 3 b in the United States.

If the rank-order of social bias for each of the populations is correlated with the percentage of an age group for each country in each of the populations, it becomes apparent that the lower the percentage of an age group in the preuniversity year, the greater the social bias. The correlation in Population 3 a is 0.66, and in 3 b is 0.78.

There is, therefore, bias both in the mathematics group and in the nonmathematics group. It is reasonable to expect that it would be stronger in the latter program, which in many cases represents the "classical" program.

It would be interesting to discover to what extent general selectivity is correlated with the age at which students are selected. It seems reasonable to expect that students of 10 to 12 years of age could not be committed to the same extent to their future educational and vocational careers as are students of 16 to 18. Systems with an early selection for academic secondary education might therefore show a higher degree of social bias than systems in which the students are kept together in roughly the same program until the age of 15 or later. West Germany, with the age of 10, has the earliest selection. Then come England, Finland, and France with age 11. Scotland, Belgium, Israel, and the Netherlands select at age 12, Japan at age 15, Sweden at age 16, and the United States, finally, at the ages of 17 to 18. The rank correlations between this variable and social bias are -0.84 for Population 3 a and -0.70 for Population 3 b.

Social bias to a very high extent goes together with age of selection. The earlier selection takes place, the stronger the bias.

Conclusion

The general hypothesis that there is to a greater or lesser extent a difference in social class composition between students at the 13-year-old level and those in the preuniversity year is confirmed for all countries except for United States students enrolled in the preuniversity nonmathematics program. These results confirm previous studies within particular nations. The social bias is in favor of the middle classes and against the working and farming classes, but it varies from country to country and between students enrolled in terminal mathematics and nonmathematics programs. The bias is stronger in the mathematics program in Finland, Japan, Scotland, and the United States. The reverse is true for Belgium, England, Germany and Sweden.

It was further hypothesized that social class bias is greatest in these countries which have the smallest proportion of an age group reaching the last grade of the academic preuniversity school. This proved to be true.

Finally, it was hypothesized that social bias would be related to the age at which selection occurred. It was found that this was true for both mathematics and nonmathematics program students.

Mathematical Achievement and School Retentivity

The hypotheses in this part deal mainly with the differences between countries in the proportions of students who continue at school to the preuniversity stage. In the grade in which 13-year-olds are normally to be found in the various systems of education represented in this study, there is still approximately 100 percent of the age group in school. After this point drop-out begins, and the rate varies greatly among the systems, as can be seen from Chapter 13 in Volume I. By the preuniversity year the proportion of students leaving school has been so different that, while 70 percent of an age group are still in school in the United States, in the Netherlands only 8 percent remain. Data were collected about the actual number of students in each year group still in full-time schooling, as well as the actual number of students in each grade group. The heads of the national centers were asked to estimate (1) the percentage of an age group in school at the preuniversity level and (2) the proportion who were specializing in mathematics (enrolled in terminal mathematics-science).[1] It would seem that in some national centers approximations were made to the nearest whole number whereas in others the proportion was calculated to the first decimal place. Despite these differences, the figures supplied are used in this analysis.

TABLE 3.34. *Percentage of an Age Group in the Preuniversity Year Enrolled in Terminal Mathematics-Science and in Full-Time Schooling.*

Country	Percentage of Age Group in Terminal Mathematics Programs	Percentage of Age Group in Terminal Nonmathematics Programs	Percentage of Age Group Enrolled in Full-Time Schooling
United States	18	52	70
Japan	8	49	57
Sweden	16	7	23
Australia	14	9	23
Scotland	5.4	12.6	18
Finland	7	7	14
Belgium	4	9	13
England	5	7	12
Germany	4.7	6.5	11
France	5	6	11
The Netherlands	5	3	8
Israel	7	—	—

[1] The operational distinctions in each country between the mathematics and nonmathematics students are given in Chapter 13 of Volume I.

TABLE 3.35. *Indices of Retentivity, Income, Industrialization, and Comprehensive Education.*

Country	Retentivity (Percentages of Age Group)			Income (Dollars)	Industrialization (Kilos)	Comprehensiveness (Percentages of Age Group)
	Total	3 a	3 b			
Australia	23	14	9	1,445	4,213	70
Belgium	13	4	9	1,224	4,668	0
England	12	5	7	1,280+	5,090+	9
Finland	14	7	7	1,081	2,072	0
France	11	5	—	1,282	2,788	0
Germany	11	4.7	6.5	1,322	4,121	0
Israel	—	7	—	1,002	1,473	96
Japan	57	8	49	524	1,532	100
The Netherlands	8	5	3	1,036	2,896	0
Scotland	18	5.4	12.6	1,280−	5,090−	44
Sweden	23	16	7	1,817	3,950	64
United States	70	18	52	2,652	8,507	92

This property of retaining students has been named retentivity. It has been referred to in Chapter 13 of Volume I, where the term used was attrition, and was used for hypothesis 08, where the percentages of the age group or cohort were shown in Table 3.32. For convenience of reference the figures are repeated in Table 3.34, in which there are also given the percentages of the age group in the mathematics section (Population 3 a) and the nonmathematics section (Population 3 b), respectively. These percentages we use as indices of retentivity.

It might be thought that the retentivity of a country would be strongly related to its general affluence or to the degree of industrialization in it. The first of these two has been measured by the average income per head as given by the United Nations *Statistical Year Book* for 1964. The second has been measured by the energy production and consumption per head as recorded in the same volume, and both are shown in Table 3.35.

A third measure which may be associated with retentivity is the degree to which each country has adopted a comprehensive system of education. This has been assessed by the percentage of students in the younger and complete age group (Population 1 a) attending schools of this type. This measure is shown in the last column of Table 3.35, and rank correlations have been calculated between the three measures of reten-

TABLE 3.36. *Mean Mathematics Score and Percent of Age Group in Population for Populations 3a and 3b.*

Country	Population 3 a		Population 3 b	
	Mean Math. Score	Percent of Age Group in Population (3 a)	Mean Math. Score	Percent of Age Group in Population (3 b)
Australia	21.6	14	—	—
Belgium	34.6	4	24.2	9
England	35.2	5	21.4	7
Finland	25.3	7	22.5	7
France	33.4	5	—	—
Germany	28.8	4.7	27.7	6.5
Israel	36.4	7	—	—
Japan	31.4	8	25.3	49
The Netherlands	31.9	5	—	—
Scotland	25.5	5.4	20.7	12.6
Sweden	27.3	16	12.6	7
United States	13.8	18	8.3	52
Rank Correlation	−0.62		−0.36	

tivity, i.e., those for the total preuniversity population and for Populations 3 a and 3 b, and each of the other variables.

The United Nations data show England and Scotland under one heading, "United Kingdom". On internal evidence, England has been placed slightly above the United Kingdom average, Scotland slightly below, and France has been given a place between the two on the income scale.

The rank correlations of each of the three retentivity indices with income are 0.32 (total), 0.31 (3 a), and 0.10 (3 b), respectively. With the index of industrialization they are 0.17, −0.03, and 0.30. With the extent to which pupils are being educated in comprehensive schools, they are 0.89, 0.76, and 0,73, respectively. There is, therefore, a much stronger association between retentivity and the provision of comprehensive schools than there is between retentivity and either average income or degree of industrialization.[1]

[1] The age at which compulsion to attend school ceases might also be expected to be associated with retentivity. These ages are given in Table 13.1 of Volume I. The rank correlations with the three retentivity indices are 0.25 (total), 0.36 (3 a) and 0.16 (3 b).

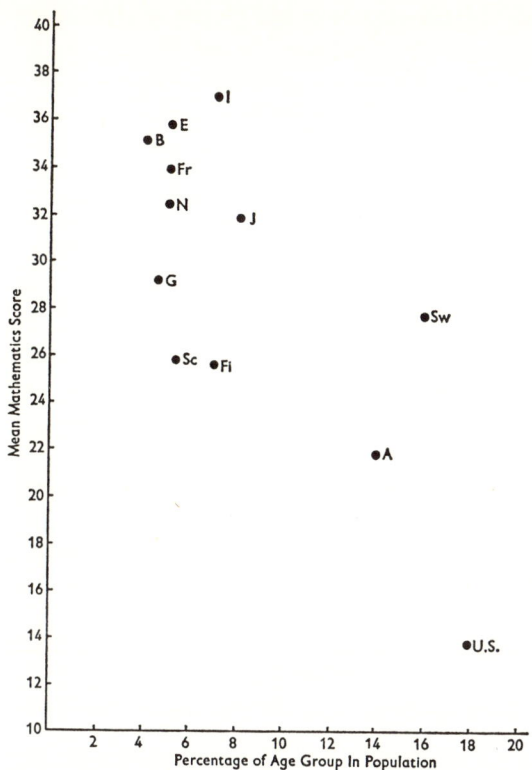

Figure 3.5. Relation of mathematics score to percentage of age group in population, by country (Population 3 a).

Does More Mean Worse?

The first hypothesis to be tested states: *The average level of mathematics achievement in both of the terminal groups will be lower in countries with larger percentages of the age group still in school* (Hypothesis 09).

The data for testing the hypothesis are shown in Table 3.36.

The fall in score as the percentage of the age group retained increases is clearly discernible in both populations (see Figures 3.5 and 3.6). The rank-order correlations between the mean score and the percentages of the age group in each population are −.62 and −.36 for Populations 3 a and 3 b, respectively. The difference in correlations might be expected, since the effect on mathematics score is likely to be less where the additional students are in the nonmathematical section of the preuniversity population.

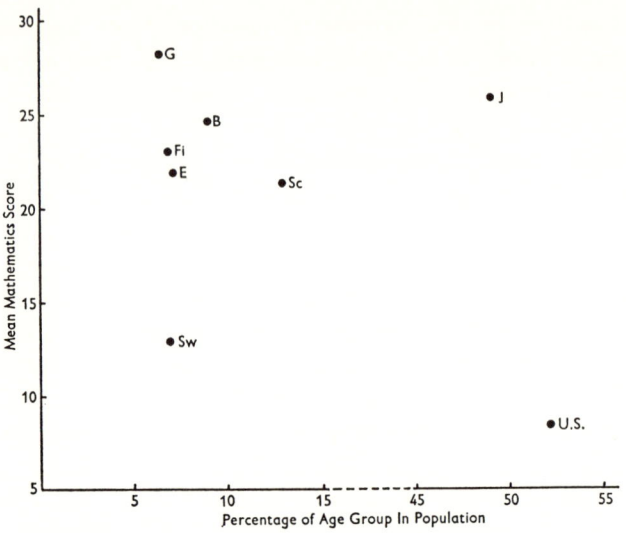

Figure 3.6. Relation of mathematics score to percentage of age group in population, by country (Population 3 b).

CONCLUSION

These results confirm the hypothesis. In this sense, "more means worse"; countries which retain larger percentages of an age group to the preuniversity stage produce *on average* lower standards of achievement than do those countries retaining smaller percentages.

Does More Mean Worse at All Levels?

INTRODUCTION

The above results immediately raise the question: Is this a wide-spread deterioration in achievement or is it attributable to the fact that the additional students are not of the same caliber as the smaller groups? The hypothesis now to be dealt with states: *When equal proportions of the total age group are compared, countries will not differ in the terminal level of mathematics achievement* (Hypothesis 10). It is, in fact, a statement that the lowering in achievement is not a decline in the achievements of the best students, but rather an addition to their number of a group of inferior students.

Subgroups of students were selected from each country's total sample to represent a constant proportion of the age group. The country with the lowest proportion of an age group in a terminal population determined the proportion of the age group which would be selected from all the countries in that terminal population. Since Belgium had the lowest percent (four) of its age group enrolled in terminal mathematics programs, 4 percent of the age group was sampled in each country in Population 3 a. Similarly, the Netherlands, with 3 percent of its age group enrolled in terminal nonmathematics programs, determined the proportion of the age group to be sampled in each country in Population 3 b.[1]

The procedure for the selection of students was accomplished by selecting students from the top part of each mathematics score distribution for each country so as to represent a constant proportion of each country's corresponding age group. In the case of Belgium at the terminal mathematics level, this involved the entire score distribution. In the other countries, this was determined by dividing the percent of the age group to be selected, for example, 4 percent at the 3 a level, by the percent of the age group enrolled in full-time schooling at that level. The resulting figure was the proportion of cases in the mathematics score distribution to be used in the comparison.

This procedure was carried out country by country for each of the two terminal populations. Means and standard deviations for the selected groups were then computed for each country in each population.

Objections might be raised to this procedure on the grounds that it gives an advantage to the countries with greater retentivity. No selection procedure has complete validity, and a country with 4 percent in its terminal population will certainly not have the best 4 percent, although it is likely that most of the students will be correctly placed in that group. Our procedure of waiting until after the test to pick the best 4 percent from the countries with greater retentivity will therefore produce a higher scoring group than would otherwise be the case. (There is also the question of errors of measurement; the true scores of none of those students are known.)

The reply to the first of these objections is that our hypothesis refers to the *actual* products of each country on the tests. From its actual 4 percent, the country with lower retentivity produces a certain level of score. From its larger 14, 16, or 18 percent, another country can extract a best-scoring subgroup of 4 percent producing another mean score. How do these levels compare? If they are approximately the same, more does not mean worse as far as the top level of mathematical achievement is concerned.

---

[1] Later, the Netherlands data for Population 3 b were rejected but the cut-off point of the top 3 percent was kept.

RESULTS

The results for the 3 a population from each country are presented in Table 3.37 and Figure 3.7. The means for the total or intact samples range from 13.8 for the United States to 36.4 for Israel. This range of 22.6 score points is statistically significant. The results for the selected groups, which are representative of the upper 4 percent of the *age group,* are quite different. The mean mathematics score ranges from 29.4 for Scotland to 43.9 for Japan, a range of 14.5. The means for the selected groups are therefore much less variable than those for the intact samples; this finding is shown diagrammatically in Figure 3.7. However, the reduction in variability from the intact samples to the selected groups is not great enough to justify acceptance of the hypothesis that the countries would not differ in the level of mathematics achievement at the end of secondary school when equal proportions of the age group are compared.

It is interesting to note the changes which occur as one moves from the total sample to the selected group representing the upper 4 percent of the age group. The rank correlation between the two sets of means is +.45. This indicates that one can predict only moderately well the

TABLE 3.37. *Means and Standard Deviations of Mathematics Test Scores for Total Sample and Equivalent Proportion of the Age Group in Each Country at the Terminal Mathematics Level.*

Country	All Terminal Mathematics Students			Selected Terminal Mathematics Students Representing Upper Four Percent of Age Group		
	M.	S.D.	N.	M.	S.D.	N.
Australia	21.6	10.5	1,089	33.7	6.6	350
Belgium	34.6	12.6	519	34.6	12.6	519
England	35.2	12.6	967	39.4	9.8	783
Finland	25.3	9.6	369	32.1	6.5	211
France	33.4	10.8	212	37.0	8.0	174
Germany	28.8	9.8	649	31.5	7.8	557
Israel	36.4	8.6	146	41.7	4.4	89
Japan	31.4	14.8	818	43.9	7.7	413
The Netherlands	31.9	8.1	462	34.7	6.2	373
Scotland	25.5	10.4	1,422	29.4	9.8	1,048
Sweden	27.3	11.9	776	43.7	6.2	177
United States	13.8	12.6	1,568	33.0	8.9	345
Range	22.6			14.5		

mathematics performance of the upper 4 percent of the *age group in each country* from the mathematics performance of the students in full-time schooling. This is, of course, due to the fact that the countries exercise different policies with regard to selection of students for retention in school. In examining the reasons for the moderateness of the correlation between the two sets of means, one notes a number of interesting shifts in relative position among countries as one moves from the intact sample to the selected sample for the *age group*. Downward shifts occur in Belgium, which moves from third to seventh place; Germany, which moves from seventh to eleventh place; and Scotland, which moves from ninth to last place. Upward shifts occur in Japan, which moves from sixth to first place; Sweden which moves from eighth to second place; and the United States, which moves from twelfth to ninth place. In general, those countries with the least restrictive policies as to who will continue in school show the greatest upward shifts, while those countries with stricter selection policies and practices show the greatest downward shifts in relative position.

The consequences of these shifts can be seen most clearly in Figure 3.7. When equal proportions are compared, the countries vary less in mathematics achievements. This would support the proposition that countries do *not* differ considerably in the proportions of students talented in mathematics, but that the differences in selection policies and practices cloud the picture. It is at least a feasible suggestion that the "cream" of mathematical talent is distributed equally over the various countries and that it is only the procedures for diluting the cream that vary from country to country.

These findings can have important implications for educational policy and practice. The view that the lowering of selection barriers would lead to a decline in achievements and, especially, a reduction in achievement among the cream of a nation's talent is questioned by these data. The results indicate that the most talented students continue to achieve at a high level, even when as much as 70 percent of the age group is enrolled in full-time schooling.

The performances of preuniversity students in the Japanese and Swedish systems are two excellent examples which refute the contention, and the performance of the students in the United States, while decidedly low for the total sample, compares quite favorably with most other countries when equal proportions of the age groups are compared. Thus, it would seem that a nation need not fear for its most talented students when it contemplates the expansion of educational opportunity at the secondary school level.

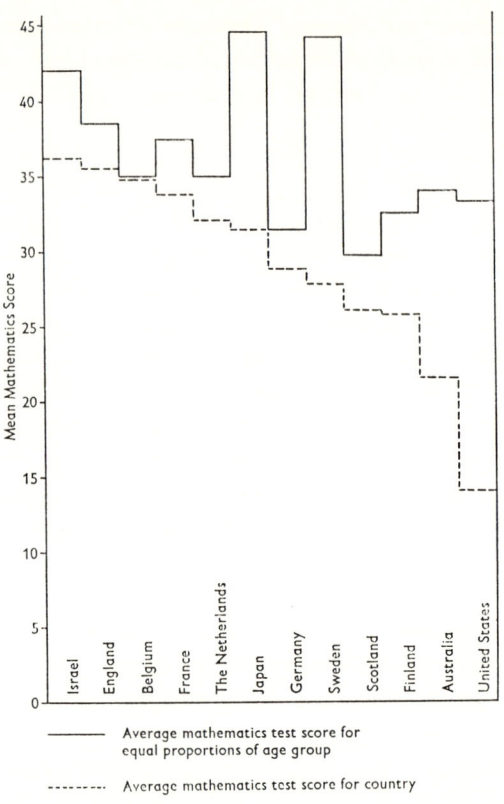

Figure 3.7. Mean mathematics test scores (1) for the total sample and (2) for equal proportions of the age group in each country for terminal mathematics population.

When one turns to the 3 b populations, the results are similar (Table 3.38 and Figure 3.8). The mean scores for the total 3 b sample range from 8.3 in the United States to 27.7 in the Federal Republic of Germany. For the selected group which represents the upper 3 percent of the age group in each country, the means range from 21.8 in Sweden to 34.9 in Belgium, with Japan's 51.7 in isolation. It can readily be seen that the shift from the total sample to the selected group results in a substantial increase in average performance and in reduced variability among countries. For the selected group, 6 of the 8 countries have means which lie between 29 and 35, a fairly narrow range for a 58-item test. The two major exceptions are Japan and Sweden. Japan is at the top both of the 3 a and the 3 b adjusted scores. Sweden is well above the average on the 3 a adjusted scores and well below the average on the 3 b adjusted scores. There would appear to be some mechanism in

TABLE 3.38. *Means and Standard Deviations of Mathematics Test Scores of Total Sample and Equivalent Proportion of the Age Group in Each Country at the Terminal Nonmathematics Level.*

Country	All Terminal Nonmathematics Students			Selected Terminal Nonmathematics Students Representing Upper Three Percent of Age Group		
	M.	S.D.	N.	M.	S.D.	N.
Belgium	24.2	9.5	1,004	34.9	4.9	254
England	21.4	10.0	1,782	30.2	6.0	796
Finland	22.5	8.3	399	29.9	5.6	181
Germany	27.7	7.6	643	34.2	4.0	301
Japan	25.3	14.3	4,372	51.7	2.2	272
Scotland	20.7	9.5	2,123	32.7	4.1	553
Sweden	12.6	6.2	222	21.8	3.8	40
United States	8.3	9.1	2,042	30.7	3.6	118
Range	19.4	(Excluding Japan)		13.1		

the Swedish system leading to a much greater differentiation than is found in other countries between the mathematical achievements of the best terminal mathematics students and those of the best terminal nonmathematics students. The total performance in Sweden matches that of other countries. For all 8 countries, the hypothesis that there will be little variability in mathematics performance when equal proportions of the age group in each country are compared is refuted. However, in 6 countries the hypothesis receives substantial support.

When one compares the performance of the total sample of fulltime students with the selected group representing the upper 3 percent of the age group for the various countries, a number of interesting shifts in relative position occur. Upward shifts occur in Belgium, which moves from fifth to second; Scotland, which moves from eighth to fifth; and the United States, which moves from tenth to sixth. These shifts are reflected in the rank-order correlation between the two sets of means, which is +.54. Again, this indicates a moderate relationship between the mathematics test performance for the total sample and the selected group representing the upper 3 percent of the age group in each country.

As in the case of the 3 a populations, the consequences for the more talented students seem to be slight when selection barriers are lowered. There is no evidence to suggest that a general deterioration sets in when the doors of the school are opened to a large proportion of the age

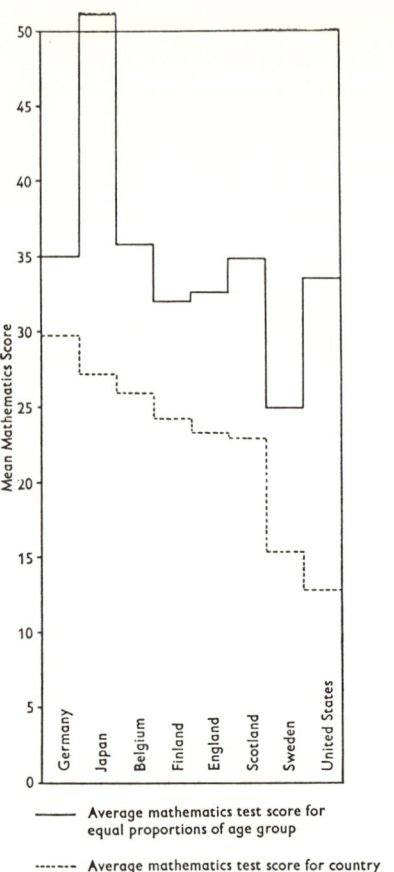

Figure 3.8. Mean mathematics test scores (1) for the total sample and (2) for equal proportions of the age group in each country for terminal nonmathematics population.

group (in the United States, over half of the age group is included in this population).

In addition to the findings relating to retentivity, important evidence relating to the issue of specialized versus comprehensive schools can be examined in this study. Table 3.39 sets forth the rank-order positions for the three countries in the study which have comprehensive systems of education, that is, Japan, Sweden, and the United States. For the sample representing equal proportions of the age group in the two terminal populations, the countries with comprehensive systems of education range from first ranked (at both of the terminal levels) to last place at the terminal nonmathematics level. One can only conclude

TABLE 3.39. *Rank Orders of Mean Mathematics Scores for Countries with Comprehensive School Systems.*

Country	Terminal Mathematics Population		Terminal Nonmathematics Population	
	Total Sample	Selected Group	Total Sample	Selected Group
Japan	6	1	3	1
Sweden	8	2	9	10
United States	12	9	10	6

from these data that there is no clear-cut relationship between the type of school organization and selection policies on the one hand and the educational achievement of the most talented students as indicated by mathematics test performance on the other. It would appear that proponents of any one type of school organization and selection policy have no reason to base their claims for their particular organizational pattern on increased achievement of the very able students under the two types of organization.

CONCLUSION

When equal proportions of an age group are compared, countries differ much less among themselves than they do when the proportions actually tested in Populations 3 a and 3 b are included in the reckoning. The range of mean scores is reduced to about 60 percent of its original value in Population 3 a and to about 67 percent of its value for Population 3 b. The reduction is not sufficiently large to warrant the statement that the remaining differences are negligible and that the top levels of achievement in all the countries in the study are identical. It does, however, suggest that the differences between countries in mean score found in both Population 3 a and 3 b are attributable, not to a general lowering of achievement throughout the population in countries retaining greater percentages of the age group to the preuniversity year, but to a preservation of a high level of achievement among the best students coupled with a dipping into lower levels of mathematical ability to provide the balance of the larger percentages retained. More in the school population does not mean worse, as far as the achievements of the best 3 or 4 percent of the whole age group are concerned.

Does More Mean Worse by International Standards?

INTRODUCTION

Another method of examining this problem is to fix a set of international standards and find what proportions of its preuniversity students each country has been able to bring to each of these standards. The advantage of this method is that it enables us to examine not only what is achieved by the best students but also the levels of achievement of the less able in each country. The hypothesis to be tested is: *In countries retaining larger proportions of an age group in school, higher levels of mathematical achievement will be attained by a smaller proportion of those still in school but by a larger proportion of the total age group* (Hypothesis 11).

The various levels of achievement have been defined for each population as the scores obtained by the top 5, 10, 15 percent, and so on, of all the students tested in all the countries. There are many differences among these preuniversity populations. There is a wide variation in the social class composition of this group, as was dealt with in some detail in the discussion of Hypothesis 08. A second major disparity is in the mean age which ranges from 17 years 6 months in Scotland to 19 years 10 months in the Federal Republic of Germany. A third variation is in average number of subjects studied; again, this ranges from 3 in England to 9 or more in Belgium, France, the Netherlands, and Japan.

With all these differences in mind, one might query whether it is justifiable to use combined distributions of scores from all countries as a base from which to derive percentiles for international comparisons. The reply would be that, whatever the national populations that contributed to produce them, the scores marked by the 95th, 90th, and 85th percentiles of the combined distributions denote fixed points which can be used for at least some comparisons. For example, the 95th percentile for Population 3 a is the score exceeded by only the best 5 per cent of the combined preuniversity populations for that level. If this 5 percent were composed of exactly 5 percent from each of the national preuniversity populations, we should conclude that, in this respect at least, all the participating countries were equal. If the 5 percent international elite is not so composed, the question arises whether the differences are attributable in part at least to the varying percentage of the age group still at school. This is the main point to be tested in this hypothesis. It is, however, of interest to consider the size of the elite in any country in relation to the age group from which it is drawn and the data have been analyzed from this point of view also.

RESULTS

Table 3.40 presents for each country the percentage of those students in Population 3 a reaching the various *international* percentile levels. For example, 36 percent of the 3 a population in the United States reached the 25th percentile levels, as compared with 97 percent of the 3 a population in the Netherlands. First decimal places have been added to some entries to increase the precision of the rank correlations. These rank-order correlations between the percentage of an age group in Population 3 a (that is, column 1 in Table 3.40) and the percentage of that population reaching each percentile level are shown in the last row of the table.

The negative correlations indicate that the smaller the proportion of the total age group taking the mathematics program at the preuniversity stage, the larger will be the proportions of those taking the program who reach given levels of performance. Thus, those who maintain that increasing the intake will lower the "standards" have a point, particularly in terms of the bottom half of those taken in. However, it is of interest that the effect at the upper end of the distribution is weaker. The between country ranges of percentages scoring above various international percentile points are very large, ranging from 61 percent at the 25th and 50th percentiles to 19.9 percent at the 95th percentile (see

TABLE 3.40. *Percentage of Preuniversity Mathematics Students Reaching Given Standards (Population 3a).*

Country	Retentivity	International Percentiles					
		25th	50th	75th	85th	90th	95th
Australia	14	67	37	10	5	3	1.1
Belgium	4	90	70	44	30	23	21.0
England	5	94	79	50	34	26	12.0
France	5	92	69	39	29.2	22	9.0
Germany	4.7	90	63	26	11	7	2.0
Finland	7	81	48	18	6	3.4	1.2
Japan	8	82	63	43	29.4	21	10.0
The Netherlands	5	97	77	35	14	5	1.3
Scotland	5.4	83	44	16	9	6	3.7
Sweden	16	81	53	26	13	8	3.1
United States	18	36	18	9	7	4.5	3.6
Range		61	61	41	29	23	19.9
Rank Correlation with Column 1		−.61	−.72	−.47	−.59	−.52	−.35

Table 3.40). Of those countries where only 4 or 5 percent of an age group are enrolled in the mathematics program, Belgium and England are outstanding, particularly in the top international quartile. It is remarkable that 21 percent of Belgian students achieve scores above the 95th percentile (as, for example, compared with 12 percent in England) when it is remembered that Belgian students are studying an average of six more subjects than English students. The Netherlands, on the other hand, has a high proportion of students up to the 50th international percentile, but a rapid fall then occurs. The United States is consistently lower than Sweden (except at the 95th percentile), whereas Japan is consistently higher than Scotland except at the 25th percentile.

If there were no relation between the degree of retention and the scores made by the pupils retained, we might expect that each country would have 5 percent of their 3a population above the 95th percentile, 10 percent above the 90th percentile, etc. It will be seen from Table 3.40 that this is not the case. Countries with a higher rate of retention bring less than 5 percent to the 95th percentile. Although, in general, the less the intake the better the performance, there are some interesting differences among countries with similar enrollments. Scotland, England, France, the Netherlands, Germany, and Belgium all have similar sizes of intake but differ considerably in the proportions of the enrollment they bring into the international top three percentile levels.

Although the suggestion that "more means worse" has been seen to have some justification, in particular in the bottom half of the distribution, it is more meaningful to see whether the size of the "elite" group (as a proportion of the total age group) can be increased by increasing the size of the intake. If the numbers reaching particular percentile levels are calculated as percentages of the *whole* age group, some differences may become apparent. These percentages are presented in Table 3.41.

The rank-order correlations between the percent of an age group enrolled in the mathematics-science program and the percent of the whole age group reaching various percentile levels are given in the last row of Table 3.41.

These positive correlations indicate that the higher the enrollment is as a percentage of the total age group, then the higher is the percentage of the whole age group reaching various international percentile levels. The greatest changes from Table 3.40 to Table 3.41 occur in Sweden, United States and Japan, all three countries with a more comprehensive system at the secondary level. Thus, it is possible to increase the size of the elite group (as a percentage of the total age

TABLE 3.41. *Percentage of Age Group Reaching Given Standards (Population 3a).*

Country	Retentivity	International Percentiles					
		25th	50th	75th	85th	90th	95th
Australia	14	9.4	5.2	1.4	.7	.42	.15
Belgium	4	3.6	2.8	1.8	1.2	.92	.84
England	5	4.7	3.9	2.5	1.7	1.30	.60
Finland	7	5.7	3.4	1.3	.4	.24	.08
France	5	4.6	3.4	1.9	1.5	1.10	.45
Germany	4.7	4.2	3.0	1.2	.5	.32	.09
Japan	8	6.6	5.0	3.4	2.3	1.68	.80
The Netherlands	5	4.8	3.8	1.7	.7	.25	.06
Scotland	5.4	4.5	2.4	.8	.5	.32	.19
Sweden	16	13.0	8.5	4.2	2.1	1.28	.50
United States	18	6.5	3.2	1.6	1.3	.81	.65
Range		9.4	6.1	3.4	1.9	1.44	.78
Rank Correlation with Column 1		+.89	+.55	+.15	+.25	+.14	+.10

group), but only to a small extent. The increase of mathematical "yield" in the bottom half of the distribution is considerable, but this is not surprising since many more students have been given the opportunity who did not previously have it. Again, the between-country range varies from 9.4 percent at the 25th percentile to .78 percent at the 95th percentile. The percentage of the whole group reaching particular international percentile levels is obviously a function of size of enrollment to a large degree at lower levels, although less so at the top levels. It is perhaps not without significance that students reaching the 99th percentile are drawn only from the United States, Sweden, and England (.18, .16, and .05 percents, respectively, of their respective total age groups).

If the "mathematical yield" of this population can be described as "how many get how far", Sweden, the United States, Australia, and Japan have the largest yields. It is possible to represent the "yield" diagrammatically by plotting the cumulative percentile frequencies for each country, as in Figure 3.9.

The convergence of the curves for the different countries at the upper end of the score scale shows that, in terms of the performance of the elite group (that is, the top 10 and 5 percent international group), the size of the enrollment has only a weak effect. It is Japan, Sweden, England, and Belgium which are performing well.

Similar information for Population 3b is given in Tables 3.42 and

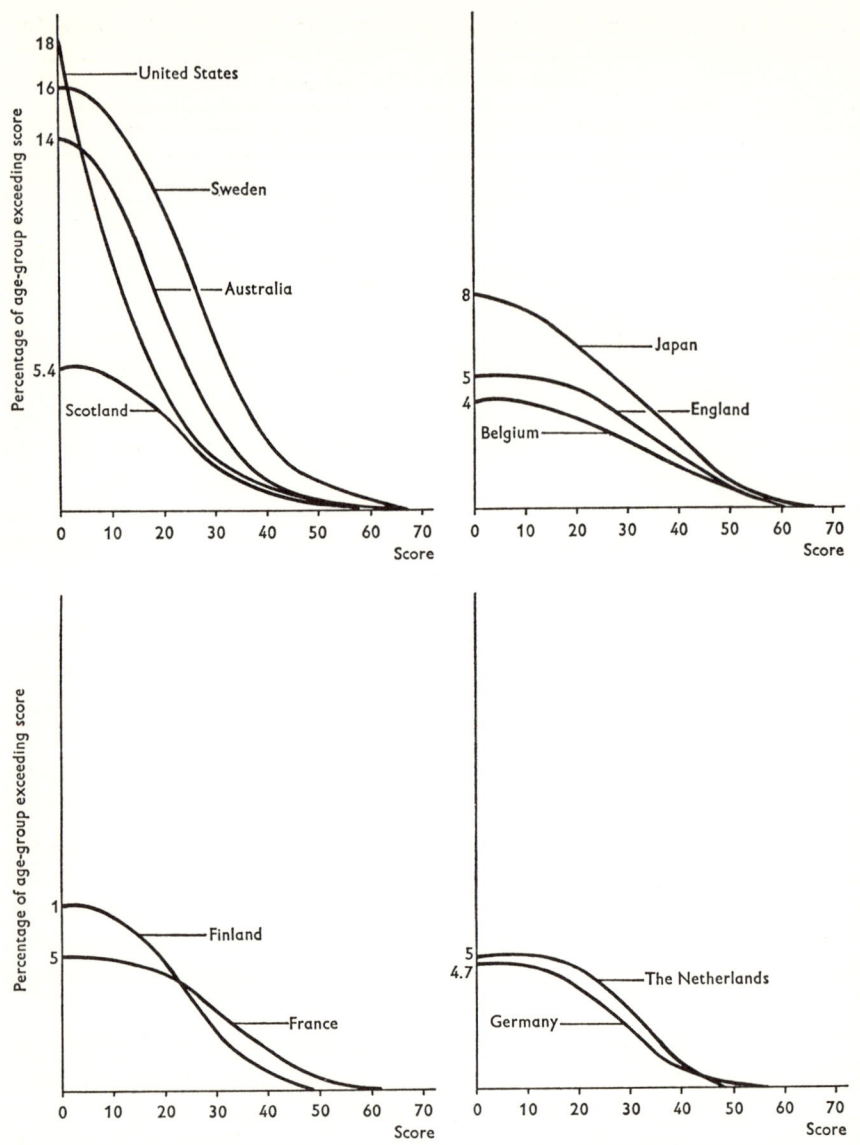

Figure 3.9. Cumulative percentile frequencies (smoothed) for Population 3 a.

3.43. The results agree closely with those obtained for Population 3 a. There is a negative relationship (except at the 95th percentile) between the percentage still at school and the percentage of that population reaching various international percentile levels. The small size of the negative correlations for the 75th, 85th, and 90th percentiles and the

TABLE 3.42. *Percentage of Preuniversity Nonmathematics Students Reaching Given Standards (Population 3b).*

Country	Retentivity	International Percentiles					
		25th	50th	75th	85th	90th	95th
Belgium	9	93	63	27	15	8	2
England	7	84	53	20	10	5	2
Finland	7	90	57	17	10	5	1
Germany	6.5	99	81	37	20	8	1
Japan	49	81	60	38	28	21	12
Scotland	12.6	82	50	18	7	3	1
Sweden	7	56	10	2	0	0	0
United States	52	30	12	3	2	1	1
Range		69	71	36	28	21	12
Rank Correlation with Column 1		−.99	−.28	−.02	−.09	−.06	+.38

TABLE 3.43. *Percentage of Age Group Reaching Given Standards (Population 3b).*

Country	Retentivity	International Percentiles					
		25th	50th	75th	85th	90th	95th
Belgium	9	8.4	5.7	2.4	1.3	.72	.18
England	7	5.9	3.7	1.4	.7	.35	.14
Finland	7	6.3	3.9	1.2	.7	.32	.07
Germany	6.5	6.4	5.3	2.4	1.3	.52	.06
Japan	49	39.7	29.4	18.6	13.7	10.3	5.9
Scotland	12.6	10.3	6.3	2.3	.88	.38	.13
Sweden	7	3.9	.7	.14	0	0	0
United States	52	15.2	6.2	1.6	1.0	.52	.52
Range		35.8	28.7	18.6	13.7	10.3	5.9
Rank Correlation with Column 1		.81	.95	.34	.40	.53	.81

positive correlation at the 95th percentile indicate that at these levels, the degree of retentivity is irrelevant or, at the top level, favorable for high scorers. Again, as with 3 a, if the numbers reaching the various percentiles are calculated as proportions of the total age group, there are positive correlations.

CONCLUSION

The proportions of the *in-school* population reaching various international percentile levels are *negatively* related to the proportion still at school. The proportions of the *total age group* reaching various international levels are *positively* related to the proportion still at school. When an intake is increased in size, the performance of the additional students tends to be of lower quality. From the evidence studied here, however, it would seem fairly certain that countries can increase their total "mathematical yield" of an age group by having larger intakes. It also seems from this evidence that the performance of high ability students is unlikely to be affected by increasing the intake.

In Population 3 a, Belgium, England, and Japan have a consistently high performance of all students. Sweden and Japan demonstrate very well that increasing the size of the intake does not necessarily mean lowering standards. Sweden has an intake approximately three times as large, for example, as that of England, and yet approximately the same proportions of the total age group are still reaching the 90th and 95th percentiles. Again, although systems with smaller intakes bring these students to higher levels, this might be expected when the selection processes and smaller numbers are considered. What is more important, however, is the proportion of the total age group reaching particular levels. Here, the size of intake may have an important effect at the lower levels (see Table 3.41, Sweden at 25 percent level), and at the top levels it is possible for countries with large intakes (e.g. the United States and Sweden) to bring high proportions of an age group to the 90th and 95th percentiles. At the 95 percent level, Finland, the Netherlands, and Germany show percentages well below those of other countries. Germany is particularly surprising considering its high selectivity. From Table 3.40 it appears that the weaker half of the United States group are much below the standards of other countries.

For Population 3 b, Japan, Belgium, and Germany perform well, whereas Sweden and the United States perform relatively poorly. It must be remembered that in Germany the 3 b group have all studied mathematics up to the end of the penultimate preuniversity year (that is the *Unterprima*).

It is therefore possible to have higher yields even when the intake size is larger. The problem of the appropriateness of the "acquired" yield of the mathematics-science program group poses another problem. This would require research being undertaken to discover the "required" yield of each society, which would then be matched against the "acquired" yield. It would seem, however, that the educational policy makers in those countries where the mathematics-science program intake is small might well reconsider the aims of their secondary school systems in terms of the yield (both total and elite) of the mathematics-science group. It is the performance in relation to the age group as a whole which is important. The data obtained in this study make it clear that it is possible to have both a high overall yield and an undiminished elite yield.

An Attempt to Construct a Model of the Effects of Selection

The table of means and variances of the mathematics scores in the various countries (Table 3.44), with their different proportions of students in the terminal mathematics-science courses, suggests that it might be possible to construct a model to represent the situation. The basic idea underlying the model is that each country has the same distribution of mathematical ability in the complete age group and that the differences in means and variances found at the terminal mathematics stage are a result of selection procedures. This is obviously a gross oversimplification of the situation, but the function of a model is to represent the situation with the minimum of assumptions. If the data do not fit the model, the basic assumptions must be modified or added to, but the principle of Occam's razor is the one on which the model-maker must operate.

The means and variances are shown in columns 2 and 3 of Table

TABLE 3.44. *Data for Constructing a Model (Population 3a).*

(1) Country	(2) Mean Score	(3) Variances of Scores		(4) Percentage of Age Group in Program	(5) Expected Mean in Standard Scores	(6) Expected Variance in Standard Scores
		3 a	1 a			
Australia	21.6	110	196	14	1.59	.189
Belgium	34.6	159	225	4	2.15	.132
England	35.2	159	289	5	2.06	.138
Finland	25.3	92	98	7	1.92	.152
France	33.4	117	154	5	2.06	.138
Germany	28.8	96	137 (1b)	4.7	2.09	.137
Israel	36.4	74	216 (1b)	7	1.92	.152
Japan	31.4	219	286	8	1.86	.158
The Netherlands	31.9	66	253	5	2.06	.138
Scotland	25.5	108	213	5.4	2.02	.139
Sweden	27.3	142	117	16	1.52	.200
United States	13.8	159	177	18	1.46	.208
Japan A	47.7	79	286	1	2.67	.097

3.44, and the percentage of the age group in the terminal mathematics-science course is shown in column 4. "Japan A" is a highly selected group, composed of students following an academic course in a comprehensive school, which has been added to the data to test the model at the extreme point. The data for columns 1 to 4 are those already used in earlier parts of this volume; the method of calculating the entries in columns 5 and 6 is explained below.

The most simple assumptions are that

1. the scores in each country would be normally distributed over the whole age group if all in the age group had taken the tests;
2. these hypothetical distributions are identical for all countries;
3. those in the terminal mathematical population are the best mathematics students in the age group in each country.

On these assumptions we can calculate the expected mean scores and variances of the groups forming the selected portions of the age group. These are given in standard scores in columns 5 and 6 of Table 3.44.

The formulae are

$$\text{Mean} = y/q; \text{ variance} = 1 - (y/q)\{(y/q) - k\}$$

where q = proportion selected
y = ordinate of normal curve at point of cut-off
k = point of cut off

The actual means and expected means are plotted against each other in Figure 3.10. The rank correlation between actual and expected means is .74. For such sweeping assumptions as have been made, the agreement between theory and results is moderately good.

The fit of actual variance to expected variance is much less satisfactory. The large variances of Belgium, England, and especially Japan do not fit in with any line that could be drawn to fit the other observations. This may be partly due to the relatively large variances which two of these countries (England and Japan) had already shown at the 13-year stage. The model may therefore be further complicated by using the ratio of the 3 a variance to the 1 a variance as the variable to be plotted against the expected variance. This has been done in Figure 3.11. The trend is evident although there are considerable deviations.

One defect of the model is that it assumes that retentivity has operated with perfect efficiency to select the best mathematics scores for Population 3 a (Assumption 3). We have assumed that the 5 percent of the

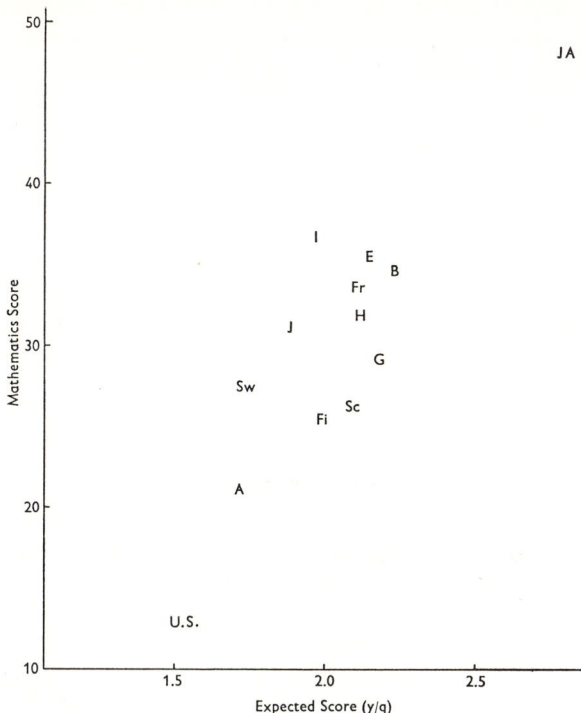

Figure 3.10. Mean score in mathematics made by Population 3 a in different countries plotted against the score expected on basis of the model.

age group following the preuniversity mathematics course is also the 5 percent who would make the highest scores if the mathematics tests were applied to the whole age group (compare Figure 3.12). This is certainly not the case in the real situation. The model can be modified to assume a correlation r between the variable operating to select the population (assumed to be normally distributed) and the mathematics score obtained. The diagram illustrating the calculations would then be Figure 3.13. If r is assumed to be the same for all countries, no difference is made to Figures 3.10 and 3.11 for the expected mean becomes $r\, y/q$, and the expected loss of variance becomes

$$r^2\, (y/q)\{(y/q) - k\}$$

that is, the shapes of the graphs are unchanged. But it is certain that r varies from country to country. It is an inviting thought that the fit might be improved by the insertion of appropriate values for r in the different countries, but we have little idea of the differences between

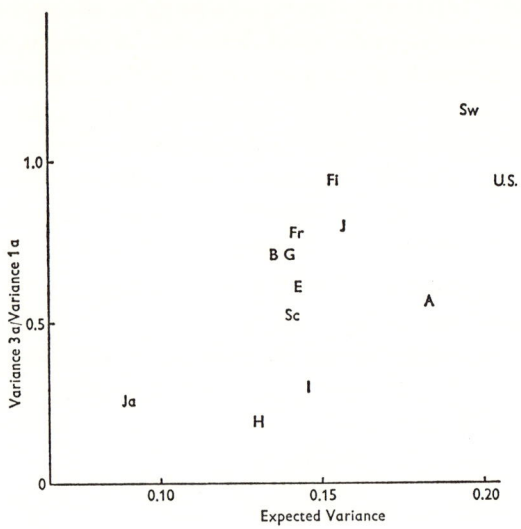

Figure 3.11. Ratio of variances in Populations 3 a and 1 a in different countries plotted against the variances expected on the basis of the model.

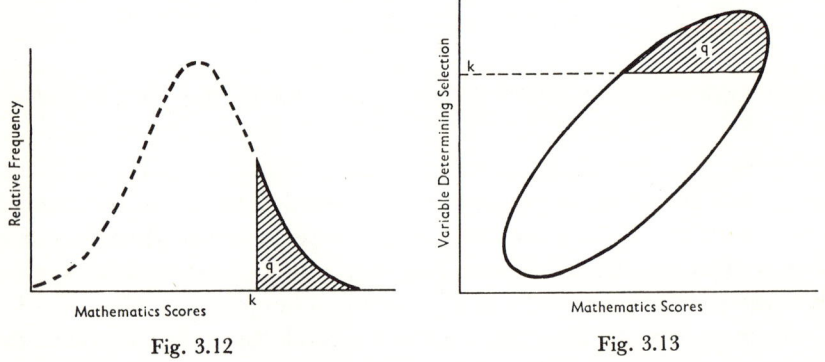

Figure 3.12. Model A. Hypothetical distribution of mathematics scores for whole age group with score k cutting off Population 3 a to form proportion q of the age group.

Figure 3.13. Model B. The correlation surface relating mathematics score to selection variable, with k cutting off proportion q from the rest.

countries in the magnitude of r, although it may be assumed that there is a fairly strong association between enrollment in a mathematics program at the terminal stage and proficiency in mathematics.

It is clear from an inspection of the Population 1 a scores that it is extremely unlikely that assumption 2 will be satisfied. There are already fairly large differences among countries at the earlier age, where the whole age group is under consideration. Defects of the data have no doubt also increased the difficulty of testing the model: the percentages of the age groups listed in column 4 of Table 3.44 are not sufficiently well scattered to give a good test.

The fact that the data, in spite of these defects, provide distributions of the types depicted in Figures 3.10 and 3.11 supports the idea of a model of this kind with its implications that

1. the pool of mathematical ability is identical in the countries represented in the study;
2. countries retaining larger percentages of an age group in the preuniversity stage are doing so by drawing on the lower levels of ability;
3. the differences in mean score at the preuniversity level can to a great extent be explained in this way.

Summary and Conclusions

At the beginning of this chapter it was anticipated that the conclusions reached after testing the various hypotheses would be drawn together to present a picture of the effects which changes in school organization might have on attainment in mathematics. Some of these conclusions indicate changes that can be made with the likelihood of satisfactory results so far as mathematics is concerned. Even where the conclusions are negative in the sense that they report no change in attainments when a certain variable is altered, they are of interest to those whose duty it is to take decisions on school organization or to administer the decisions made by others.

Changing the age of entry to school is likely to make no substantial change in mathematics score. Those countries which have an entry age of 5 produce poorer scores in mathematics at age 13 than do countries with an entrance age of 6. Delaying the age of entry to 7 is associated with even lower scores at 13, but the whole pattern suggests that the important variable is not the age of entry. Even the children of parents in the lower occupational groups do not seem to profit substantially from an earlier entry to schools as now constituted.

Delaying the age for completing secondary school education prior to transfer to university does not appear to pay dividends as far as mathematics achievement is concerned. The best scores among the preuniversity mathematics students are made by those countries presenting students between the ages of 18 and 18 $^1/_2$; countries presenting students aged on the average between 19 and 20 do not do so well. When the scores are adjusted for differences in the proportions still at school, it is found that the gains between the 1 a and the 3 a stages are directly related to the time interval between the two stages, the rate of gain being the same in practically all of the countries. In the other preuniversity population there is a negative correlation between the gains in score and the time interval between the two stages, when the appropriate adjustments have been made for retentivity.

To the teacher, an obvious way of raising standards seems to be to reduce the size of classes. Here, the results in our investigation are conflicting. At the higher levels, smaller classes are associated with superior attainments. At the lower levels, the trend is reversed. There are so many complicating factors in this study that it is almost impossible to separate out the effect of class size from the others, especially since most of them are not under control. It is, at any rate, apparent that merely reducing the size of classes is not likely to increase mathematical attainment significantly.

A reduction in the number of subjects studied at the preuniversity stage is not necessarily accompanied by an increase in mathematical attainments but rather by a reduction in the age of the group.

The relation between size of school and score varies with the age of the student and the type of school. For the younger students, aged about 13, schools with enrollments exceeding 800 make better scores than do smaller schools. At the terminal stage where the students are from 17 to 20 years of age, there are differences between comprehensive and selective academic schools. In comprehensive schools, the larger the school, the higher the score. In selective academic schools, those with enrollments between 700 and 1,100 achieve higher scores than any other group, including the larger schools with enrollments over 1,100. But these international averages conceal national differences which are occasionally significant.

The main difference among the countries taking part in the IEA study is in the use made of specialized schools as contrasted with comprehensive schools. This is an area in which the countries could learn much from each other. It has been possible to show how students fare under the different systems as far as mathematics achievement is concerned.

The first finding is that 13-year-old students following academic courses in specialized schools attain a higher level and show slightly less variability than do students following similar courses in comprehensive schools. At the preuniversity stage their superiority has vanished, and there is no significant difference between the scores of the two groups. On the other hand, 13-year-old students following general courses do better in comprehensive schools than do students following similar courses in schools not containing academic pupils.

If interest in mathematics rather than attainments be considered, we find that students in specialized academic schools show the greatest interest, and those in specialized nonacademic schools show the least, comprehensive schools occupying the middle position. When the score of the specialized academic schools is averaged with those of specialized nonacademic schools, the average is found to be practically the same as that for comprehensive schools. In other words, the total products of the two systems are almost identical. If, however, we assess countries on a specialist-comprehensive scale and compare the assessments with the average interest score we find that the countries towards the comprehensive end of the scale are also those high on the scale of student interest.

The question of specialized and comprehensive school organization is closely linked with that of socio-economic bias in the preuniversity year. The study shows that this bias exists in all countries in the sense that the preuniversity-year group differs from the 13-year-old group in having a higher proportion of students whose fathers have upper- or middle-class occupations. But the degree of bias is markedly greater in countries operating a selective system and reaches a very high level in some of these countries. It is also closely related to the age at which selection occurs, being greater at the younger ages.

An important question, closely related to the one just discussed, is that of the retentivity of the school system. The countries in the inquiry differed substantially in the extent to which students remained in the school system until the preuniversity year. It has been shown that countries retaining higher proportions of an age group show lower scores where the preuniversity population of students is concerned. But it has been shown that this appears to be due not to a lowering of the standards of the best students, but rather to a dipping into lower levels of ability to provide the additional students. If this is accepted as a reasonable interpretation of the results, it is possible to provide a model of how the process operates, and this has been done on pages 135-139.

An alternative way of examining the situation is to estimate what proportion of an age group in each country reaches a high international standard and to relate it to the proportion of the age group retained in the school system. The results show that the two proportions are positively related, although the correlations are not high. In other words, the higher the proportion of students retained in the school system, the higher the proportion of the age group making high scores is likely to be.

To sum up the last two paragraphs: "More means worse" only in the sense that the average score of the expanded group is likely to be lower than that of the original smaller group. From the evidence of this inquiry it does not seem likely that the mathematical attainments of the most able students will be affected; on the contrary, the total yield of advanced students is likely to be increased.

Countries seeking an increase in the supply of mathematicians would therefore be wise to look closely at the structure of their secondary school system, at the proportions entering courses, and at the ages at which final choices are being made. These seem to be the important factors as far as school organization is concerned, but this aspect, as the reader will recognize, is only one facet of a complex problem.

Chapter 4

Problems Related to the Curriculum and Instructional Methods

Introduction

A number of problems that arise in connection with the curriculum content and instructional methods were investigated by the IEA. This chapter comments on the background of the questions that were raised and the hypotheses that were formulated to sharpen the focus for the study. It also reports the findings and discusses the conclusions and implications that seem warranted.

Persons responsible for the evaluation of curriculums and instructional methods are of course first of all concerned with mathematics achievement as revealed by total scores on the mathematics test and with the subscores under various classifications. These were reported in Chapter 1 in this volume. The present chapter deals for the most part with the exploration of *relationships* between achievement and certain other factors—for example, the use of "discovery" approaches to learning, the amount of time spent on homework, and the strength of the association between achievement and interest in mathematics and between achievement and certain attitudes toward mathematics.

Administrators of schools are responsible for many decisions that affect, directly or indirectly, the achievement of students. Teachers vary as to their training and their perceptions of the limits within which they have freedom to act. These factors are commonly believed to influence not only achievement but also students' interest and attitude toward school in general and toward particular subjects and their goals in life. Some students who are thus influenced strive for better achievement. Others become relatively unresponsive to the efforts of the teachers to help them learn, and in many instances drop out of school soon after the law permits them to do so. By gathering data bearing on this

---

This chapter was written by Professor M. L. Hartung, who drew extensively upon reports concerning individual hypotheses from the following group of persons: B. S. Bloom, A. W. Foshay, S. H. Hilding, M. Kojima, G. Mialaret, K. M. Miller, G. Ögren, J. A. Pidgeon, G. F. Peaker, M. I. Takala, D. A. Walker.

complex of interrelationships from various countries it was hoped that generalizations could be made, some of which would be useful in making these decisions.

In the following pages the questions studied are considered in three groups, as follows:

1. Those dealing primarily with *student* achievement in relation to certain methods of teaching, student interests, and attitudes.
2. Those that are concerned with the *teachers'* training and their perception of the situation they face.
3. Those that are related to *school* decisions that are administrative in type—as, for example, achievement as related to the number of hours per week allocated to mathematics.

Achievement in Relation to Interests and Attitudes

Formulation of Hypotheses

Inquiry-Centered Methods

Many discussions of curriculum issues assume implicitly or explicitly that education is better when it seeks to develop the "higher" as well as the "lower" mental processes. One process commonly accepted as "higher" is the discovery and formulation of generalizations from particular instances. Another "higher" process is the recognition of the applicability of a known generalization or principle in a situation new to the learner. In contrast, the mere recall of information or the exhibition of a skill acquired by repetitive practice are commonly accepted as involving only "lower" mental processes.

Research in the United States (summarized by Bloom, 1954; Chausow, 1955; Dressel and Mayhew, 1954; Ginther, 1964; and Sheehan, 1965) supports the thesis that knowledge objectives and other objectives requiring little more than the remembering and comprehension of subject matter can be learned as a result of a great variety of teaching methods and learning experiences. The lower mental processes can be learned equally well in large or small classes, in lectures or by discussion, through teaching by television or films or the use of programed learning materials or teaching by regular classroom procedures. Some students can learn the lower mental processes through independent study as well as through the more conventional classroom procedures. Basically, all that seems to be required for the development of lower mental processes is an attentive and well-motivated learner and a set of learning activities in which an accurate version of a piece of information is communicated to the student by means of the printed page, the spoken word, or through pictures or illustrations.

The research referred to in the preceding paragraph makes it clear that the achievement of complex types of critical thinking (higher mental processes) is not likely to be attained by simple lecture methods or by merely telling the students what they are to do or how they are to do it. Demonstrations of appropriate problem-solving processes are not very effective in bringing about actual problem-solving competence. The general import of these studies is that the more complex and higher categories of the cognitive domain require far more sophisticated types of learning experiences than the simple communication of a correct version of an idea or event to the student. Much more motivation is required, much more activity and participation on the part of the learner are necessary, and more opportunities must be available to help the individual gain insight into the processes he uses or misuses.

In the actual classroom, it is possible for teachers in general and teachers of mathematics in particular to adopt rather different approaches. On the one hand, they may place emphasis on drill and rote learning and use a "textbook" approach. On the other hand, they may stress the need for students to understand principles and ideas and to employ "discovery" methods. In practice, of course, in every mathematics class there is more or less emphasis on factual and rote learning on the one hand or learning by "discovery" and active participation methods on the other.

Early in this study it was hypothesized, therefore, that mathematics learning would be greatest in classrooms that emphasized the use of inquiry approaches and understanding and that it would be least in classrooms that emphasized drill and memorization. The original formulation of the hypothesis and the preliminary analysis of the data was in terms of scores on the items classified as calling for the higher mental processes. When it became apparent that the coefficient of correlation between total score and score on the "higher" mental processes was high (see Chapter 1, Volume II, page 36 and Table 1.11), the formulation was revised and the final analysis was made in terms of the total mathematics scores. Recognizing, however, that student performance on the mathematics tests would clearly be dependent upon the level of instruction offered, the revised formal statement of the hypothesis was as follows: *When level of mathematics instruction is held constant, inquiry-centered approaches to learning will produce higher and less variable scores in mathematics than will more traditional approaches* (Hypothesis 12).

It was also anticipated that the opportunity for learning the specific mathematics content involved in each test item would influence the total achievement score (see Chapter 1, Volume II, page 37). In the *teacher questionnaire* the teachers were therefore asked to indicate for each item whether or not their students had an opportunity to learn the mathematics involved. These ratings made it possible to formulate an additional hypothesis concerning the effect of inquiry-centered approaches. This was stated as follows: *When the opportunity to learn mathematics is held constant, scores on the mathematics test will be related to student descriptions of mathematics teaching and school learning* (Hypothesis 13).

As formulated, this hypothesis refers to "student descriptions of mathematics teaching and learning", which is essentially the name of the instrument used for gathering the data on the sort of teaching approach which, in the students' view, characterized the classroom. The analysis, however, is in terms of whether there was "little" or "much" inquiry-centered activity in the learning situation.

Interest and Attitudes

It is usually taken for granted that high achievement in mathematics is associated with high interest in the subject. Previous studies in various countries participating in the present project have already shown that there is a positive correlation between interest and achievement in mathematics. There are, however, remarkable variations in the size of these coefficients depending, for example, on the particular population sampled and on the tests and other instruments used. The IEA study provided an opportunity to investigate this relationship cross-nationally, using a considerable range of grade levels.

Also, it was believed that scores on the mathematics test would correlate positively with certain attitudes toward the subject. Several of these attitudes or beliefs are often cited in discussions of the objectives of the "New Mathematics" movement. For example, it was expected that the students with high achievement would tend to believe that mathematics is an "open" system—a subject, that is, in the process of development and one that allows for a number of ways of viewing and solving problems. This may be contrasted with the not uncommon view that mathematics is a fixed system governed by rigid rules which the student must learn and apply.

Further, the prediction was made that beliefs concerning the difficulties in learning mathematics would be related to achievement. In particular, students with high achievement scores would tend to consider mathematics as a subject everyone can learn. In addition, it was presumed that the student performing well in mathematics would believe that a good knowledge of mathematics on the part of citizens is of great importance for a nation's development and that more of the most able people should be encouraged to become mathematicians and mathematics teachers. The hypothesis concerned with student's interest and attitudes therefore ran as follows: *Total mathematics scores will be related to students' interest in and attitude toward mathematics. In particular, students with higher achievement scores will:*

1. *Have greater interest in mathematics.*
2. *Have greater interest in taking more mathematics.*
3. *View mathematics as an open system.*
4. *View mathematics as a subject most students can learn.*
5. *View mathematics as an increasingly central subject for occupations and for the development of society* (Hypothesis 14).

Interests and Other Factors

Availability of data bearing on student interest in mathematics and also on many other factors invites study of possible interrelationships among these factors. Among the many possibilities three were selected that have some relevance to the curriculum and instructional methods.

First, what is the relation between interest in mathematics and the student's views about the sort of mathematics teaching he has experienced? In particular, is high interest associated with an inquiry-centered approach to learning?

Second, what is the relation between interest in mathematics and students' attitudes toward mathematics? In particular, is high interest associated with the view that mathematics is an open system?

Third, is there a relation between interest in mathematics and parental status and occupation? In particular, is high interest associated with high social-economic status (as judged by the father's occupation), and is interest higher when the father's occupation is of a scientific or technological nature?

The data from the IEA study provided an opportunity to investigate these questions on an international basis. The hypothesis thus ran as follows: *Students will have more interest in mathematics when:*

1. *They describe mathematics teaching as emphasizing inquiry.*
2. *They view mathematics as an open system.*
3. *Their father's occupation is a high status one* (Hypothesis 15).

Findings

Inquiry-Centered Methods

Hypothesis 12 is concerned with the relation between mathematics achievement and inquiry-centered approaches to learning. Before turning to the data, a very brief description will be given of the instrument used to discover the approach used in the various classrooms and schools. The data were collected by means of a student attitude scale incorporated in the student opinion booklet. Each student was asked to mark "agree", "disagree", or "uncertain" to a series of statements intended to describe his mathematics class or his school. This scale, "Description of Mathematics Teaching and School Learning", is described in more detail in Chapter 6 of Volume 1 and contains items such as:

> My mathematics teacher does not like pupils to ask questions after he has given an explanation.
>
> Much of our classroom work is discussing ideas and problems with the teacher and the other pupils.

High scores indicate that the students perceive the teaching they receive as promoting inquiry, independent study and student activity.

Since it is likely that students may differ in their description of the same teacher, class, or school for many reasons, including their own biases and competence, it was decided that the best estimate of the approach used would be the average of the descriptions from all the students in a particular class. Thus, the "Description of Mathematics Teaching and School Learning" score, assigned to any one student for purposes of Hypothesis 12, is the mean score of the students in his school class who participated in the IEA study.

In Figure 4.1 below, the correlations by countries and by populations of the total scores with the scale, "Description of Mathematics Teaching and School Learning" are presented (see also Chapter 6, Tables 6.6–6.9).

The relationship between achievement in mathematics and the students' description of their class and school is generally very low. Although the coefficients in Populations 1 a and 1 b are generally positive, they are very small. Only in the case of three countries (Finland, the Netherlands, and Sweden) do we find correlations of .11 or higher. While correlations of about .08 or higher are statistically significant, it seems evident that the relationship has little practical or scientific significance.

At the terminal levels of secondary education, the correlations are also small, and they are characteristically negative. In only two countries (England and the Federal Republic of Germany) are there correla-

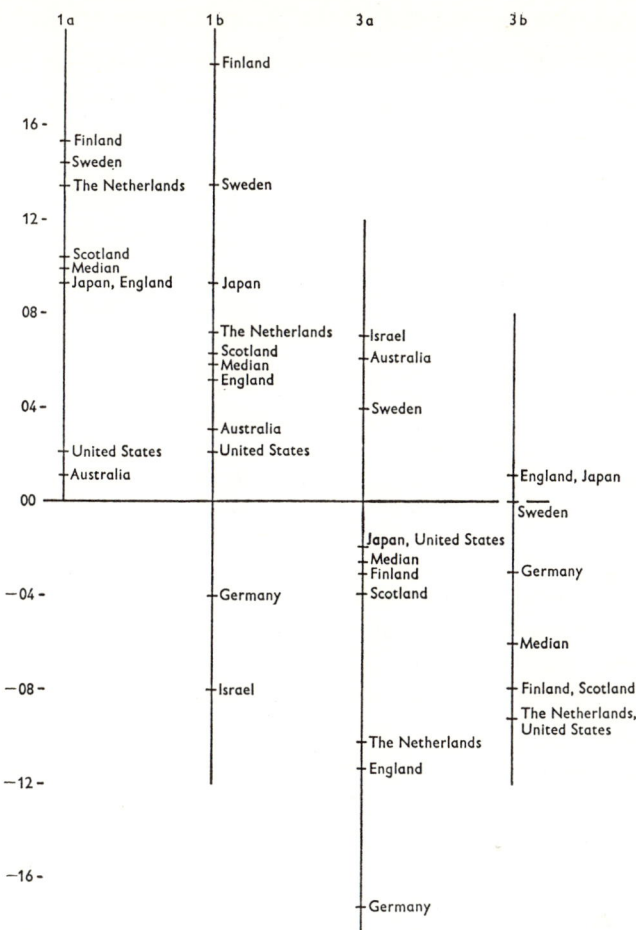

Figure 4.1. Correlations between the scale *Descriptions of mathematics teaching and school learning* and *Total mathematics scores* for Populations 1 a, 1 b, 3 a and 3 b.

tions of .11 or higher, and these too are negative. Again, the low levels of relationship make them of dubious value for practical or scientific purposes. The fact that the majority of correlations are negative suggests that as far as the cognitive achievement measured in this study is concerned, inquiry-centered approaches to teaching older groups of students (ages 17 to 20) clearly did not result in higher scores than those obtained under more restricted and directive approaches.

The correlations within countries may mask possible differences between countries. However, Table 4.1 shows that there is little difference among the countries with regard to either the mean of student descrip-

TABLE 4.1. *Means, Standard Deviations, and Number of Cases of the Scale: Description of Mathematics Teaching and School Learning, for Populations 1a and 3a.*

Country	Population 1 a			Population 3 a		
	M.	S.D.	N.	M.	S.D.	N.
Australia	23.4	3.5	2,917	22.7	3.3	1,089
England	23.8	3.8	2,949	22.9	3.6	967
Finland	22.5	0.5	747	19.7	0.3	369
Germany	—	—	—	22.8	0.5	649
Israel	—	—	—	25.2	3.9	146
Japan	24.8	3.3	2,050	25.1	3.1	819
The Netherlands	24.6	3.5	429	24.5	3.4	462
Scotland	23.7	3.9	5,256	23.5	3.6	1,422
Sweden	23.4	3.6	2,554	25.0	3.7	776
United States	23.8	3.8	6,231	24.7	4.0	1,568
All Countries	23.8	3.7	23,132	23.8	3.6	8,266

tions of mathematics teaching and school learning or the variance in these descriptions. And, in fact, the rank-order correlation for the eight countries concerned is only .27.

As an alternative approach in the investigation of the effect on achievement of the methods of instruction and approach to school learning, the scores of the scale "Description of Mathematics Teaching and School Learning" were divided into three groups labeled "little", "some", "much". "Little" represents the extreme of classes which emphasize drill, memorization, and a relatively rigid system of instruction. These divisions were made on all countries combined.

In considering the results to be obtained by this method of analysis, it was anticipated that student performance on the mathematics test would be influenced both by the level of instruction and by the student's opportunity to learn mathematics items. Therefore, separate analyses were carried out. In the first, the mean total mathematics scores for those students in each country who came into either the "much" or the "little" groups with respect to their scores on the "Description of Mathematics Teaching and School Learning" scale were calculated, holding *Level of Instruction* constant. In the second, a similar set of mean total mathematics scores were calculated, but this time the *Student Opportunity to Learn* the mathematics items was held constant.

Dealing first with the analysis holding *Level of Instruction* constant, Table 4.2 gives the all countries means, standard deviations, and numbers of cases for Populations 1 a, 1 b, and 3 a. Population 3 b is not

TABLE 4.2. *Mean Total Mathematics Scores, Adjusted for Level of Instruction with S.D.'s and N.'s for Students in "Little" and "Much" Inquiry Groups, for Populations 1a, 1b, and 3a.*

	Little Inquiry			Much Inquiry			Difference of Means
	M.	S.D.	N.	M.	S.D.	N.	Much–Little
Population 1a	19.6	12.9	2,931	21.7	13.4	5,454	2.1
Population 1b	22.0	13.3	3,466	23.4	13.8	8,006	1.4
Population 3a	27.4	11.0	926	26.6	10.6	2,005	–0.8

included, as only four countries were represented. The differences between the means for the "little" and "much" groups in Populations 1 a and 1 b are statistically significant, suggesting that for the younger age groups included in this study a rigid drill and memorization approach to teaching and learning is less effective in terms of mastery of the subject than inquiry-centered approaches. This confirms the results of the correlational analysis given earlier and offers support for the hypothesis. The contention in the hypothesis, however, that there would be less variation in achievement in the "much" inquiry group is not supported by the evidence. If anything, in these two populations there is rather more variation in achievement in the "much" group than in the "little" group. In Population 3 a, the difference between the means of the two groups is in the opposite direction, although it is nonsignificant, and the variability of the "much" inquiry group is slightly less than that for the "little" inquiry group. The results for the separate countries are shown in Table 4.3.

The results of the second analysis, in which the *Student Opportunity to Learn* was held constant, are given in Table 4.4 and for separate countries in Table 4.5, and repeat in all respects the results of the first analysis.

As stated, therefore, Hypotheses 12 and 13 are partly confirmed. A relationship does exist for the younger students of Population 1 a with high mathematics scores being accompanied by the students' view of their teaching as inquiry-centered. For the preuniversity students of Population 3 a, however, no significant relationship was found, although there appears a tendency for high scores to be associated with the students' view of their teaching as being more formal and prescribed than inquiry-centered.

The central thesis that the nature of the learning situation affects

TABLE 4.3. *Mean Total Mathematics Scores (Adjusted for Level of Instruction) Related to Inquiry-Centered Approaches to Learning for Various Countries.*

	Little Inquiry			Much Inquiry		
Country	M.	S.D.	N.	M.	S.D.	N.
Population 1 a						
Australia	18.6	12.5	501	19.3	12.5	813
England	16.5	16.1	511	21.0	16.7	917
Japan	27.7	18.2	197	32.5	16.7	881
The Netherlands	25.5	9.1	61	24.1	10.9	178
Sweden	14.1	9.5	529	17.1	10.6	713
United States	15.5	12.1	1,132	16.1	12.8	1,951
Population 3 a						
Australia	21.0	8.8	251	22.3	8.6	209
England	36.4	13.1	247	33.6	12.6	226
Japan	32.3	14.3	69	31.3	14.1	369
The Netherlands	33.1	6.7	65	30.7	7.4	203
Sweden	26.9	10.3	88	27.8	10.2	354
United States	14.7	12.8	206	13.9	10.9	644

the learning outcomes as measured by achievement and other tests is not strongly supported by the attempt to relate inquiry-centered teaching and learning to mathematics achievement scores. Questions about the validity of the scale "Description of Mathematics Teaching and School Learning" may be raised. Clearly, more precise and valid methods of measuring inquiry approaches to teaching can be developed, and this is one possible avenue for future research. However, it may be that the difficulty is not in the instrumentation but is in the tenuous relation between method of teaching and cognitive achievement. Learning theory may help in the future to understand the dynamic relations between methods of teaching and learning outcomes, but at present it is only possible to speculate about the chain of relations between methods and student achievement in mathematics.

Interests and Attitudes

Hypothesis 14 is concerned with the relation between mathematics achievement and interest and between achievement and certain attitudes. Data for the investigation of interest were collected by the student questionnaire. Responses to six different questions were used to produce a derived "interest index". (Details as to the method of arriving at this index are given in Chapter 12, Volume I.) For example, if a student

TABLE 4.4. *Mean Total Mathematics Scores, Adjusted for Student Opportunity to Learn, with S.D.'s and N.'s for Students in "Little" and "Much" Inquiry Groups, for Populations 1a, 1b, and 3a.*

	Little Inquiry			Much Inquiry			Difference of Means
	M.	S.D.	N.	M.	S.D.	N.	Much–Little
tion 1a	20.8	13.1	2,792	23.0	13.2	5,376	**2.2**
tion 1b	24.0	13.3	3,400	25.4	13.7	7,030	**1.4**
tion 3a	28.9	12.0	936	27.4	10.4	1,904	−1.5

TABLE 4.5. *Mean Total Mathematics Scores (Adjusted for Student Opportunity to Learn) Related to Inquiry-Centered Approaches to Learning for Various Countries.*

Country	Little Inquiry			Much Inquiry		
	M.	S.D.	N.	M.	S.D.	N.
			Population 1 a			
England	27.2	14.3	436	28.7	14.9	928
Japan	27.9	17.8	192	32.4	16.5	840
Scotland	19.0	11.1	972	20.7	11.0	1,656
Sweden	12.8	9.4	416	16.1	11.3	492
United States	17.1	12.8	776	17.3	12.9	1,460
			Population 3 a			
Australia	21.1	9.6	196	21.1	8.4	232
England	36.7	13.0	168	32.2	11.5	152
Japan	33.2	11.8	64	32.0	12.4	352
Scotland	26.8	10.7	268	25.4	10.5	408
Sweden	26.7	11.6	60	28.6	11.2	252
United States	17.8	12.2	180	16.8	11.6	508

indicated on the questionnaire that he "wishes to take additional mathematics courses" he was awarded one point.

The measures of student attitude were obtained from student responses to attitude scales constructed specifically for this study. (The procedure followed in the construction on these scales, the items, and data about them are also reported in Chapter 6 of Volume I.)

The coefficients of correlation between the mathematics scores and the interest or attitude variable studied, based on the total distribution for all countries, are given in Table 4.6. It is immediately evident that interest in mathematics is positively related to achievement, and that

TABLE 4.6. *Correlations with Total Mathematics Scores for Interest and a Number of Attitudes over All Countries, Populations 1a, 1b, 3a, and 3b.*

Variable	1a	1b	3a	3b
Interest	.27	.30	.34	.31
Desires More Mathematics	.23	.25	.19	.19
Mathematics as a Process	−.08	−.09	−.24	−.21
Difficulty of Mathematics	.00	−.02	−.07	.00
Importance of Mathematics	.04	.08	.09	.13

the coefficients are statistically significant at each level of instruction. As might be expected, the largest coefficient (.34) is found for the population that is still taking mathematics courses at the terminal level of secondary education. The coefficients for "wishes to take more mathematics" are also positive but somewhat smaller for all populations. It is perhaps not a surprise to discover that the coefficients are smaller for the upper levels, either because students by then have already had considerably more mathematics, or possibly because with greater maturity the interests of the group have tended to become more diverse and specialized.

The coefficients of correlation between achievement and attitudes are small in general. Achievement in mathematics is positively, but weakly, correlated with student belief about the importance of mathematics to society, the largest coefficient (.13) occurring in the case of the more mature terminal nonmathematics population. On the other hand, achievement scores tend to correlate negatively with attitude toward mathematics as a process and toward the difficulty of learning mathematics, which is contrary to the original hypothesis. The negative coefficients between achievement and the belief that mathematics is an "open" system occur in both the younger and older populations, and they are greater in absolute value for the older students (−.24 and −.21).

On the basis of these results, it must be concluded that these affective outcomes of teaching are not consistently related to the cognitive outcomes measured except in regard to interests. Moreover, at each instructional level, students with high achievement in mathematics tend to view it as a "closed" system. Questions immediately arise as to whether there are significant national differences in interest and in attitude toward mathematics. In some countries, for example, the nature of the instruction may emphasize the difficulty of mathematics more than it

TABLE 4.7. *Means, Standard Deviations, and Number of Cases of Scores on the Scale of Interest in Mathematics for Populations 1a, 3a, and 3b.*

Country	Population 1 a			Population 3 a			Population 3 b		
	M.	S.D.	N.	M.	S.D.	N.	M.	S.D.	N.
Australia	5.9	1.8	2,917	6.7	2.1	1,089	—	—	—
Belgium	5.7	1.8	1,686	6.0	2.1	519	4.5	2.1	1,004
England	5.7	1.6	2,949	7.3	1.9	967	5.1	1.3	1,782
Finland	6.2	1.6	747	6.2	1.9	369	4.8	1.9	399
France	5.5	1.9	2,409	6.9	1.9	222	—	—	—
Germany	—	—	—	6.6	1.5	649	5.7	1.5	643
Israel	—	—	—	7.4	1.6	146	—	—	—
Japan	6.1	1.6	2,050	6.3	2.2	818	5.2	2.0	4,372
The Netherlands	5.4	1.7	429	6.4	1.8	462	—	—	—
Scotland	5.3	1.8	5,256	5.8	2.2	1,422	4.4	1.9	2,123
Sweden	5.8	1.4	2,554	6.5	1.4	776	5.8	0.9	222
United States	6.2	1.7	6,231	6.4	2.3	1,568	4.5	1.8	2,042
All Countries	5.8	1.7	27,228	6.4	2.1	9,007	4.9	1.9	12,586

does in other countries. When the data are pooled, such national differences may be obscured. National means and standard deviations of scores on these attitudes for each instructional level were reported in Chapter 1, Tables 1.16–1.19. National means and standard deviations of scores on the "Interest Scale" are given for three populations in Table 4.7. The data for Population 1 b, which overlaps Popuation 1 a, are not given in this table.

It is perhaps of some value to compare the interest measures for Population 3 a (students who are taking mathematics at the terminal level) and Population 3 b (students who are *not* taking mathematics). In every case, the difference of the means is significant at the 5 percent level. Thus, if "higher mathematics achievement" is interpreted as meaning taking *courses* at higher grade levels, rather than in terms of scores on mathematics tests, these data provide some basis for the conclusion that higher interest is associated with "higher achievement". Table 4.8 reports the correlations between national mean scores in mathematics and interest in mathematics and also "wishes more mathematics" for each population. The rank-order correlation coefficients between national mean scores in mathematics and national mean attitude measures were given in Chapter 1 but are repeated here for convenience of reference. Also included in Table 4.8 are the correlations between mean scores in mathematics and the mean score on the scale

TABLE 4.8. *Correlations Between National Mean Scores in Mathematics and Means of Interest and Various Attitude Scales for Populations 1a, 1b, 3a, and 3b.*[a]

Variable	Population 1 a	Population 1 b	Population 3 a	Population 3 b
Interest	.17	.30	.28	.32
Desires more Mathematics	−.32	−.46	.47	−.23
Mathematics as a Process	−.78	−.64	−.54	−.61
Mathematics Difficulty	−.64	−.45	−.39	−.23
Importance to Society	.28	.27	.57	.88
Mathematics Teaching	.70	.15	−.51	−.59
Views of School Learning	.60	.23	.10	.56

[a] Coefficients for Populations 1a and 3a are Pearson r's taken from Table 2.1. Correlations for Populations 1b and 3b are rank-order coefficients taken, for the most part, from Chapter 1, Table 1.20.

"Views About Mathematics Teaching" and the scale "Views About School and School Learning", some of which are not elsewhere reported separately.

The correlations between mean achievement and interest are positive; students in countries with high mean achievement are more interested in mathematics. However, in countries where the mean achievement is high, the younger students, as well as the terminal nonmathematics students, do not want to take more mathematics.

The rank-difference correlations between achievement and attitudes indicate a consistent pattern. In countries with high achievement in mathematics, the students tend to consider mathematics a "closed" system and a difficult subject. These relationships can be illustrated by references to particular countries. For example, in Japan where the achievement is relatively high (see Chapter 1, Table 1.1) the students in Population 1 a tend to consider mathematics as a "closed" system and a difficult subject (see Chapter 1, Table 1.16). On the contrary, in the United States, where the achievement scores are relatively low, the students tend to perceive mathematics as an "open" system and a comparatively easy subject.

In the countries with high achievement in mathematics, the preuniversity students tend to consider mathematics an important subject for the society (coefficients +.57 and +.88). This result may be connected with that concerning the students' "wishes to take more mathematics". In the countries with high achievement, the younger populations have less tendency to want more mathematics and at the same time they do not consider mathematics, on the average, to be very important. On the contrary, the preuniversity Population 3 a in countries with high achievement wish to study more mathematics and they consider mathematics to be very important to modern society and a subject for the most able people.

In summary, then, the hypothesis concerning interest was confirmed for all populations. The coefficients were, however, somewhat lower than might have been expected. In the case of the attitude scales, some results were contrary to the original hypothesis. The within-country correlations between achievement and certain attitudes (mathematics as an "open" system, the difficulty of mathematics) were usually negative and very low. On the other hand, the between-country correlations were always negative, and for the younger populations, at least, they were surprisingly high. Therefore, it may be concluded that the negative relationship is related to the national characteristics of learning situations. It may be that pressure toward high achievement tends to create an atmosphere in which mathematics is considered a difficult subject for the majority of students, and mathematics is more often seen as a fixed system of rules. This result cannot be explained by referring to national differences in selectivity because a similar viewpoint was found in the younger as well as the older populations. It should, however, be pointed out that the correlations between attitudes and achievement are quite low within single countries and, therefore, the interpretation may not hold for individual teachers or schools. The negative correlations for the younger populations between mean achievement and desire to take more mathematics in spite of the positive correlations for these populations in the case of attitude toward the importance of mathematics may also be a result of a pressure toward high achievement in mathematics.

A more detailed study of attitudes may show whether this interpretation is sufficient. It is possible that the attitude toward mathematics as an "open" system can be divided into two or three more or less independent continua, in which the understanding of the deductive nature of mathematics and of the more attitudinal aspects (creative and noncreative, changing and rigid) are separated.

Interests and Other Factors

Hypothesis 15 is concerned with the relationships between students' interests in mathematics on the one hand and with certain selected variables on the other. These were (1) the extent to which students view their mathematics teaching as involving a process of inquiry, (2) the

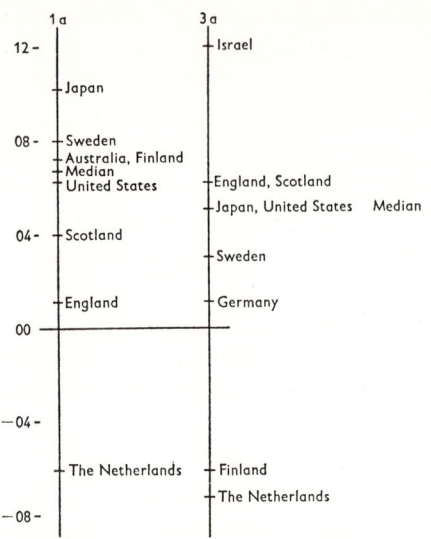

Figure 4.2. Correlations between *Interest in mathematics* and *Student's views about mathematics teaching* for Populations 1 a and 3 a.

extent to which students view mathematics as an open system, and (3) the occupational status of the father. Data for all of these investigations were collected by means of the student questionnaire and the student opinion booklet. The scales for the first two are described in Volume I Chapter 6, and in Chapter 1 of this volume.

The correlations between students' interests and views about mathematics teaching for Populations 1 a and 3 a are given in Figure 4.2. For most countries, the coefficients are positive in both populations.

TABLE 4.9. *Means, Standard Deviations, and Number of Cases of Interest Related to Extent that Learning Is Viewed as Inquiry Centered for Population 1a.*

Country	Little Inquiry			Much Inquiry			Difference of Means
	M.	S.D.	N.	M.	S.D.	N.	
Australia	5.4	1.5	140	6.1	1.8	528	.7
England	5.4	1.4	168	5.9	1.6	519	.5
The Netherlands	4.9	1.6	17	5.5	1.7	84	.6
Scotland	4.9	1.6	255	5.4	1.9	951	.5
Sweden	5.5	1.2	100	5.9	1.4	379	.4
United States	5.7	1.6	630	6.5	1.8	888	.8
All Countries	5.3	1.5	219	5.9	1.7	558	.6

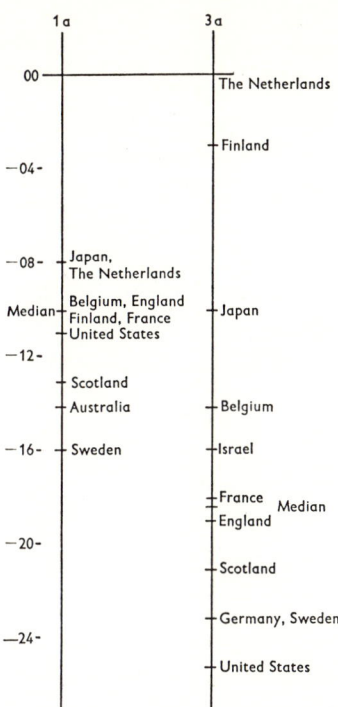

Figure 4.3. Correlations between *Interest in mathematics* and *Student's views about mathematics as an open system* for Populations 1 a and 3 a.

but few reach statistical significance. In order to examine further the relationship between these two variables, the students in Population 1 a were sorted into groups according to whether they viewed their mathematics teaching as having "little", "some", or "much" emphasis on inquiry methods, and the mean interest scores of these groups were then compared. As the results in Table 4.9 show, the difference between the groups, in each country concerned in the analysis, was significant. Thus, there is some evidence to support the hypothesis that students will have more interest in mathematics when they view their mathematics teaching as employing inquiry, but the relationship is fairly weak.

The correlations between students' interest and their views of mathematics as an open or closed system are given in Figure 4.3. For both Populations 1 a and 3 a the coefficients are all negative and tend to be somewhat larger for the older students. This means that in both populations "higher interest in mathematics" is associated with the view that mathematics is "closed", that is, given once and for all time. This

159

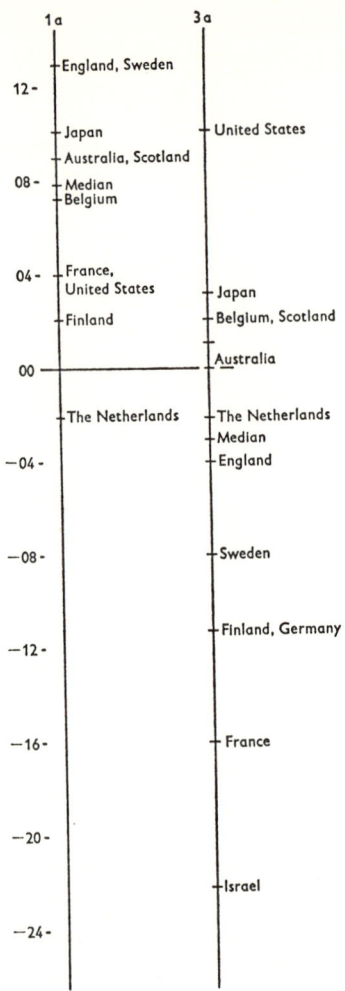

Figure 4.4. Correlations between *Interest in mathematics* and *Status of father's occupation* for Populations 1 a and 3 a.

result is contrary to the hypothesis and perhaps would tend to conflict with the opinions of many teachers who are attempting to teach in such a way as to develop the attitude that mathematics is an open system. It may well be, however, that many students are attracted to mathematics simply because they *do* view it as closed and "fixed". They may in fact then find little in their learning of the subject in the secondary school to change their view, so that by the preuniversity year the interest of those students who are continuing with the subject is strengthened.

TABLE 4.10. *Means, Standard Deviations, and Number of Cases of Interest in Mathematics in Relation to Scientific or Nonscientific Nature of Father's Occupation for Population 1a.*

Country	Scientific			Nonscientific			Difference of Means
	M.	S.D.	N.	M.	S.D.	N.	
Australia	6.0	1.9	614	5.9	1.8	115	0.1
Belgium	5.7	1.9	32	5.7	1.8	390	0.0
England	6.0	1.6	99	5.7	1.6	638	0.3
Finland	6.3	1.7	50	6.2	1.6	136	0.1
France	5.9	1.9	56	5.5	1.8	546	0.2
Japan	6.5	1.5	102	6.0	1.5	411	0.5
The Netherlands	5.3	1.8	11	5.4	1.7	96	−0.1
Scotland	5.5	1.8	156	5.2	1.8	1,158	0.3
Sweden	6.2	1.3	115	5.8	1.4	523	0.4
United States	6.2	1.7	222	6.2	1.7	1,335	0.0
All Countries	5.9	1.7	146	5.8	1.7	535	0.1

The relationship between interest in mathematics and the status of father's occupation was also investigated. The correlations between these two variables are shown in Figure 4.4. For Population 1 a, higher interest does tend to be associated with higher status, but the relationship is weak. In Population 3 a, however, although the overall correlation is zero, in a number of countries relatively large negative coefficients appear, indicating that in these countries there is a tendency for the students with fathers in lower status occupations to have a greater interest in mathematics. Perhaps the explanation is that students from the lower status groups who have survived to Population 3 a must in any case be more strongly motivated than students coming from higher status families and hence demonstrate a greater interest. Furthermore, in some countries a higher proportion of students from lower status groups tend to go from secondary school into technological occupations. An additional analysis was made in Population 1 a to discover whether the scientific or nonscientific nature of the father's occupation had any relationship with the student's interest in mathematics (Table 4.10). Although there is a tendency for interest to be a little higher when the father's occupation is of a scientific nature, the relationship over all countries is not significant.

Finally, knowledge of the relationships between different populations on some of these factors may be useful. Table 4.11 gives rank-order

TABLE 4.11. *Rank Correlation Coefficients of National Mean Scores on Interest and Selected Attitudes Between Different Populations.*

Variable	Between		
	1 a and 3 a	1 a and 3 b	3 a and 3 b
Interest	.05	.25	.67
Wishes to Take More Mathematics	.36	.35	.59
Mathematics as a Process	.52	.70	.76
Difficulty of Mathematics	−.13	.10	.90
Mathematics Importance	.61	.55	.77
Views About Mathematics Teaching	.60	.57	.83
Views About School Learning	.48	.71	.95

correlation coefficients between different populations of the national mean scores on interests and selected attitudes. It will be noted immediately that many of these coefficients are substantial, particularly those between Populations 3 a and 3 b. They indicate that, with the exception of views about the difficulty of mathematics, these affective aspects associated with the learning of mathematics are rather stable. They support the well-known statement that changes in interest and attitude are quite difficult to bring about and that learning experiences especially designed for this purpose are generally necessary before dramatic changes can be expected.

Achievement in Relation to Teachers' Perceptions and Training

This section discusses certain questions related to the perception held by the teachers of their instructional situation. It also includes discussions of the relation between scores on the mathematics test and certain aspects of teacher training.

Formulation of Hypotheses

Achievement and Opportunity to Learn

One of the factors which may influence scores on an achievement examination is whether or not the students have had an opportunity study a particular topic or learn how to solve a particular type problem presented by the test. If they have not had such an opportunity they might in some cases transfer learning from related topics to produce a solution, but certainly their chance of responding correctly to the test item would be reduced.

Questions related to this factor arose in connection with the pilot investigation referred to in Chapter 1 of Volume I and reported as one of the results of that project.[1] Some teachers helping with the pilot project expressed the opinion that their students had been unable to answer some of the questions because they had not been given any instruction in those parts of mathematics. This difficulty is bound to arise in any test which is administered to pupils in different countries with different educational systems and different curricula. Scottish teachers were therefore asked to estimate the opportunity which their students had been given to become acquainted with the material covered by the item. It was found that the teachers' ratings were not related in a statistically significant way to the scores made by the pupils, a conclusion which appeared rather surprising. The factor found to be associated with success on most of the items studied was the ability of the class, again assessed in a very rough way. It was decided that a similar investigation on a larger scale and covering several countries should be incorporated in the present inquiry.

Teachers assisting in the IEA investigation were asked to indicate to what extent the test items were appropriate for their students. This information is based on the *perception* of the teacher as to the appropriateness of the items. The hypothesis that was formulated in this case was as follows: *Total mathematics scores will be related to the teachers' perception of the students' opportunity to learn the mathematics involved in the test items* (Hypothesis 16).

Achievement and National Emphasis

Another factor that may influence the performance of a teacher, and indirectly of the students, is the teachers' perception as to the relative emphasis put upon different topics in the curriculum of their country. The topics included and the teaching methods employed in the mathematics education within most of the countries in this study depend mainly on three factors: (1) the curriculum as set down by the school authorities; (2) the textbooks; (3) the teachers' use of these books. In many countries the curriculum is formulated only in a general way without any specific prescriptions. In some countries, for example in

[1] An analysis of the reactions of Scottish teachers and pupils to items in the geography, mathematics, and science tests. See A. W. Foshay (ed.), *Educational Achievement of Thirteen-Year-Olds in Twelve Countries,* UNESCO Institute for Education, Hamburg, 1962.

England, there is no officially defined curriculum, and what is taught is left to the teacher or to each school to decide, although their decisions are influenced in no minor degree by the examination system.

In all cases, however, what is treated in mathematics classes in any school is largely governed by the textbooks available, and the authors of such books are themselves clearly influenced by the requirements of the curricula. In some countries (for example, Sweden) regulations govern the teachers' freedom of choice among textbooks, for a book must be approved for use in the national school system by a special board. There are instances when the books of well-known textbook authors have failed to receive this approval.

In these circumstances one would anticipate finding only relatively small variation within countries in the perception of different teachers concerning the appropriate topics to be emphasized. On the other hand, one may expect to find considerable variation between countries as to the relative national emphasis on a given topic. Moreover, the achievement subscores on various topics in each country should vary with the relative national emphasis.

There are several values to be derived from investigating these questions. In the first place, curriculum authorities need to know whether teachers' *perceptions* of the importance of topics correspond to that desired by authorities, be they official or professional. In the second place, if there is in fact no relationship or a very weak one between teachers' perceptions of importance and the actual achievement of students, the situation certainly calls for further study. Perhaps teachers' perceptions in this respect need to be sharpened or perhaps the national emphasis is unrealistic when judged by teachers in the light of their experience. Finally, if there is in fact a strong relationship between teacher's perceptions of national emphasis and students' achievement on particular topics, some countries may want to reconsider their distribution of emphasis in order to change the profile of achievement among topics.

The formal statement of the hypothesis bearing on these questions follows: *The profile of test performance in each country will be related to the national emphasis on each topic (as reported in teachers' ratings) in each school program* (Hypothesis 17).

Teacher Perceptions of Freedom

A third factor that may influence the performance of a teacher is his perception of the amount of restraint or freedom under which he carries on his task. Is student performance better when teachers feel that they must follow directives as to curriculum and instruction quite closely, or is it better when they believe that they have some freedom of choice as to what is taught or how it is taught or both?

Mathematics has a long history as a school subject, and it has been one of the most stable parts of the curriculum until recently. The basic elements of arithmetic and mathematics have been known and used for a long time. Strong traditions have developed as to what is to be taught and how it is to be taught. Communication between countries has brought about a certain international uniformity in content and methods in this field.

Recently, however, there has been a strong movement to introduce new content and methods into mathematical education. As teachers become aware of the ferment now going on, some of them feel restrictions on their freedom to try out new ideas. Others may have no such feeling and may, in fact, want such restraint. The question that arises is how differing perceptions by teachers of the amount of freedom that exists are related to their students' achievement. Because it was thought that there might be considerable variation between countries in the perception of freedom and constraint, it seemed opportune to take advantage of the international scope of the IEA study to gather data bearing on this question. The formal statement of the hypothesis was: *When level of instruction*[1] *is held constant, the total mathematics score will be higher in schools where teachers feel themselves to have greater freedom in determining what will be taught and how it will be taught* (Hypothesis 18).

Achievement, Interests, and Attitudes of Students in Relation to Recent In-Service Training of Teachers

When achievement in mathematics is under discussion, usually there are comments sooner or later about the training of the teachers. The length and kind of their training, and the recency of preservice training or of in-service supplementation are assumed to influence the achievement of their students. One question is whether or not there is a difference between the achievement of students whose teachers have had recent in-service training in mathematics and students whose teachers have not had recent in-service training. If there is substantial evidence that such training is associated with higher achievement, school authorities would have reason to encourage or demand the provision of opportunities for in-service training and the participation of teachers in it.

---

[1] "Level of instruction" is defined by a scale described in Volume I, p. 131.

There is considerable variation at present among countries in the nature and amount of in-service training in mathematics available to teachers. Moreover, since many of the existing in-service training programs are unofficial and participation is voluntary, the motivation which leads teachers to participate may vary. In some cases (as in England) it is the teachers with the best previous training who tend to participate. They are doubtless motivated mainly by a desire to do a better job of teaching. In contrast, although the programs in the United States are for the most part "unofficial" (that is, not required by the state for maintenance of a certificate to teach), participation in certain forms of in-service training is commonly rewarded by an increase in salary. In these circumstances, many poorly trained teachers may seek to improve themselves as teachers as well as financially.

Whatever the reason may be for participation by teachers it is clear that appreciable amounts of time and money are being spent on such programs. Is there, then, any evidence that in-service *mathematics* training produces measurable increases in the achievement of students who are subsequently taught by these teachers? Several hypotheses were formulated, as follows: *Students whose teachers have had recent (within 5 years) in-service training in mathematics will*

1. *Have higher total mathematics scores than students studying under teachers who have not had such training.*
2. *Be more inclined to view mathematics as an open system.*
3. *Have greater interest in mathematics.*
4. *Describe mathematics teaching as emphasizing inquiry.*
5. *View mathematics as increasingly central for occupations and the development of society* (Hypothesis 19).

Amount and Type of Preservice Training

Clearly, study of the relation of student achievement to the nature and duration of teachers' preservice training is important also, and interest in this question has been growing in many countries in recent years. During the generation since 1940, the shortage of qualified teachers has led to the issuance of emergency certificates in some countries. In mathematics especially, it has led to the employment of teachers who have studied mathematics but little else of an academic sort—for example, some men from the military. Others who were originally prepared to teach at the primary school level have been drawn into secondary schools despite relatively slight preparation in mathematics.

More recently, the function of the specialized teacher-training institution has come under review. Should these institutions not broaden their functions and their curricula? Do they not inevitably depress the quality of people entering teaching because they accept students who would not qualify for entrance to the university?

The experience of educationalists suggests that the answers to these questions are less obvious than they appear. Some will argue that the good quality of university graduates is a consequence less of the university curriculum than of its admission policies—if only good students are permitted to enter, it should not be surprising that good graduates emerge. The teacher-training institutions, on the other hand, also have good graduates despite their far less selective entrance policies.

What of the others—those teachers who have attended both teacher-training institutions and universities, who have come into teaching from institutions other than the two that supply most teachers, or who have come from institutions (such as the military) having little connection with the formal educational structure? The hypothesis dealing with these questions was formulated as follows: *When the level of teachers' training is held constant, mathematics achievement will be directly related to the amount and the quality of the preservice training of the teachers* (Hypothesis 20).

Findings

Achievement and Opportunity to Learn

Hypothesis 16 involves the relationship between mathematics scores and the teacher's perception of students opportunity to learn the specific mathematics content involved in the test items. Question 18 of the teacher questionnaire was as follows:

Attached to this questionnaire are the sets of tests that are being given to some of your students. Also attached is the special answer sheet on which they are recording their answers by blackening the appropriate response position. The response positions for the questions in each test are located in blocks on the answer sheet, e.g., T A, T 3, T 6.

To have information available concerning the appropriateness of each item for your students, you are now asked to rate the questions as to whether or not the topic any particular question deals with has been covered by the students *to whom you teach mathematics and who are taking this set of tests.* Even if you are not sure, please make an estimate according to the scale given below.

Please examine each question in turn and indicate in the way described below, whether, in your opinion

A. All or most (at least 75%) of this group of students have had an opportunity to to learn this type of problem.

B. Some (25% to 75%) of this group of students have had an opportunity to learn this type of problem.

C. Few or none (under 25%) of this group of students have had an opportunity to learn this type of problem.

Please indicate your rating (A, B, or C) by blackening the appropriate response position for each question on the attached answer sheet.

These ratings were scaled by assigning the value 87.5 (midway between 75 and 100) to rating A, 50 to rating B, and 12.5 to rating C. The ratings given by a teacher to each of the items in the tests taken by his students were averaged, unrated items being excluded from the calculations. For each teacher there was thus a mean rating and the mean score made by his pupils on the tests rated.

The agreement between these ratings and the scores in mathematics was estimated by calculating the coefficients of correlation between these variables. This was done for all four populations, and the results are shown in Figure 4.5 (see also Chapter 6, pp. 269–274).

There was a small but statistically significant positive correlation between the scores and the teachers' ratings of opportunity to learn the topics. There was, however, much variation between countries and between populations within countries in the size of these coefficients. Two countries have no significant coefficients for Populations 1 a and 1 b, and two other countries have correlation coefficients exceeding .50.

One reason for at least some of the coefficients being small is the homogeneity of the teachers' ratings in these countries. The larger coefficients obtained in England and Scotland may well be, as the author of Chapter 1 suggested, an indication of the wide variety of mathematics courses offered to the younger students (Populations 1 a and 1 b) in these countries. Support for this idea is given by the appearance of much smaller coefficients at the higher levels in both countries, where the curricula are much less variable.

These considerations suggest that it would be profitable to examine the between-countries correlations for these variables. It is found that they are .64, .73, .80, and .40 for Populations 1 a, 1 b, 3 a, and 3 b, respectively. In other words, students have scored higher marks in countries where the tests have been considered by the teachers to be more appropriate to the experience of their students.

The average ratings and scores are of interest in themselves and are shown in Table 4.12. The ratings range from a low of 37.4 for Sweden's Populations 1 a and 1 b to a high of 75.2 for France's Population 3 a.

The general conclusion is that a considerable amount of the variation between countries in mathematics score can be attributed to the differences between students' opportunities to learn the material which

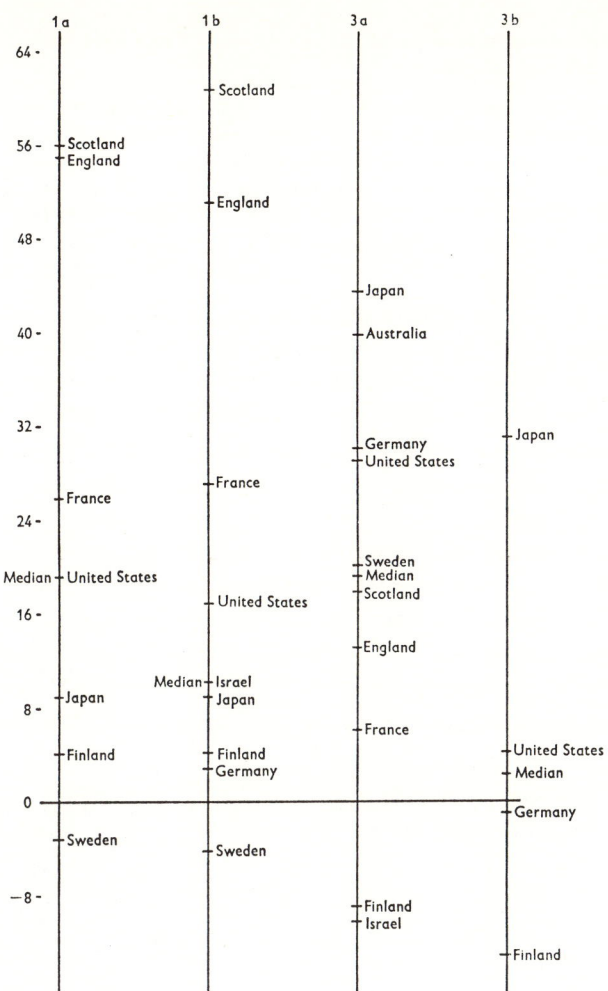

Figure 4.5. Correlations between *Total mathematics score* and *Teacher's ratings of opportunities to learn*.

was tested. This applies in particular to Populations 1 a and 1 b. In Sweden and Finland, for example, there has been a tendency among curriculum experts in recent years to defer many of the more "abstract" topics to a later grade level. Thus, in these countries where the entry to school takes place at 7, only arithmetic is studied before the age of 14, except for the relatively small group which transfers from the elementary to the secondary academic school before that age. Within countries and within schools the effect is often masked by the homogeneity of the groups.

TABLE 4.12. *National Means of Teachers' Ratings of Opportunity to Learn and of Mathematics Scores.*

Country	Population							
	1 a		1 b		3 a		3 b	
	Ratings	Scores	Ratings	Scores	Ratings	Scores	Ratings	Scores
Australia	—	20.2	—	18.9	56.7	21.6	—	—
England	60.4	19.3	60.4	23.8	66.6	35.2	—	21.4
Finland	47.4	24.1	47.4	26.4	65.2	25.3	74.1	22.5
France	49.9	18.3	50.9	21.0	75.2	33.4	—	—
Germany	—	—	58.1	25.5	67.5	28.8	73.5	27.7
Israel	65.7	—	65.6	32.3	63.4	36.4	—	—
Japan	63.1	31.2	63.1	31.2	63.5	31.4	74.3	25.3
Scotland	51.3	19.1	51.1	22.3	58.3	25.5	—	20.7
Sweden	37.4	15.7	37.4	15.3	54.8	27.3	—	12.6
United States	47.5	16.2	50.9	17.8	50.3	13.8	50.3	8.3
All Countries	54.5	19.8	55.9	23.0	58.7	26.1	68.3	19.6
Range	28.3	15.5	28.2	17.0	24.9	22.6	24.0	19.4
Correlation Between Ratings and Scores by Countries	.64		.73		.80		.40	

Achievement and National Emphasis

Hypothesis 17 is concerned with the relation between achievement and national emphasis in mathematics education. In particular, it calls for a study of the *profiles* of emphasis on various topics. The teachers reported their judgment of the relative emphasis on certain topics. It was recognized that by making comparative studies of the textbooks in use it would be possible to get one sort of information concerning the emphasis on different topics that prevails in the various countries. However, such studies would involve a larger investigation than has been aimed at in the present research. Although they would no doubt give a clearly defined picture of the mathematical education in different countries (since most teachers, in their instruction, follow the textbooks closely, in some cases against their own will—compare Hypothesis 18), this method of determining emphasis was not feasible.

The same basic data used in the study of Hypothesis 16 were used for Hypothesis 17, but the data were treated differently. Responses to Question 18 of the teacher questionnaire (TCH 1) were used to compute an "index of emphasis". For each country, each population, each school program, and each topic the "index of emphasis" was calculated in percentages. In this procedure the three-step scale of the teachers' rating

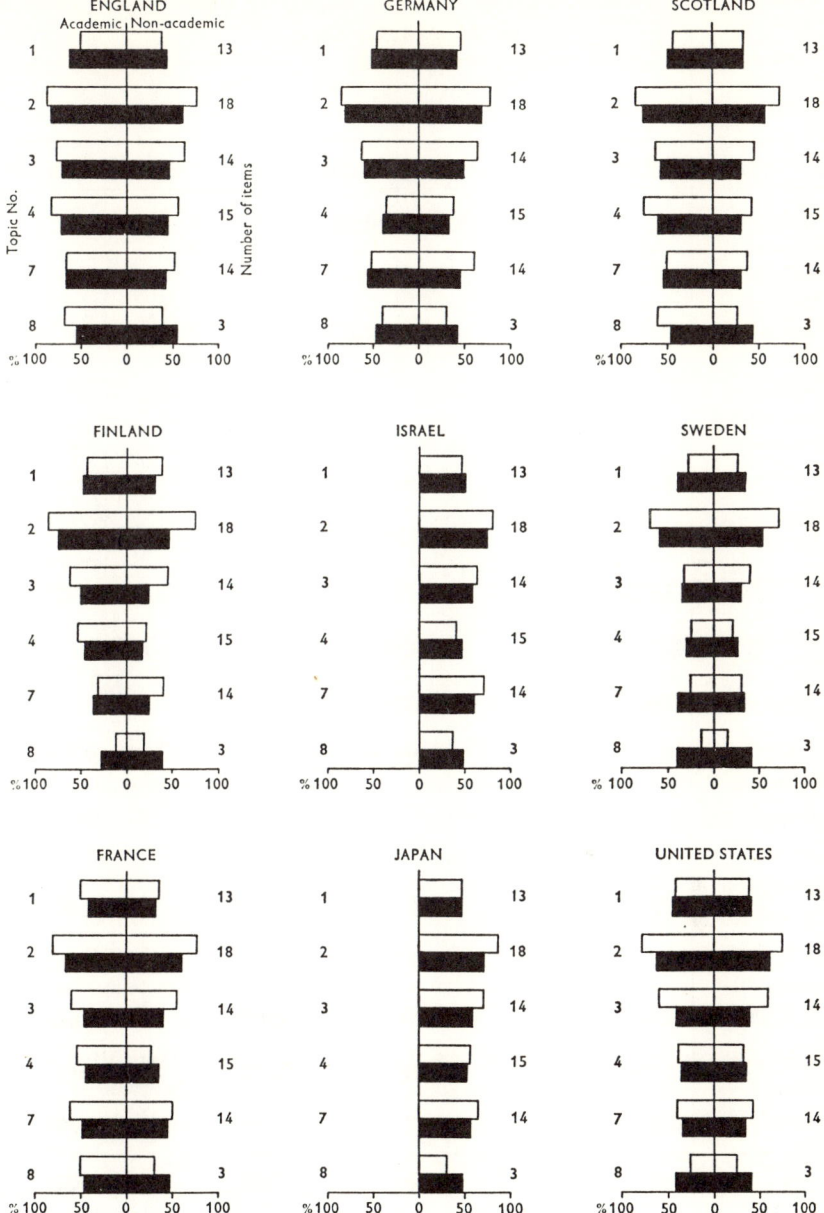

Figure 4.6. National profiles of percent emphasis on selected topics in academic and nonacademic courses, Population 1 b. Topics: 1 = New Mathematics, 2 = Basic Arithmetic, 3 = Advanced Arithmetic, 4 = Elementary Algebra, 7 = Intuitive Geometry, 8 = Demonstrative Geometry. □, Estimation; ■, performance.

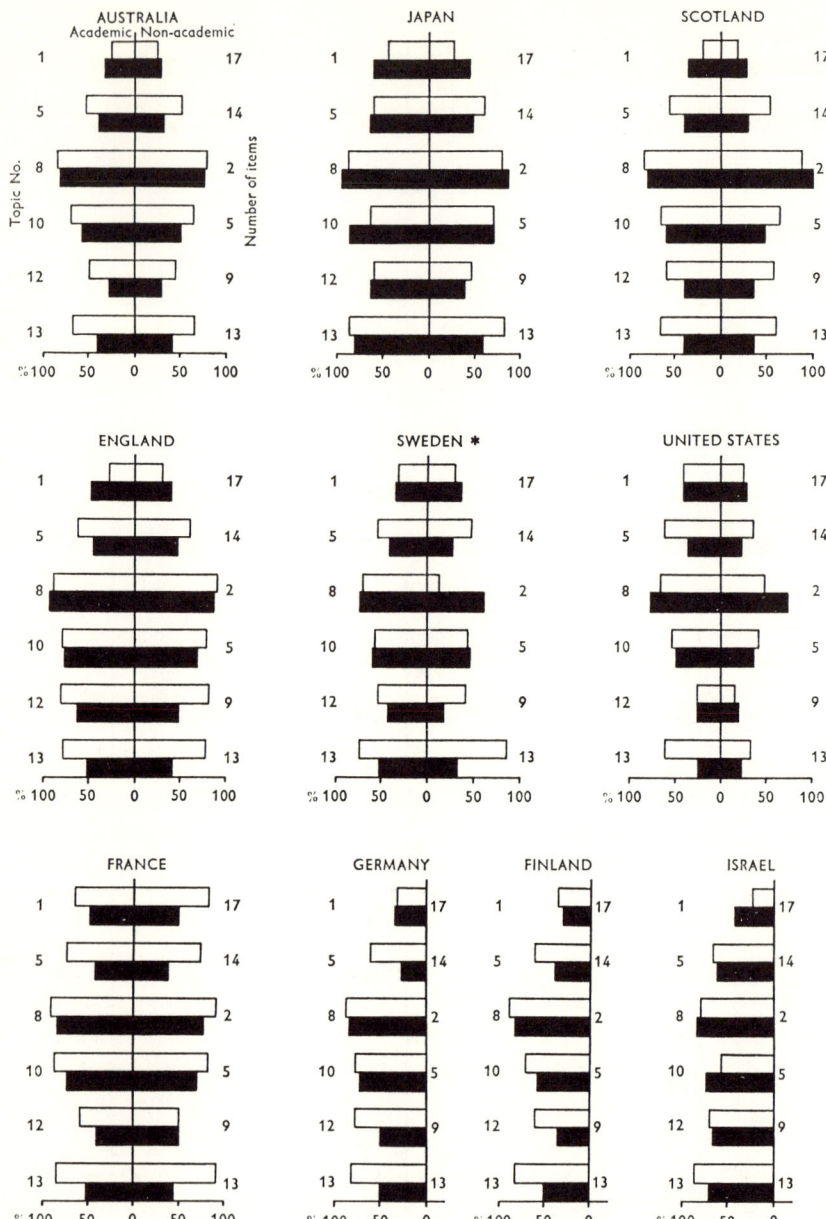

Figure 4.7. National profiles of percent emphasis on selected topics in academic and nonacademic courses, Population 3 a. Topics: 1 = New Mathematics, 5 = Intermediate Algebra, 8 = Demonstrative Geometry, 10 = Analytic Geometry, 12 = Calculus, 13 = Other Analysis. □, Estimation; ■, performance.

* "Academic" indicates students, who plan to go to university, "non-academic" are the rest.

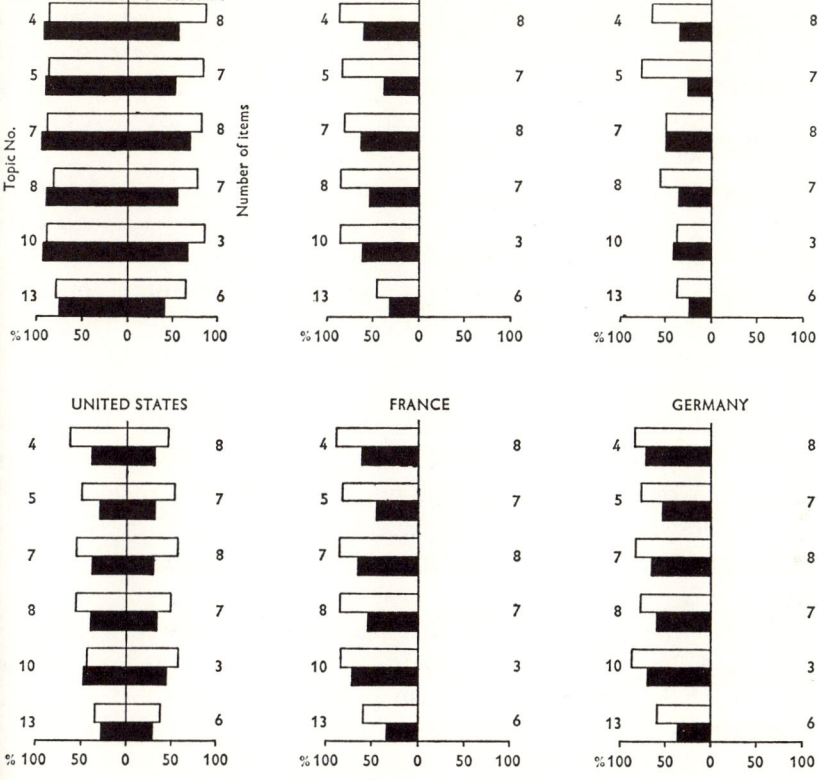

Figure 4.8. National profiles of percent emphasis on selected topics in academic and nonacademic courses, Population 3 b. Topics: 4 = Elementary Algebra, 5 = Intermediate Algebra, 7 = Intuitive Geometry, 8 = Demonstrative Geometry, 10 = Analytic Geometry, 13 = Analysis. □, Estimation; ■, performance.

was translated to 87.5, 50.0, and 12.5 percent for each rating of each item in the group of items that formed a topic. The result of the inquiry is shown in the diagrams of Figures 4.6–4.8, where the white bars represent, in percent, the emphasis on each topic as estimated by the teachers in the way indicated above.

The students themselves indicated on their questionnaire to which course of study (academic, vocational, or general) they belonged (see page 91). Only two types of school programs, that is, academic and nonacademic (vocational and general) were used. It was expected that a certain difference in emphasis between a purely theoretical program and the two other programs would appear.

For the lowest age groups only the data for Population 1 b, where the characteristic common factor is the grade rather than the age, have been used. The Populations 1 a and 1 b do in fact have a large number of students in common, but for an investigation of the curriculum at this level, by means of teachers' ratings, more precise information can be expected by concentrating on a certain grade to which most of that age group belong.

As was mentioned earlier, the investigation of the distribution of emphasis on various topics could have been carried out through comparative study of the textbooks. The kind of investigation that was carried out, however, gave a unique opportunity for finding out how the distribution of emphasis, as estimated by teachers, compared with the actual result in the test. In other words, it enabled a comparison to be drawn between theory and reality, between hypothetical emphasis and final outcome. Both had the same source, the actual items used in the test. Thus, in the diagrams of Figures 4.6–4.8, there is below each white bar a black one which indicates the percent of students giving a correct answer on the items representing the topic in question. For each population, six characteristic classifications of subject matter were chosen.

In Population 1 b there is a striking relation between the teachers' ratings and the students' achievement, and this holds true for all countries. The only noticeable discrepancy occurs for "demonstrative geometry" where, as can be seen, the test result is roughly the same in all countries, despite the different emphasis which this topic was perceived as having in the various countries.

For Population 3 a, the hypothesis is not supported as conclusively as for Population 1 b. In some countries the relation between the emphasis in the teaching and the outcome is fairly strong (for example, Australia, England, Israel, and Japan); in other instances, the hypothesis seems to be disproved. In certain countries the outcome appears to be systematically lower than the teachers' perceptions of emphasis (for example, Finland and France).

The last-mentioned finding seems to be a general one for Population 3 b. A reasonable explanation is that, for students who do not specialize in mathematics, the opportunity of studying problems of certain type is no guarantee of their actual learning because mathematical studies at this level have become much more complex than Population 1 b.

In general, the conclusion seems warranted that the profile of achievement of characteristic classifications of subject matter does follow the national emphasis as judged by the teachers.

Teacher Perceptions of Freedom

Hypothesis 18 concerns the relationship between student achievement, as measured by the total mathematics score, and teacher freedom, as perceived by the teachers. The teacher questionnaire was used to obtain the data on the teachers' perceptions of freedom. The index of "Degree of Freedom Given Teachers" is a composite score, derived from four questions, relevant to four aspects of constraint: (1) by a syllabus or curriculum that must be followed, (2) by a textbook that must be used, (3) by prescriptions as to teaching methods, and (4) by examinations that students are required to take. The scale has a possible range from 10 to 20. The maximum of 20 signifies complete freedom from any sense of constraint and the minimum of 10 indicates a sense of hampering constraint on all four counts. Details concerning the composition of the index are given in Chapter 12 of Volume I.

In addition to the quantitative data, comments came from some countries. From France came the following remark as to syllabus and textbooks: "Teachers have to follow a fixed program but remain free to choose their books and their methods."

An analysis of answers from other countries to the same question reveals that in Scotland it is not clear who was thought to restrict the freedom, the Scottish Education Department or the principal teacher of the subject in the school, both receiving some blame.

The comment from the United States read:

The autonomy of the teacher does not vary mainly with the type of school. The main factor is administrative "tightness" of school, which varies enormously. In better schools and schools with better teachers and in schools with separate mathematics departments, teachers have much independence in textbooks, ways of teaching, etc.

England commented:

The individual teacher is free to choose how the subject will be taught and which textbooks will be used in all the above school categories. Freedom in what is taught depends on the extent to which students will take external examinations, e.g. G.C.E. "O" and "A" Level. The examining bodies determine the syllabus although the teacher is free to submit an individual syllabus for acceptance by the examining body. The general position as regards the last two facts is that the nine examining bodies in England and Wales vary fairly little in mathematics syllabuses and that individual teachers rarely submit their own syllabuses.

Australian teachers seem to be under similar restrictions due to external examinations as are their English colleagues. There are, however, other factors that limit the freedom of the teachers; for example, only one prescribed textbook.

Teachers in Japan are able to express their opinions in regard to the

TABLE 4.13. *National Means, Standard Deviations, and Number of Cases for Scores on Degree of Freedom Given Teachers, for Populations 1b and 3a.*

Country	Population 1 b			Population 3 a		
	M.	S.D.	N.	M.	S.D.	N.
Australia	15.6	3.3	514	15.4	3.0	125
Belgium	12.4	2.2	61	13.8	2.1	23
England	16.3	3.4	994	16.0	3.2	184
Finland	16.6	1.3	77	16.5	1.3	22
France	16.8	3.2	203	14.7	2.7	14
Germany	16.7	2.8	184	16.0	3.2	37
Israel	17.9	2.8	118	13.9	5.2	5
Japan	16.9	1.6	409	16.9	1.6	140
The Netherlands	18.1	2.7	58	16.7	2.4	31
Scotland	16.8	3.4	234	15.7	3.7	126
Sweden	19.0	2.1	134	18.9	1.9	48
United States	17.7	2.9	727	17.2	3.3	217
All Countries	16.7	3.1	3,713	16.4	3.1	972

textbooks adopted but the decision is made *en bloc* by the local board of education.

An analysis of the data by means of the multiple regression model is reported in Tables 6.6–6.9. In Chapter 6 it is reported that the contribution made by the variable "Degree of Freedom Given Teachers" to the total variance for all countries is negligible. Examination of the correlation coefficients for individual countries reveals that the coefficients for France and Populations 1 a and 1 b are positive and comparatively large (.17 and .26), but otherwise they are mostly small and tend to be negative. In Population 3 a the coefficient for Finland is relatively large but negative. Such exceptions to the general trend of zero correlations are perhaps due to conditions within the individual countries, but from these data as a whole there appears to be no appreciable relationship between student achievement and the degree of freedom perceived by teachers.

The national means of the scale for "Degree of Freedom Given Teachers" are reported in Table 4.13 for Populations 1 b and 3 a. The means for Population 1 a are, in most cases, identical with those for Population 1 b and, similarly, the means for Population 3 b do not differ appreciably from those for Population 3 a.

The coefficients of correlation between the national means and the means of achievement scores are −.50 and −.45 for Populations 1 a and

3 a, respectively. (See Table 2.1.) Between countries, therefore, high mathematics achievements is associated with a greater perception of *restriction* of their freedom on the part of teachers. This is most evident in the case of Belgium, where the perception of freedom is relatively low yet achievement is high, and in the cases of Sweden and the United States, where teachers think they have considerable freedom but achievement is low. Hypothesis 18, which was that "the total mathematics score will be higher in schools where teachers are given the greatest freedom in determining what will be taught and how it will be taught" is not supported by the data. The conclusion is that higher mean scores tend to occur in countries where teachers think they are given less freedom.

Achievement, Interests, and Attitudes in Relation to In-Service Training of Teachers

The hypotheses in this group state that students whose teachers have had recent in-service training in mathematics will have higher total mathematics scores and both greater interest in and more favorable attitudes toward mathematics. The hypothesis as originally formulated restricted the study to the effect of in-service training in *new* mathematics. A question was included in the teacher questionnaire (TCH 1) to ascertain whether or not the teachers had participated in such training. The question was: Have you attended any lecture/seminar course dealing with "New Mathematics"?

There is little doubt that the term "new" in this question did not mean the same thing to all respondents. When the data were examined it was found that the results were for the most part insignificant. The hypotheses dealing with in-service training in "New Mathematics" were therefore abandoned. The only results that will be reported are for recent in-service training of any kind.

The correlation coefficients between achievement in mathematics and recent in-service training are reported in Tables 6.6 to 6.9. For Population 1 a, the largest coefficient is .12, which was found for both The Netherlands and Scotland. However, the correlation for Finland is $-.15$, and the mean is .05. For Population 3 a, coefficients of .18 and .16 were found for the United States and Finland, but there are five negative coefficients and the mean is only .02.

In the case of the other variables related to in-service training in Hypothesis 19, the results of correlations within countries were, in general, not significant and will not be reported. A few exceptions stand out—for example, the coefficient between in-service training and interest

TABLE 4.14. *Correlation Coefficients Between National Mean on In-Service Training and Five Other Variables for Populations 1a and 3a.*

	Population	
	1 a	3 a
Total Mathematics Score	−.58	−.54
Tendency to View Mathematics as an Open System	−.21	.78
Interest in Mathematics	.21	−.11
Description of Mathematics Teaching as Emphasizing Inquiry	−.57	−.03
View of Place of Mathematics in Society	.59	.43

is −.12 for Population 1 a in Finland, and is .13 for Population 3 a in the United States.

Since interesting differences between countries may be masked by the foregoing analysis, the correlations between national means and the six variables under consideration are reported in Table 4.14. It is immediately evident that in most cases these coefficients are substantially larger than those for the individual countries reported above.

For Population 1 a there are large coefficients that indicate that increased in-service training is accompanied by *lower* achievement scores, a tendency *not* to describe mathematics learning as emphasizing inquiry, and the view that mathematics is important to society. For Population 3 a there are large coefficients that indicate that increased in-service training is again accompanied by *lower* achievement. For this population, however, the view that mathematics is an open system accompanies increased in-service training, as does a strong belief that mathematics is important for society.

It is surprising that such relatively large coefficients emerge in view of the very rough measure of in-service training that was obtained. Nothing is known about the type of training, its intensity, or its quality. The high negative correlation coefficients with the mathematics score may be explained by assuming that more in-service training is taking place where it is most needed—with teachers of lower scoring students. This assumption, however, is a dubious one.

Amount and Type of Pre-service Training

Hypothesis 20 concerns the relationship between mathematics achievement and the length and type of the teachers' postsecondary education.

In all countries, a university education has the greatest prestige, and

It is natural that noneducators question the adequacy of the preparation offered by any other postsecondary institution. If this view is valid, students in the classes of university-trained teachers should receive higher scores. Moreover, in many schools and in several of the countries represented in the present study, only the more able secondary-school students are placed in the classes of university-trained teachers. Frequently, only university-trained teachers teach the most advanced mathematics classes. It follows that if there is no significant difference between the scores of students in classes of university-trained teachers and nonuniversity-trained teachers, or where the nonuniversity teachers' students receive significantly higher scores, something is happening that contradicts popular assumptions on this issue.

Data to test this hypothesis were gathered by means of the teacher questionnaire. The teachers were classified according to their postsecondary school education. If all of their training was in a teacher-training institution, they were put in the teacher training group. If their training was (all, finally, or mostly) in a university, they were put in the university group. In all other cases they were classified as "other", for example, a retired military officer who received his advanced education under military auspices.

Teachers were further classified according to whether they had little (3 years or less), some (4 years), or much (more than 4 years) postsecondary education. Unfortunately, in a great many instances there was either no response from the teacher on this question, or the number of teachers in the category was so small that the mean score had no significance. Tables 4.15 and 4.16 give the results for Populations 1 a and 3 a. Data for other populations were as a rule scanty and scattered; those for Population 1 b, when they are available, do not differ markedly from those of Population 1 a. In several instances where data for Population 1 a were not available, but those for Population 1 b were, the latter were substituted. Also, to simplify the tables, the "some" category was eliminated except in several instances where it substituted for "little".

In general, the data seem to support the hypothesis that the more training a teacher has received, the better will be the achievement of his students, but this does not hold in all cases. For Population 1 a, the All Countries means are identical for students of teachers who received "much" and "little" training at teacher-training institutions. However, the All Countries mean for university-trained teachers who have had "much" training is substantially greater than the mean for those who have had "little" training. For Population 3 a, the means for uni-

TABLE 4.15. *Means of Mathematics Scores, Standard Deviations, and Number of Cases Related to the Type of Training of Teachers and to Whether the Teachers Had Little (L) or Much (M) Training.*

Population 1 a.

Country		Teacher Training			University			Other	
		M.	S.D.	N.	M.	S.D.	N.	M.	S.D.
Australia	L	18.9	13.9	1,115	17.4	13.2	199	10.0	8.2
	M	18.5	8.8	181	23.6	13.5	814	10.0	8.6
Belgium	L	23.6	13.1	642	—	—	—	—	—
	M	19.4	8.8	52	40.2	9.9	87	—	—
England	L	12.6	12.9	1,168	19.6	12.8	17	33.4	11.2
	M	14.5	12.3	128	31.6	16.6	608	20.4	14.2
Finland	L	20.8	9.9	70	—	—	—	21.3	9.7
	M	14.9	9.9	25	—	—	—	21.8	8.3
France	L	19.9	12.7	710	25.0	12.2	85	19.4	13.3
	M	14.7	10.9	23	27.3	10.5	88	18.8	8.5
Germany (Level 1b)	L	24.1	11.7	1,114	16.3	5.3	89	24.4	9.7
	M	27.7	10.1	482	32.5	9.3	744	26.5	9.0
Israel (Level 1b)	L	31.3	12.9	718	35.2	11.0	192	33.8	11.8
	M	32.7	12.0	183	38.6	10.3	350	26.3	10.1
Japan	L	30.7	16.2	372	26.0	13.1	45	29.0	16.1
	M	31.8	16.8	506	32.6	17.5	270	29.4	13.5
Scotland	L	17.0	10.8	51	10.7	10.8	117	8.9	5.7
	M	18.4	12.1	572	19.6	14.0	3,030	29.6	11.7
United States	L	12.3	8.3	41	13.8	10.4	146	—	—
	M	18.3	13.6	950	17.1	13.0	2,964	—	—
All Countries	L	21.1	12.3	6,001	20.5	11.1	890	22.5	10.7
	M	22.1	10.7	3,102	29.2	12.7	8,955	22.8	10.5

versity-trained teachers and "other" teachers who have had "much" training are also greater.

University-trained teachers may also be compared with those who had a different kind of training. For Population 1 a, the mean for those who had "much" university training is significantly greater than the mean for those who had "much" work at teacher-training institutions. A different conclusion, however, seems to hold for Population 3 a, where students taught by teachers who were educated at teacher-training institutions achieve significantly better than students taught by university-trained teachers, regardless of whether either group of teachers had "little" or "much" training. The same finding holds for students trained

TABLE 4.16. *Means of Mathematics Scores, Standard Deviations, and Number of Cases Related to the Type of Training of Teachers and to Whether the Teachers Had Little (L) or Much (M) Training.*

Population 3 a.

		Teacher Training			University			Other		
try		M.	S.D.	N.	M.	S.D.	N.	M.	S.D.	N.
alia	L	16.4	20.5	29	16.3	9.6	36	—	—	—
	M	20.2	10.3	49	22.7	9.8	771	—	—	—
nd	S[a]	23.8	9.4	32	37.5[a]	11.7	118	39.1[a]	12.6	69
	M	28.6	9.4	26	36.0	11.4	628	38.8	8.4	42
	L	37.7	6.8	28	16.3	8.4	33	21.5[a]	11.1	111
	M	32.8	12.9	202	35.3	12.1	220	32.0	9.0	24
d States	L	—	—	—	15.8	7.8	4	—	—	—
	M	—	—	—	16.5	13.0	926	—	—	—
untries	L	26.0	12.2	89	21.4	9.4	191	30.3	11.8	180
	M	27.2	10.9	277	24.8	11.6	2,545	35.4	8.7	66

[a] Some.

in "other" ways. Thus, for Population 3 a, the view that students in classes of university-trained teachers should receive higher scores is contradicted by these data.

Some comments on individual countries may be warranted. Thus, for Australia at Population 1 a, there is a significant difference in favor of "much" university training over both "little" and much "other" training. For Belgium, there is a significant difference (exceeding 20) between "much" university training and "much" teacher training.

In England, there is a significant difference between "little" teacher training and "little" other training and, surprisingly, the scores of students taught by the "other" group are superior. There is also a significant difference between the scores of students taught by teachers with "much" university training and "much" teacher training.

In conclusion, the hypothesis that mathematics achievement would be directly related to the amount of the preservice training of the teachers was, in general, confirmed. But the view that the students of university-trained teachers would achieve higher mean scores than the students of teachers trained at teacher-training institutions was weakly supported for Population 1 a and contradicted for Population 3 a.

School Decisions Relative to Curriculum

Formulation of Hypotheses

Time Spent on School Work

One of the important decisions made by school authorities concerns the amount of time students are required to spend on school work. In this chapter we are not concerned with differences among countries in the number of years of schooling required. Instead, questions are raised as to the effect on achievement in mathematics of variations among countries in the number of hours of schooling per week required for all subjects and also for mathematics in particular. Related to these questions are two others concerning the number of hours per week devoted to homework in all subjects and to the study of mathematics in particular.

It is natural to suppose that the more hours per week students spend in school, the more they learn. Within a given country, the variation between schools in time required tends to be small because of standardizing regulations, the power of national tradition, and similar factors. Between countries, however, such forces have not been as uniform, and it was anticipated that considerable variation would be found. The question was asked, therefore, whether higher national achievement is related to the expenditure of time in school.

A little reflection on this question suggests that it is naïve and superficial to believe that the expenditure of more time per week in school may be associated very closely with higher achievement in mathematics. It seems much more likely that the *way* this time is used—the *kind* of experience the student has—is a more important consideration. In particular, the amount of time devoted to *mathematics* instruction ought to be more closely associated with mathematics achievement than is total time spent in school.

Questions about the influence of homework on achievement lead to a similar line of thought. For example, is achievement in mathematics associated with the number of hours per week the students spend in doing homework in all subjects? One reason for thinking so might be that there is a tendency for good students in mathematics to be good students in general. If "good" students spend more time on homework in general than "poor" students, the existence of such a relation might be expected. But "good" students also tend to be bright students who might be expected to expend fewer hours on homework.

Again, presumably, it is the sort of experience the students have in connection with homework that is important. Homework frequently consists of an effort to store information in the memory, or in practice to improve a skill, rather than being work which demands and cultivates the higher mental processes. If this is the case, the correlation between achievement in the lower mental processes and time spent on homework ought to be higher than the correlation for achievement in the higher mental processes.

The hypotheses formulated around these questions were as follows:

When level of instruction is held constant, the total mathematics score will not be related to the number of hours of schooling per week (Hypothesis 21).

When the level of instruction is held constant, the achievement of students will be related to the number of hours per week given to mathematics instruction (Hypothesis 22).

When the level of instruction is held constant, the achievement of students will be related to the number of hours per week devoted to mathematics homework (Hypothesis 23).

When the level of instruction is held constant, the relationship between the number of hours per week devoted to all school homework and the lower mental processes scores will be higher than the relationship between the number of hours per week devoted to all school homework and the higher mental process scores (Hypothesis 24).

Participation in Special Opportunities or "New Mathematics"

Another set of questions is related to the distribution of the time available. In some schools, for example, special opportunities for instruction in mathematics are offered to at least some of the students. These opportunities sometimes fall outside the regular schedule but involve some sort of adjustment of the total time available. These special opportunities may consist of voluntary participation in extra classes, attending lectures by visiting mathematicians, participating in a mathematics club, working on individual projects or problems, and other activities of this sort. The question arises as to whether there is any measurable relation between participation in such activities and achievement.

In the United States, reports have recently appeared on a few studies comparing achievement in "New Mathematics" with achievement in traditional mathematics. In some cases (see Lucow, 1963, and Ruddell, 1962, and Tredway and Hollister, 1963) the "New Mathematics" turned out to have the desired after-effect, but in other cases (see Banghart, et al., 1963, and Brownell, 1963, and Peck, 1963) no significant differences were found.

As was pointed out in Volume I, Chapter 4, the problem of whether or not to include in the IEA mathematics test items that involve content of the "New Mathematics" type was discussed at an early stage of the test construction process. The decision was to include some items of this type. Until now there have been no cross-national studies of the effect of instruction in "New Mathematics" upon mathematics achievement in general. Since the question is a live and at present a controversial one, it seemed opportune to use this international study to throw additional light on the subject.

The hypotheses concerning these relations were as follows:

Mathematics achievement will be highest among students who have participated in special opportunities offered students (Hypothesis 25).

Students who have had courses in "New Mathematics" will have higher scores than other students on items in traditional mathematics (Hypothesis 26).

Findings

Time Spent on Schooling

Hypotheses 21–24 form a group linked by the fact that the independent variables in these are the time allowances for all schooling, all homework, mathematics instruction, and mathematics homework. Relevant data were collected by means of the student questionnaires. Since the analyses were largely correlational, they will be fully dealt with in the analysis reported in Chapter 6 of this volume. The main points to emerge from that analysis concerning the group of hypotheses are the importance of all homework in the younger groups of students and of mathematics homework in Population 3 a.

The greatest variation between countries in the amount of time given to mathematics instruction occurs in Population 3 a. The mean number of hours per week given to mathematics instruction of this population in the various countries is reported in Table 4.17, which for convenience also includes the national mean total mathematics score. The correlation between these measures is .13. This is perhaps smaller than might have been expected, but it should be remembered that time spent on other subjects may be more or less directly helpful to mathematics and that there are other ways of spending time than that which is given directly to learning mathematics. For example, for England, the measure of time does not include the very considerable length of time during which applied mathematics is taught, because this subject, which usually goes under the heading of mechanics or physics in other countries, was not tested. Whatever it is called, time spent on it is likely to improve proficiency in mathematics.

TABLE 4.17. *Means, Standard Deviations, and Number of Cases of Time Given Mathematics Instruction and of Total Mathematics Scores, by Countries, for Population 3a.*

Country	Time (Hour/Week)			Total Mathematics Score		
	M.	S.D.	N.	M.	S.D.	N.
Australia	6.9	1.6	1,089	21.6	10.5	1,089
Belgium	7.4	1.1	519	34.6	12.6	519
England	4.3	1.3	964	35.2	12.6	967
Finland	4.0	0.0	369	25.3	9.6	369
France	8.9	0.5	222	33.4	10.8	222
Germany	4.2	0.5	649	28.8	9.8	649
Israel	4.7	0.3	146	36.4	8.6	146
Japan	5.5	1.1	817	31.4	14.8	818
The Netherlands	5.1	0.3	462	31.9	8.1	462
Scotland	6.2	1.5	1,417	25.5	10.4	1,422
Sweden	4.6	1.6	752	27.3	11.9	776
United States	5.0	0.9	1,557	13.8	12.6	1,568
All Countries	5.5	1.7	8,963	26.1	13.8	9,007

It was stated in Chapter 1 of this volume that the two scores for lower and higher mental processes were hardly different enough to make their separate analysis promising. This has been found to be true in most instances in this study. Hypothesis 24, however, depended upon the distinction being made and hence the appropriate analysis was carried out. With level of instruction held constant, the partial regression coefficients for all homework with scores on the lower and higher mental processes are given in Table 4.18.

For Populations 1a and 1b the majority of the coefficients are positive with a tendency for them to be slightly higher for the lower mental processes than for the higher. While this offers some support for the hypothesis, it must be noted that the differences between the coefficients within countries are rather small, a result to be expected from the relatively high correlation found between the two process scores.

For Populations 3a and 3b, the within-countries coefficients have positive and negative signs in about equal number and in only a few instances are they statistically significant. Over all countries, however, there is again a tendency for all homework to correlate more highly with the lower mental process score than with the higher.

One interesting feature of Table 4.18 is the variation between coun-

TABLE 4.18. *Partial Regression Coefficients for All Homework with Mathematics Scores on Lower Mental Process (LMP) and on Higher Mental Process (HMP), with Level of Instruction Held Constant.*

Country	Population 1 a		Population 1 b		Population 3 a		Populati
	LMP	HMP	LMP	HMP	LMP	HMP	LMP
Australia	.16	.16	.14	.15	.03	.00	—
Belgium	.15	.09	.12	.05	.04	.04	−.05
England	.29	.27	.29	.26	−.06	−.10	.00
Finland	−.01	.01	.02	−.02	−.05	−.11	−.01
France	.02	−.02	−.01	−.05	.04	−.03	.06
Germany	—	—	.11	−.02	−.08	−.10	−.07
Israel	—	—	.14	.12	.07	−.19	—
Japan	.10	.10	.10	.10	.13	.07	.15
The Netherlands	.04	−.05	.03	−.01	−.02	−.13	−.09
Scotland	.20	.15	.18	.15	−.02	−.01	−.04
Sweden	.03	.02	.03	.00	−.05	.00	.01
United States	.13	.09	.11	.09	.16	.12	.16
Signs +	9	8	11	7	6	3	4
Signs −	1	2	1	4	6	7	5
Range	.30	.32	.30	.31	.24	.31	.25
Mean	.11	.08	.11	.07	.02	−.04	.01
All Countries	**.17**	**.13**	**.21**	**.18**	**.18**	**.06**	**.14**

tries. In Finland, France, the Netherlands, and Sweden, the relationship between all homework and both lower and higher mental process score is virtually nonexistent in all populations, while in England and Scotland, and to some extent in Australia and Belgium, it is relatively substantial with the younger students but disappears with the older ones. In Japan and the United States, on the other hand, a significant positive relationship persists across all populations, with some tendency to become stronger with the older student.

From the regression analysis (see Chapter 6) it can be seen that when all countries are considered, achievement in mathematics has little relationship to the number of hours per week of schooling. Thus, Hypothesis 21 was supported in three of the four populations. For Population 3 b, the relationship between these variables is somewhat stronger. For all populations, students with higher achievement scores actually tend to have fewer hours of schooling per week.

The analysis has also shown that achievement in mathematics bears a slight relationship to the number of hours per week devoted specifi-

TABLE 4.19. *Means, Standard Deviations, and Number of Cases of Hours per Week for Mathematics Homework (MH) and All Homework (AH), and Percent of MH to AH by Populations.*

Country	Math. Homework			All Homework			Percent MH/AH
	M.	S.D.	N.	M.	S.D.	N.	
			Population 1 a				
Australia	2.4	1.4	268	6.1	3.8	2,788	39
Belgium	3.6	2.5	1,621	11.4	7.3	1,584	32
England	1.7	0.9	1,972	5.4	3.1	2,042	31
Finland	2.8	1.6	730	11.0	6.9	732	26
France	3.4	1.9	2,104	9.1	5.7	2,214	37
Germany[a]	3.4	1.9	4,249	9.0	4.4	4,217	38
Israel[a]	4.4	2.6	3,145	14.0	7.0	3,096	31
Japan	3.0	1.8	2,030	8.3	5.5	2,021	36
The Netherlands	2.6	1.8	289	9.0	6.9	348	29
Scotland	2.3	1.7	3,115	4.8	3.7	3,621	48
Sweden	1.9	1.3	2,131	6.1	3.9	2,093	31
United States	3.1	2.5	5,479	6.9	5.5	5,437	45
All Countries	2.7	2.0	22,089	7.1	5.4	22,881	38
			Population 3 a				
Australia	6.1	3.3	1,079	19.7	7.0	1,085	31
Belgium	8.7	4.6	504	18.8	8.3	500	46
England	4.1	1.9	946	14.3	4.9	960	29
Finland	6.6	3.5	359	22.1	8.0	354	30
France	9.6	3.5	213	22.0	6.3	209	44
Germany	5.1	2.9	571	14.6	6.9	573	35
Israel	7.5	3.7	140	24.0	6.9	140	31
Japan	5.2	4.3	767	13.8	10.4	761	38
The Netherlands	5.7	3.4	456	20.0	6.1	450	28
Scotland	4.1	2.3	1,349	12.1	5.4	1,379	34
Sweden	4.9	2.9	681	17.0	6.5	714	29
United States	4.1	2.4	1,474	11.4	6.7	1,524	36
All Countries	5.2	3.4	8,537	15.5	7.9	8,648	34

[a] The data for Israel and Germany are those for Population 1b. In most cases, the data for the two populations for other countries did not differ substantially.

cally to mathematics instruction. Hypothesis 22 received very slight support in three of the four populations. In this case, it is Population 3 a that is the exception; the relationship is a little stronger.

The results for Hypothesis 23, which concerns mathematics homework, are substantially similar to those for Hypothesis 22. Except for

TABLE 4.20. *Coefficients of Correlation Between National Mean Mathematics Scores and National Means of Several Time Variables for Populations 1a and 3a.*

	Population	
Variables	1 a	3 a
Mathematics Homework	.44	.35
All Homework	.61	.45
No. of Hours per week of Mathematics Instruction	−.01	.18
No. of Hours per week of Schooling	−.18	.31

Population 3 a, the number of hours per week spent on mathematics homework makes no appreciable difference in achievement.

Because the data may have some interest for mathematics teachers and persons interested in the curriculum as a whole, the means of the number of hours per week spent on mathematics homework and on homework for all subjects are reported in Table 4.19 for Populations 1 a and 3 a. It is immediately obvious that for all countries taken together, the time spent by Population 3 a is about double that spent by Population 1 a. A bit more than one third of all homework time, on the average, is spent on mathematics. It is to be noted also that the variance in total time spent by students seems to be considerable.

It is of some interest to investigate the relationships between the national means on the achievement test and the mean number of hours devoted to mathematics homework, all homework, to mathematics instruction, and to all schooling. The coefficients of correlation for Populations 1 a and 3 a are given in Table 4.20.

The coefficients are in most cases substantially greater than those based on the total sample for all countries which are reported in Tables 6.10 and 6.12. It is apparent that national characteristics concerning the requirement of homework both in all subjects and in mathematics are involved. The coefficients are positive and fairly high, reaching .61 in the case of the relationship between mean mathematics achievement and mean number of hours per week spent in all homework in Population 1 a. On the other hand, for this population the national mean number of hours per week devoted to mathematics instruction is unrelated to achievement and is negatively related to the number of hours devoted to schooling in all subjects. A possible explanation may be that the poorer students are required to spend more

TABLE 4.21. *Percent of Participation in Special Opportunities Related to Difference Between Mean Achievement of Students Who Participated and Students Who Did Not.*

Population 3 a.

Country	Percent of Participation	Math. Mean Score Difference
Australia	7	1.7
Belgium	61	−1.5
England	12	**5.0**
Finland	13	**4.0**
France	7	2.2
Germany	12	**5.2**
Israel	13	**4.8**
Japan	2	**8.8**
The Netherlands	32	**4.4**
Sweden	5	0.6
United States	28	**5.1**
All Countries	17	**3.6**

time in school. For Population 3 a the relationship is positive, however, and is somewhat greater for hours devoted to schooling than to mathematics instruction.

Participation in Special Opportunities

Hypothesis 25 concerns the relationship between achievement and participation in special opportunities offered students. The relevant data were again collected by the student questionnaire. Students were asked whether they had participated in extra activities in mathematics and to describe the nature of these activities. Table 4.21 reports both the percent of participation and the amount by which the mean scores of participants exceeded those of the remainder. More than half of the individual differences reach significance, and the mean over countries is about 8 times its standard error. Since general experience suggests that there will be a positive difference, the question is really one of estimation rather than significance. The mean difference is rather more than a quarter of the standard deviation per student, and it is therefore quite considerable. The explanation for this may be not that students secure better scores because they participate in special opportunities but that teachers do in fact provide special opportunities on a fairly large scale and that students who do participate (perhaps a majority of whom are the more able students) do achieve higher scores.

Courses in "New Mathematics"

It will be recalled that Hypothesis 26 concerns the relation of exposure to "New Mathematics" and achievement in traditional mathematics. To collect data for study of this hypothesis the students were given an opportunity to indicate on the student questionnaire whether they had taken any courses in "New Mathematics".

The list of courses from which the students had to choose when making their responses was worked out in different ways in different countries. As an example, the following excerpts from the list used in Scotland are cited:

69 Modern Arithmetic (e.g., binary numbers)
79 Modern Algebra (e.g., sets)
89 Modern Geometry or Trigonometry (e.g., vectors)
99 Modern Mathematics (e.g., groups, matrices)

In this version of the student questionnaire the nature of the "New Mathematics" was indicated by examples in order that the student would be able to recognize whether he had taken such work, even though it had not been presented to him as "new". In other countries the distinction was not defined so explicitly.

By some mischance the data processing of the material collected in the investigation indicated that no single student in any country had had any course in "New Mathematics". Of course, this is not consistent with the facts, since some students certainly had taken courses of new mathematics and, even if many of them were not aware of this fact, at least some must have known it. One explanation, beside the simple one that in some countries (for example, Sweden) the experimental courses were not listed at all on the student's questionnaire, lies in the coding of the collected data. For example, in order to facilitate the handling of data, the numbers of the two highest courses only in the above list were coded. Thus, a student may have reported himself having had courses Numbers 69, 79, 80 (plane geometry), and 85 (trigonometry). In the data processing, however, this student would be accounted for only under courses 80 and 85, neither of which is of the "New Mathematics" type. As one can see from the example, numerals with the unit's digit 9 were used exclusively for the "New Mathematics" courses.

This mishap in the data collection was very unfortunate. The hypothesis dealing with the "New Mathematics" courses is interesting and important, and particularly so to mathematics teachers. Thus, it seems desirable to estimate in some way the population of students who have studied "New Mathematics".

One possible method was to use the information collected from the teachers who had an opportunity to report on their questionnaire whether they were now teaching "New Mathematics". There were, however, two objections to this way of defining the desired population of students. First, one might then get students who themselves had no "New Mathematics" at all, since their teacher might indeed be teaching "New Mathematics", but to some other class. Second, those students would not be included who had had a course in "New Mathematics", perhaps in the previous year, but who were now being taught by another teacher using traditional mathematics only. On these grounds it was decided that it was not sufficient to use only the information from the teachers as to whether they had taught "New Mathematics". Instead, the following procedure was used in order to estimate the desired population of students in an indirect way.

From the set of items in each group of tests three were chosen which were both characteristic of the "New Mathematics" and very simple. These items are given below.

Populations 1 a and 1 b

A-12. If $x/2 < 7$, then

        A. $x < 7/2$ B. $x < 5$ C. $x < 14$ D. $x > 5$ E. $x > 14$

A-17. Which of the following is (are) true?

    I. $(53 \times 73) \times 17 = 53 \times (73 \times 17)$
    II. $133 \times (78 + 89) = (133 \times 78) + 89$
    III. $133 \times (78 + 89) = (133 \times 78) + (133 \times 89)$

        A. I only
        B. II only
        C. III only
        D. I and II only
        E. I and III only

A-23. Which of the following equals $7 \times (3 + 9)$?

    A. $(7 \times 3) + (7 \times 9)$ D. 7×27
    B $(7 \times 9) + (3 \times 9)$ E $21 + 9$
    C. $(7 \times 3) + (3 \times 9)$

Population 3 a

5-17. Below there are several definitions of new operations named * in terms of the usual operations on real numbers. For which of the definitions is the property $y * x = x * y$ valid for all positive real numbers x and y?

A. $x * y = x/y$ B. $x * y = x - y$ C. $x * y = x(x + y)$
D. $x * y = xy/x + y$ E. $x * y = x^2 + xy^2 + y^4$

7-14. If x and y belong to the set of real numbers and sets P, Q, and R are defined as follows:

$P = \{(x, y) \mid x^2 + y^2 = 4\}$
$Q = \{(x, y) \mid x - y = 2\}$
$R = \{(x, y) \mid (x^2 + y^2 - 4)(x - y - 2) = 0\}$

which of the following is true?

A. $R = P \cap Q$ B. $R = P \cup Q$ C. $R = \{(2,0), (0,2), (-2,0), (-0,2)\}$
D. $R = \{\ \}$ (the empty set) E. $R = \{(2,0), (0, -2)\}$

9-13. Find the difference $\vec{b} - \vec{a}$ of the vectors $\vec{a} = (4,2)$ and $\vec{b} = (0,3)$

A. $(-4, -2)$ B. $(-4, 1)$ C. $(4, -1)$ D. $(4,2)$ E. $(4,5)$

Population 3 b

C-21. Which of the following numbers in base two is (are) even?

I. 110011 II. 110010 III. 110101 IV. 100100
A. I only B. III only C. I and III only
D. II and IV only E. I, III, and IV

5-17. See above, under Population 3 b.

6-6. Four persons whose names begin with different letters are placed in a row, side by side. What is the probability that they will be placed in alphabetical order from left to right?

A. 1/120 B. 1/24 C. 1/12 D. 1/6 E. 1/4

Concerning these items, it was assumed first that they are so simple that every student who had "New Mathematics" would have encountered them, regardless of the country in which the course was given; second, that very few of those students who have had only "traditional" courses in mathematics will have met these items, since they represent aspects of mathematics which the "New Mathematics" includes and which have been lacking in the traditional courses.

The teacher questionnaire called for an estimate of the proportion of students who had had an opportunity to learn the mathematics involved in each item. If the teacher, with regard to all three items which for a particular population were considered to be basic and typical (according to the list above) gave the rating A, signifying that most of his students had encountered such problems, then these students were defined as being in the desired population, that is, they were assigned to the population of students who had taken "New Mathematics" courses. It is possible, of course, that the population defined in this way was smaller than the actual population of students who had studied "New Mathematics". Some teachers, for instance, made no ratings. Also,

TABLE 4.22. *Means, Standard Deviations, and Differences of Means of Scores on Items of Traditional Type by Students Who Have Had and Have Not Had Courses in New Mathematics.*

Country	New Mathematics			No New Mathematics			Difference of Means
	M.	S.D.	N.	M.	S.D.	N.	
			Population 1 a				
England	32.2	11.9	223	15.3	14.0	2,642	**16.9**
Finland	21.1	8.4	124	19.3	8.4	653	1.8
France	25.1	10.1	126	15.6	10.2	2,261	**9.5**
Japan	28.6	13.8	141	26.8	14.5	1,914	1.8
Scotland	28.4	10.5	274	15.5	12.4	4,933	**12.9**
Sweden	18.2	7.3	620	12.4	9.1	2,382	**5.8**
United States	21.0	12.2	411	12.9	11.0	5,703	**8.1**
All Countries	24.9	10.6	1,919	16.8	11.4	20,488	**8.1**
			Population 1 b				
England	33.0	11.6	340	18.9	15.5	2,647	**14.1**
Finland	22.1	7.7	111	21.1	8.2	729	1.0
France	25.3	9.6	193	18.0	11.2	3,218	**7.3**
Germany	31.1	7.7	179	21.1	9.6	4,285	**10.0**
Israel	31.3	10.0	250	27.2	12.3	2,979	**4.1**
Japan	28.6	13.8	133	26.8	14.5	1,715	1.8
Scotland	33.8	9.0	427	17.9	13.1	5,235	**15.9**
Sweden	17.6	7.4	171	12.1	9.1	2,641	**5.5**
United States	21.6	11.7	527	14.4	10.9	5,925	**7.2**
All Countries	27.2	9.8	2,329	19.7	11.6	29,374	**7.5**

a student may have had a course of "New Mathematics" before he entered the class in which he was at the time enrolled. However, in order to get information, this definition was chosen.

After this division of the total population of students into two groups, their scores on all items in their tests which were not of the "New Mathematics" type were obtained. The results are shown in Table 4.22. For both Populations 1 a and 1 b the all countries mean of the students who have had "New Mathematics" exceeds the mean of those who have not by a statistically significant amount. Almost all the differences for individual countries are significant. A possible explanation is that the "New Mathematics", with its emphasis on the fundamental structural features of elementary mathematics, gives the student a more solid knowledge which he can then use in different kinds of traditional mathematics. For the terminal Populations 3 a and 3 b the data were

sparse and the results were inconclusive. However, the useable d[ata]
support Hypothesis 26. In view of the rather limited basis which cou[ld]
be devised for distinguishing students who had "New Mathemati[cs]"
from students who had not, the conclusion may be surprising. It shou[ld]
however, be recognized that if only the more able students have h[ad]
experience with "New Mathematics", this fact alone could account [for]
their superiority.

Summary and Conclusions

The hypotheses formulated by the IEA that seemed to be particula[rly]
relevant to problems of curriculum and instructional methods, the da[ta]
and the findings have been discussed in some detail in the three p[re]
ceding sections. This final section will discuss the conclusions mu[ch]
more generally, and at times, speculatively. First, a brief summary
the major conclusions will be given.

Summary of Major Findings

As before, the findings will be presented in three groups. The fi[rst]
group to be considered concerns conclusions dealing with relationshi[ps]
between *student* achievement, interests, and attitudes.

Student Achievement, Interests, and Attitudes

First, for Population 1 a, achievement, as hypothesized, does appear
be positively correlated within countries with students' views that math
matics teaching and learning is inquiry-centered; in Population 3 a th[is]
hypothesis was not supported by the data and, on the contrary, the[re]
was some indication that students who viewed mathematics learning
dominantly a process of memorization and learning to follow ru[les]
achieved higher scores than those who viewed the learning as inqui[ry]
centered. A distinction between the higher and lower mental proces[ses]
did not prove fruitful.

Second, achievement was found to be positively correlated within a[nd]
between countries at all levels with interest in mathematics, as w[as]
hypothesized. When the data for all countries were pooled, the co[ef]
ficients of correlation between achievement scores and various attitud[es]
toward mathematics were quite low and in several instances were neg[a]
tive—particularly those which involved the tendency to view mathemati[cs]
as an "open" system. On the other hand, when the coefficients of c[or]
relation between the national means scores on these variables we[re]

mputed, in a number of instances they were remarkably high (for ample, −.78 between achievement and tendency to view mathematics an "open" system for Population 1 a), which indicates that national aracteristics influence student attitudes quite strongly.

Third, the relationships between interest and various attitudes were und to be very weak, indicating that these variables are relatively dependent. However, interest was significantly greater for Population when learning was viewed as inquiry-centered. Whether the father's cupation was of a scientific or nonscientific nature made little difence in the interest scores.

In general, the hypotheses concerning student relationships among hievement, interests, and attitudes were supported, especially for Population 1 a. However, in the case of several attitudes, the findings were ntrary to the hypotheses, and in most cases the coefficients of correlan were low or at best moderate, and differences between means of res pooled for all countries were small.

achers' Perceptions and Training

e second group of conclusions concerns *teachers'* perceptions about dent opportunity to learn, national emphasis on topics, and their n freedom in teaching. Also in this group are conclusions relative student achievement, interests, and attitudes in relation to teachers' service training and achievement in relation to the length and type their preservice training.

First, within countries the correlations between achievement scores d teachers' perceptions of students' opportunities to learn the mathetics involved in the test items were always positive and usually subntial. Between countries they were large (.62 and .80 for 1 a and 3 a). e conclusion is that a considerable amount of the variation between intries in mathematics score can be attributed to the differences be:en students' opportunities to learn the material tested.

Second, the profile of achievement of different classifications of subject tter does follow, as hypothesized, the national emphasis as judged teachers.

Third, the relation between achievement and teachers' perception of degree of freedom that they have is negligible within countries. wever, between countries, higher mean scores tend to occur in coun:s where teachers think they have less freedom, which is contrary the hypothesis.

ourth, the hypothesis that "students of teachers who say they have had ent in-service training in mathematics will achieve better than students of teachers who have not" received only feeble support from [the] analysis within countries. However, the analysis between countries [for] Population 1 a indicated that in-service training is accompanied [by] *lower* achievement scores and the tendency *not* to describe mathemat[ics] learning as emphasizing inquiry. Thus, these parts of the hypothe[sis] were contradicted. The part of the hypothesis which asserted a posit[ive] association between in-service training and the view that mathema[tics] is important to society was confirmed. For Population 3 a, the conc[lu]sions relative to achievement and the view of the importance of mat[he]matics to society were similar, but in this case there was also a stro[ng] tendency for the view that mathematics is an open system to be p[osi]tively associated with in-service training.

Fifth, the data support the hypothesis that the more preservice tra[in]ing a teacher had received, the better was the achievement of his s[tu]dents. However, students in Population 3 a who studied under univers[ity] trained teachers achieved less than students of teachers trained in teach[er] training institutions.

School Decisions Related to Achievement

The third group of hypotheses was concerned with the effect of stud[ent] time spent on school work, student participation in "special opportu[ni]ties" in mathematics, and with the study of "New Mathematics". T[he] major conclusions were as follows.

First, when all countries are considered together, achievement [in] mathematics had little relationship to the number of hours per w[eek] of schooling, but did have a slight relationship to the number of ho[urs] per week devoted to *mathematics* instruction. Similarly, the num[ber] of hours per week spent on all homework appears to have made a [dif]ference in achievement in Populations 1 a and 1 b, but the number [of] hours spent on *mathematics* homework made little difference except [in] Population 3 a, where it did make a difference. Also, the number [of] hours spent on all homework tended to correlate more highly with [the] lower mental process score than with the higher mental process sc[ore.] The between-countries analysis reveals that the relation between achie[ve]ment at the national level and hours spent per week on mathema[tics] homework and on all homework is positive and substantial.

Second, teachers did provide special opportunities in mathema[tics] and the students who participated achieved higher mathematics sco[res.]

Third, students who had had courses in "New Mathematics" achie[ved] higher scores than other students on items in traditional mathema[tics]

There were other findings concerning particular countries and particular populations which will be left for discussion at the national level.

Some General Implications of the Findings

One of the most striking features of the results is the paucity of general (that is, both among countries and consistently in each country) relationships between mathematics achievement and the curriculum variables investigated. The correlations between achievement scores and these curriculum variables, when based on the pooled data from all countries, were usually quite low. The weakness of many of the relationships that were found limits their usefulness in curriculum planning.

One notable exception to this is the positive relation between achievement and interest in mathematics. We have not shown, however, which is cause and which is effect, but there is reason to suppose that on occasion it works both ways. If so, efforts to increase interest will often produce better achievement, and better achievement, in turn, may lead to increased interest. Also notable was the positive relation between achievement and teachers' perception of opportunity to learn the material tested.

On the other hand, when relationships between achievement and these curriculum variables are studied on a between-country basis many of the correlation coefficients are substantial. Moreover, in several cases the sign for the preuniversity students is opposite from that for the 13-year-olds. Thus, although the relationships are not, as a rule, such as to transcend national boundaries, they are obviously influenced by features of the national programs.

The results can perhaps be explained most easily by noting that within countries, and within different populations in the same country, there is considerable homogeneity in what mathematics is taught, in the methods by which it is taught, and in the perceptions of teachers as to the degree of freedom they have. For such variables as amount of time spent on mathematics instruction and homework, it is quite reasonable to expect homogeneity of response within a given country. In some cases, there is also homogeneity on an international basis. As an example, consider the similarity of the mean scores for different countries on the scale "Description of Mathematics Teaching and School Learning".

Within a given country this means that most teachers and textbooks present similar content by very similar (and, for the most part traditional) instructional methods. For example, the students are expected to learn the mathematics presented whether they are interested in it or not. Teachers like to have their students interested, but in many cases, do not go out of their way to stimulate interest. If a majority of students study mathematics not because they are deeply interested or successful in learning it, but because it is required of them, moderate and positive correlations between achievement and interest may be expected—and were in fact found.

The attitudes which were investigated by the IEA differ from the interest variable in that they tended to focus upon some of the ideas that are associated with "New Mathematics" teaching and a particular educational philosophy and learning theory. Thus, the view that mathematics teaching and school learning emphasizes inquiry would not usually be the traditional one. Again, the view that mathematics is an "open" system—with the possibility of future development, with different approaches possible to the same problem, etc.—is a view more modern than traditional at the school level. Similarly, concern with the higher mental processes, while often given lip-service in mathematical education, has not been a prominent characteristic of traditional mathematics teaching.

If the hypothesis that mathematics teaching is predominantly traditional is viewed as tenable, there is considerable evidence in this study to support it. The weak relationship between total achievement and the attitudes measured would be predicted under traditional teaching. Teachers who have not been accustomed to great freedom of choice as to curriculum and method would not be expected to be highly conscious of "restrictions" on their freedom, and there is little reason to suppose that achievement would be enhanced by greater teacher freedom, unless the teachers really want to make use of it. The international impact of the "New Mathematics" programs apparently had not been very great at the time the data were gathered.

Future international studies of the curriculum variables considered here will perhaps find fruitful a deeper investigation of the several *national* characteristics of the curriculum that will explain the relatively strong relationships between national mean achievement and the national means of the other variables. We have identified a number of variables worth deeper study. If present trends concerning content and method continue, several of our findings which are contrary to the original hypothesis (for example, the tendency of students in the pre-university population to view mathematics as a fixed system), are particularly worthy of study.

Chapter 5

Social Factors in Education

Introduction

The schools and the educational system have rarely in the history of man been completely separated from the social and cultural system. It is true that critics of the schools have frequently pointed to the slow response of the educational system to changes in the society. Evidence of the disparity between the social changes and the educational changes has been pointed out by scholars within some of the nations in this study as well as by some of the students of comparative education.

The hypotheses and questions raised in Chapter 5 all have to do with social, economic, and cultural factors as they relate to schooling and mathematics achievement. While it is likely that these factors are less directly related to schooling and achievement than the more direct influences of curriculum and teaching methods, we are attempting to understand possible chains of relations between these characteristics and specific school learning. In this study, where we are limited to psychometric and survey methods, we can only *infer* relations which should be investigated more directly by other research procedures.

For some of the hypotheses tested in this chapter we are attempting to determine whether particular relationships found in one country are repeated in the other countries and to seek some explanation for the exceptions which are found. These relationships are likely to be ones that have previously been investigated, and we are here attempting to determine how general the relationship is.

In other hypotheses, we are attempting to probe further and to determine how a simple and relatively clear relationship under one kind of analysis becomes more complex and meaningful as we search for interrelations among a set of variables.

This chapter was edited by Professor B. S. Bloom on the basis of reports concerning individual hypotheses from the following persons: L. M. Björkquist, A. W. Foshay, M. Groen, T. Harada, T. Husén, J. Keeves, K. Miller, G. Mialaret, C. Rapaport, S. Sakakibara, M. Smilansky, M. Takala, R. L. Thorndike, and S. Wiegersma.

In still other hypotheses, we seek to establish a series of connecting links among variables which will help us understand something of the process by which particular influences on the student and his learning come about.

Finally, by comparing some of the within-country differences with the between-country differences, we hope to secure a new view of certain educational phenomena which are related to social and cultural variables.

We are somewhat restricted in the variables available in this study that are pertinent to the relationship between education and the social system. The clearest social variable is the socio-economic status of students and we have studied this in some detail as it relates to schooling, achievement, interests, attitudes, and aspirations. Then we consider the urban-rural differences in mathematics achievement as a way of understanding how the quality of education is influenced by geographic (and cultural) factors. We next turn to a consideration of the way in which financial support for education influences the quality of education as reflected in mathematics achievement. Then we investigate the ways in which the roles of men and women in a society influence the learning outcomes in mathematics—regarded frequently as a "male subject". Finally, the plans and aspirations of boys and girls are studied to determine how the students' view of the future is related to their achievement in one school subject—mathematics.

Home Environment

During the past decade there have been a large number of studies which attempt to understand the links between school achievement and the home background. These studies have used observational and clinical methods to determine how the interaction between parents and children influences the child's language development, his problem-solving abilities, and his attitude toward himself, the school, and learning. While these studies have been done on relatively small samples, they are most revealing in helping us understand the relations between schools and homes. These studies confirm the earlier studies of the relation between socio-economic level (and social class) and school learning, but they do more in that they reveal something of the process underlying these relations.

What these studies help us understand is that the child has learned a great deal before he enters school and that it is the parents in the home who have been his teachers. The research of Dave (1963) and Wolf (1964) reveals the ways in which the parents influence the child's aspirations and motivation. Bernstein (1961), Hess and Shipman (1965), and Deutsch (1965) reveal the way in which the home influences language development as well as the development of cognition and thought. These studies indicate that it is the parents who stimulate the child to experience and interact with a complex environment, and it is the parents who are the models as well as stimulators of language and thought development.

Studies such as these have concentrated on the processes by which parents interact with their children and the stimulation this provides for early learning. These investigators with their careful observation and detailed securing of evidence help us become aware of the way in which variation in parents as models and parents as "teachers" account for much of the "individual differences" that teachers find in their pupils at the time of entrance to school. Unfortunately, we have in the past done little more than name these differences or accept the fact of individual variation. Only in a few instances throughout the world—Deutsch (1964), Gray and Klaus (1963), Landers (1963), Shepard (1963), and Smilansky (1964)—have the schools and teachers deliberately planned a strategy of teaching to attempt to do in the teaching in the school those things required to make up for the variation in "parents as teachers". This is not the place to enter into a discussion of the ancient nature-nurture controversy; this would not serve the purposes of our present concern. What has been pointed out or implied in the many studies which have investigated differences in the home backgrounds of children is that the quality of parents as teachers (and as models for learning) differs widely and that children have been taught well, or poorly, in the home before they enter school. No doubt also, the teachers they will have in school also differ widely in their teaching competence as well as in the models they offer for learning. In the first part of this section, we are attempting to understand how the home background relates to the learning outcomes in mathematics as well as to other variables we have included in this study. To put it in other terms, we are attempting to study the interactions between parents as teachers and teachers as teachers—primarily in relation to mathematics.

Unfortunately, in a study of more than 130,000 students in 12 countries, it is not possible to study parents as teachers with the degree of precision of such workers as Dave (1963), Wolf (1964), Bernstein (1961), Hess and Shipman (1965), Kahl (1953), and Douglas (1964). Rather, we are limited to a few variables which will in a partial way reveal some of the relationships. The variables which we have been able to secure are the parents' level of education and the father's occupation. These are useful indicators, but they reveal only a limited range of the home background. Thus, the relationships we find using such indicators are relatively weak in contrast with the more powerful relationships found with more precise measures of "parents as teachers". Thus, Dave found a correlation of +.80 between school achievement and his detailed measure of the home environment, in contrast with the usual relationship of +.30 to +.40 between school achievement and the type of socio-economic indicators we are using in this investigation. Further, more detailed studies in various socio-economic settings would have to be carried out in order to obtain a clearer picture of environmental influences.

In the hypotheses which are relevant to the home background we first point out some of the simple relationships between home background and the schools, then we attempt to introduce more complex relationships which will reveal something of the process of interaction between home background, school characteristics, and the outcomes of instruction measured by our instruments. Since the present study is of the survey type, we were unable to penetrate into the more subtle and presumably more powerful influences of the home situations upon children's success in school.

Thus, we begin by pointing out the way in which the home background is related to the kind of schooling and opportunities for learning in the various countries. This aspect of our work is quite parallel to the studies of Floud and Halsey (1961), Husén (1961), Douglas (1964), and Davie (1953). It becomes evident in this analysis across 12 countries that countries vary greatly in the extent to which the "parents as teachers" determine the schooling the child will receive as well as the "teachers as teachers" he will have.

We then move to the relationship between home background and the achievement of students in mathematics. Here we attempt to account for the differences in the relationships from one level of school to another and we attempt to explain why these relationships vary from country to country. It becomes very clear that an international study of this kind helps to put these relationships in a larger perspective than would be available within any one of the countries in this study.

We then consider the relationship between home background and level of instruction, amount of homework, interest in mathematics, and expectations for further education.

Finally, we come to more complex relationships between home background, type of schooling, and mathematics outcomes. Here we are concerned about the ways in which selection and type of schooling interact insofar as this can be revealed by our methods of investigation and our measurement devices.

We are interested in the ways in which the type of school program, interacting with socio-economic status, relates to the mathematics outcome; hence we also deal with the relation between school variability (with respect to socio-economic status) and mathematics achievement.

Place of Residence

Throughout the world the differences between urban and rural communities have been noted in many aspects of human life and behavior. We have long attributed sophistication and cultural development to the urban centers and have expected conservative views and limited intellectual development in rural areas. More recently, the movement of people from large and heterogeneous cities to the adjacent suburbs has made for smaller communities which are very homogeneous with respect to socio-economic status. Although coding difficulties in the IEA study did not enable us to differentiate between suburban and large urban communities, the effect of heterogeneity in schools (with respect to socio-economic status of the parents) has been studied in Hypothesis 28.

In this part of the study we attempt to determine whether and to what extent the quality of education is related to urban-rural differences. We are interested in learning whether the socio-cultural influence of the size of the community in which the student lives has a significant relationship with his learning of mathematics.

Financial Support for Education

The nations in this study vary in the way in which education is supported. In some countries there is little variation in the financial support from school to school—especially when educational policy and practices are determined by a centralized authority. In other countries there is a tradition of local support for education and schools may vary greatly in the support received.

We are interested in financial support for education because of widely held views that quality of education can in part be influenced by economic support for the schools. Previous research in the United States, summarized by James, Thomas, and Dyck (1963), has indicated a high degree of relationship between educational achievement and financial support for the schools. It is possible that in these studies the financial support for the schools is mainly a reflection of socio-economic characteristics of the community and the educational and occupational level of the parents.

Sex Differences

Mathematics has in many countries been viewed as a male subject. We are interested in determining the way in which cultural views of the role of men and women influence not only the taking of mathematics courses but also the mathematics achievement of boys and girls. We also attempt to determine whether sex differences are reflected in verbal as compared to computational problems in mathematics. We further attempt to understand how sex roles are related to interest in mathematics, plans to take further mathematics, and attitudes about the difficulty of learning mathematics. As a final attempt to understand sex roles in relation to mathematics, we have studied the effects of single-sex and coeducational schools on mathematics scores.

Aspirations and Mathematics Achievement

Education may be thought of as preparation for the future. Under this view, it is hypothesized that the student's present learning is likely to be influenced by his plans and his aspirations for the future. In this part, we look at the relations between the student's educational and vocational plans and desires and his mathematics test scores.

Home Background

Socio-Economic Background and Educational Selectivity

In Chapter 3 the relation between selection as an organizational feature in the school systems and the resulting social class composition of the students in the preuniversity year has been considered. This social class bias is also of interest from the societal point of view in terms of its effects in determining which members of society complete secondary education.

Another method of showing the social selectivity in the different countries is to compare the percent of students in the three highest occupations (1, 2, and 3) in Population 1 b with the percent of students in these highest occupations in Populations 3 b and 3 a. Table 5.1 shows that while all countries show some social selectivity, the bias toward the higher occupational groups is greatest in Germany, England, the Netherlands, Sweden, and France.

Still another way of showing the selectivity in the participating countries is to compare the average level of father's education in Population 1 b with that of the two preuniversity populations. This is shown in Table 5.2 where the difference between the 13-year group and the

TABLE 5.1. *Percentages of the Three Higher Occupational Status Groups in the Different Samples.*

Country	Population 1 b	Population 3 b	Population 3 a
Australia	20	—	39
Belgium	20	46	33
England	13	55	48
Finland	22	38	49
France	16	—	45
Germany	17	67	68
Israel	14	—	25
Japan	19	26	40
The Netherlands	23	—	53
Scotland	17	46	49
Sweden	22	55	52
United States	24	31	34

TABLE 5.2. *Mean Level of Father's Education.*

Country	Population 1 b	Population 3 b	Population 3 a
Australia	9.3	—	9.8
Belgium	10.4	12.2	10.9
England	10.1	11.8	11.4
Finland	7.2	8.0	8.7
France	8.1	—	11.7
Germany	8.9	13.2	12.5
Israel	9.7	—	11.0
Japan	8.9	9.4	10.9
The Netherlands	8.7	—	11.1
Scotland	9.8	10.8	11.3
Sweden	7.6	10.5	10.7
United States	11.2	11.1	11.7

preuniversity groups is greatest in Germany, France, Sweden, and the Netherlands.

In the following we will examine these background factors and their relation to mathematics achievement in more detail.

Socio-Economic Background and Mathematics Achievement

In this section we will try to relate mathematics achievement to the educational and socio-economic factors of the home background of the students. We will also try to see the extent to which these factors affect

TABLE 5.3. *Correlations Between Mathematics Achievement and Father's Educational Level and Father's Occupational Status.*

Country	Father's Educational Level			Father's Occupational Status		
	1 b	3 a	3 b	1 b	3 a	3 b
Australia	.07	.02	—	.20	.05	—
Belgium	.14	.08	−.02	.19	.04	.00
England	.25	.09	−.04	.26	.08	.01
Finland	.14	−.04	−.10	.07	−.11	−.04
France	.18	.02	—	.22	−.01	—
Germany	.13	−.02	−.06	.15	.01	−.02
Israel	.18	−.12	—	.22	.01	—
Japan	.33	.13	.13	.25	.08	.09
The Netherlands	.26	.05	—	.23	−.05	—
Scotland	.15	.04	.03	.26	.06	.03
Sweden	.15	−.02	−.02	.17	−.03	−.03
United States	.30	.32	.31	.29	.24	.26
All Countries	.16	.07	.04	.22	.11	.13
Signs	+12 / − 0	8 / 4	3 / 5	12 / 0	8 / 4	4 / 3

a The categorization of father's occupational status was used as a rough occupational scale, since the nine categories on the whole are arranged in terms of occupational status.

the kind of schooling a child gets, how they affect the expectations and desires of the students, and how students from different home backgrounds learn when they are put together in a socio-economical heterogeneous or homogeneous school system. We do not suggest that there is a simple and direct causal relationship between these factors and mathematics achievement because an index of socio-economic status is only a symptom of a very complex set of motivational and other forces determining selective factors in schooling, reality of vocational and educational goals, and pressures on children for school achievement.[1]

In Table 5.3 we show the relationship between father's educational level and occupational status on one side and the mathematics scores of the students on the other. For Population 1 b the relationship be-

[1] We have already demonstrated (in Chapter 3) the relation between socio-economic factors and selection of students for both amount of schooling and type of educational program in which the students are put. We will later demonstrate the way in which these effects are in part determined by the type of school in which the students are put.

TABLE 5.4. *Total Mathematics Test Score by Level of Father's Occupational Status, Population 1b.*

Country	Lowest Occupations			Second Lowest Occupations			Second Highest Occupations			Highest Occupations			Difference
	M.	S.D.	N.	M.	S.D.	N.	M.	S.D.	N.	M.	S.D.	N.	Highest–Lowest
Australia	16.68	11.79	1,382	19.78	11.78	469	21.53	11.95	858	23.55	12.13	260	6.87
Belgium	27.44	13.94	1,126	31.16	13.01	146	32.96	12.85	1,079	35.22	12.62	241	7.78
England	19.70	16.58	1,763	24.64	17.91	107	33.19	17.26	819	42.16	13.97	172	22.46
Finland	25.88	9.48	371	27.06	9.54	233	25.49	9.19	151	27.34	11.75	67	1.46
France	18.15	11.19	1,780	29.23	13.75	413	21.98	12.91	936	26.92	11.08	130	8.87
Germany	23.77	11.58	2,037	25.76	12.14	476	26.68	11.42	1,384	30.59	10.46	408	6.82
Israel	31.21	14.17	1,464	26.07	13.84	378	36.50	13.11	937	40.82	10.99	274	9.61
Japan	27.70	16.43	509	26.48	15.55	538	35.02	16.04	653	41.33	15.26	261	13.63
The Netherlands	18.85	11.38	595	18.35	9.63	166	23.54	12.06	489	27.94	13.06	170	9.09
Scotland	20.21	14.86	3,583	22.58	14.89	322	28.72	15.36	1,020	32.18	15.08	460	11.97
Sweden	14.28	10.08	1,237	14.27	10.68	531	16.93	10.92	733	19.88	11.89	199	5.60
United States	14.98	11.90	3,033	16.27	12.60	442	21.63	12.64	1,700	26.83	13.79	875	11.85
All Countries	21.57	12.78	18,880	23.47	12.94	4,221	27.01	12.97	10,759	31.23	12.67	3,517	9.66

[a] We are not strictly dealing with an occupational status scale. For the comparisons made in Table 5.3 we have grouped the occupations as follows. Lowest occupations comprise categories 7 and 9 (manual workers except in farming; fishing and forestry). Second lowest occupations comprise categories 5 and 8 (farming, fishing, forestry). Second highest comprise 3, 4 and 6 (mostly white collar workers). Highest occupations comprise categories 1 and 2 (professions, higher technical, executives, administrators etc.).

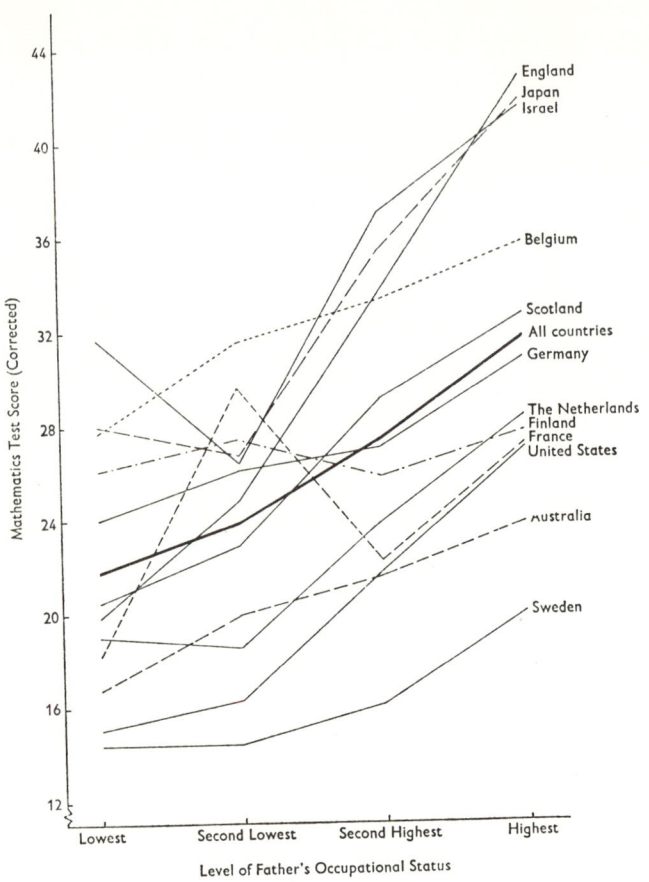

Figure 5.1. Total mathematics test score as related to level of father's occupational status, Population 1 b.

tween father's educational level and mathematics is significant in all countries, while for Population 1 a the relationship with father's occupational level is significant in all but Finland.

At the preuniversity level the correlations are smaller and frequently negative. This undoubtedly reflects the bias in the selection process (see Tables 5.1 and 5.2), such that only the higher socio-economic groups are well represented in the preuniversity level of education, except in the United States, where the correlations are larger for all populations. Put in other terms, only some of the most able students in the lower socio-economic groups are still included in the terminal levels of secondary education so that there is little relationship at these levels between socio-economic status and mathematics achievement. That these students are most able is especially clear in those countries with negative correlations, where the lower socio-economic students are in fact slightly superior in mathematics achievement to the upper socio-economic students (Finland, Sweden, and Germany).

Table 5.4 shows the mean and standard deviation of the mathematics scores for Population 1 b as related to the father's occupational status by countries as well as for all countries. We have also shown this in graphic form (Figure 5.1).

In Table 5.4 and Figure 5.1, it is evident that in almost every country the scores of the highest occupational group students are significantly higher than those of the lowest occupational group. What is most remarkable about Table 5.4 is the variability of the mean differences between highest and lowest occupations among the countries. These differences are so great that children of the highest occupational groups in some countries (Sweden, Australia, the United States and France), are below or equal to the mathematics achievement of the lowest occupational groups in other countries (Japan, Israel, and Belgium). While one may search for an explanation for these striking differences, it is likely that the curriculum and school programs account for these differences. (This has been analyzed in greater detail in Chapter 4.) We conclude that the differences in school achievement among socio-economic groups are of relative importance in the hierarchical arrangement of students within countries but that the effects of differences in the educational systems are so great that the most favored students in some countries learn (or achieve) at a lower level than the least favored students in other countries.

Socio-Economic Status, Level of Mathematics Instruction, and Further Plans for Education

Since it has been established that the home background variables are related to mathematics achievement, the question arises as to possible explanations of this relationship. We have not measured the intelligence or aptitudes of the students in all the countries and therefore cannot relate such measures to a possible complex of home background and intelligence interacting with school achievement. We will try to find other variables which may help explain how mathematics achievement is influenced by the socio-economic environment.

One possible way in which socio-economic status may influence the educational achievement of students is through the level of aspiration. Parents of higher socio-economic status do expect more education for their children than do parents of lower economic status,—see the work of Anderson (1940), Havighurst and Neugarten (1962), Kahl (1953), and Floud (1961). We will attempt to determine whether or not an intervening variable between socio-economic status and achievement is the students' motivation to seek further education.

Still another way in which socio-economic status may more directly influence mathematics achievement is through selective factors determining the level of mathematics instruction the students receive. Perhaps as a result of better preparation, motivation to do school work, and parent's concern for the child's school career, children of higher socio-economic status may secure the most advanced instruction available in the schools. We will attempt to determine whether the level of mathematics instruction is an intervening variable between parent's socio-economic status and the child's mathematics achievement. Here we are assuming that the level of instruction has some influence on the level of mathematics achievement. In Chapter 6 it is evident that the level of instruction accounts for a significant proportion of the variation in mathematics achievement in each of the populations being investigated in this study.

In Table 5.5 it is evident that the level of mathematics instruction is in part determined by socio-economic status (father's education) only in Population 1 b. While the relationship between these two variables is significant for the pooled data of all countries at this level, in only six of the individual countries is it found to be significant. Only in the United States is a significant positive relationship present in Population 3 a, and in no country is it significantly present in Population 3 b. What is likely here is that those lower socio-economic students who are included in the preuniversity classes are able to get to the same level of mathematics instruction as the higher socio-economic groups. This, as well as the reduced variation in socio-economic status of students in countries in Europe, may help to explain the low relationship between socio-economic status and mathematics achievement in Populations 3 a and 3 b. On the whole, we find that the level of mathematics instruction is a useful intervening variable for explaining how socio-economic status influences educational achievement in the data presented here for only the younger students.

In Table 5.5 it is apparent that educational motivation as measured by the student's plans for further education is affected by the socio-economic status of the parents. With few exceptions, there is a significant relationship between father's level of education and the student's plans for further education. The relationships are strongest for Population 1 b, but they are significantly high at the preuniversity

TABLE 5.5. *Correlations of Father's Level of Education with the Student's Level of Mathematics Instruction and Number of Years of Additional Education Expected.*

Country	Number of Years of Additional Education Expected			Level of Math. Instruction		
	Population 1 b	Population 3 a	Population 3 b	Population 1 b	Population 3 a	Population 3 b
Australia	.12	.03	—	.00	.00	—
Belgium	.29	.31	.24	.08	.01	.01
England	.34	.14	.11	.05	.05	−.05
Finland	.27	.19	.10	.00	.00	.02
France	.32	.27	—	.31	−.01	—
Germany	.45	.13	.18	.35	−.10	.00
Israel	.22	−.10	—	.00	.00	—
Japan	.46	.31	.33	.00	.00	.10
The Netherlands	.31	.25	—	.15	.00	—
Scotland	.23	.18	.16	.04	.04	.05
Sweden	.35	.25	.28	.13	−.08	.00
United States	.32	.25	.25	.11	.28	−.05
All Countries	.32	.19	.31	.15	.02	−.05
Signs +	12	11	8	8	4	4
Signs −	0	1	0	0	3	2

level. We cannot explain the low relationships in Australia and the negative relationship in Israel. With these two exceptions, it is clear that the educational achievement of the students in mathematics is influenced by socio-economic status through its effects on the educational plans of students. Elsewhere (see Chapters 1 and 6) we have shown the relatively high relationship between further plans for education and mathematics achievement. Thus, it is in the aspirations for the future that socio-economic differences have their clearest impact on the students.

Socio-Economic Status, Interest in Mathematics, and Amount of Homework

We have also attempted to determine whether the socio-economic status of the student influences mathematics achievement through the development of interest in mathematics. Elsewhere (see Chapter 6) we have demonstrated relatively high relationships between interest in and achievement in mathematics.

In Table 5.6 it is evident that the relationship between father's level of education and the student's interest in mathematics is slight. In

TABLE 5.6. *Correlations of Father's Level of Education with the Student's Interest in Mathematics and Number of Hours of Homework per Week.*

Country	Interest in Math.			Number of Hours of All Homework per Week		
	Population 1 b	Population 3 a	Population 3 b	Population 1 b	Population 3 a	Population 3 b
Australia	.06	−.02	—	.01	.01	—
Belgium	.07	.06	.04	.11	−.02	−.04
England	.11	−.01	−.04	.16	.06	.12
Finland	.07	−.01	.12	−.04	−.01	−.12
France	.04	−.15	.09	.17	−.05	−.10
Germany	.01	−.07	.03	.12	−.00	.04
Israel	.04	.02	—	.05	−.05	—
Japan	.15	.06	−.07	.02	−.01	.09
The Netherlands	.08	.05	−.08	.07	−.01	.04
Scotland	.03	−.03	−.04	.07	.03	.09
Sweden	.16	−.06	−.04	.08	.01	−.08
United States	.04	.11	.00	.06	.14	.10
All Countries	.07	.01	−.03	.05	−.03	.09
Signs +	12	5	4	11	5	6
Signs −	0	7	5	1	7	4

Population 1 b a few countries show significant correlations, but these are low. We conclude that interest in mathematics is not clearly influenced by the home background as measured by the father's level of education. It may be that other measures of the home background (such as father's interest in science and engineering) may be more relevant for further investigations of this problem.

In Table 5.6 we have also explored the relationship between the amount of homework the students claim they do each week and the father's level of education. Here again we generally find a weak relationship. We conclude that neither of these two variables help us in understanding the relationship between mathematics achievement and the socio-economic background of the home as measured by the father's level of education.

Socio-Economic Differences in Mathematics Achievement Within School Programs

We have already pointed to some of the relationships between socioeconomic status and the school careers of students in the study. In this section we are interested in determining the extent to which socio

TABLE 5.7. *Percent of High Occupational Status Students in Academic and General School Programs in Selected Countries, Population 1 b.*

	General Program	Academic Program
England	17	55
France	28	59
United States	36	50
Germany	35	80
Australia	35	41

economic status still has an effect on mathematics scores when we compare the students within particular types of school programs. Hypothesis 27, as originally stated by the IEA group, is *When the school program is held constant, the total mathematics test scores of pupils of lower socio-economic status will differ from those of pupils of higher socio-economic status.*

In the student questionnaires, the students were asked to determine whether they were in an academic, general, or vocational school program. Although there was some error in the students' responses, particularly in the 13-year old populations, these were not sufficiently great to create a serious problem in comparison of groups.

It is evident that by age 13, the school programs in some countries already reflect a socio-economic bias. This may be seen in the proportion of high occupational level students (children of white-collar workers, professionals, administrators, and upper-level proprietors) in the different school programs. Thus, 29 percent of the students in vocational programs are from the top occupational grouping, as compared with 35 percent of students in general programs and 50 percent in academic programs.

In five of the countries we can compare the relative proportion of high socio-economic groups in academic and general school programs. In Table 5.7 it is evident that in each of these five countries the high socio-economic students tend to be placed in the academic program in greater proportion than in the general program, with the greatest differences apparent in Germany, England, and France.

It is likely that the placement of students in the different school programs is a reflection of complicated selection mechanisms. Thus, we might expect that the more able students would be placed in the academic programs and that this selection process would reduce the difference in mathematics achievement between the socio-economic groups within each of the school programs.

TABLE 5.8. *Total Mathematics Test Scores (Corrected) as Related to Socio-Economic Background (School Program Being Held Constant).*

Population 1a.

	Father's Occupational Status													Difference
	Lowest (7 and 9)			Second Lowest (5 and 8)			Second Highest (3, 4 and 6)			Highest (1 and 2)			Highest—
Country	M.	S.D.	N.	M.	S.D.	N.	M.	S.D.	N.	M.	S.D.	N.	Lowest
I. *Academic Program*													
Australia	20.00	11.81	480	21.89	13.41	152	25.47	12.18	343	28.33	12.22	131	8.38
Belgium	32.26	11.55	286	31.86	10.58	42	33.00	11.97	497	36.90	10.48	99	4.64
England	27.32	16.23	348	35.76	12.77	31	35.61	14.12	327	38.86	12.30	116	11.54
Finland	24.73	8.62	309	25.67	7.86	184	23.81	8.81	147	23.28	11.05	70	−1.45
France	15.49	9.13	96	16.83	9.78	48	15.23	9.84	146	17.49	9.71	45	2.00
United States	14.60	11.60	1,286	14.31	11.35	149	18.80	12.50	942	24.88	13.64	540	10.28
All Countries (Average)	25.76	12.56	2,812	27.26	9.87	611	27.12	11.21	2,419	29.38	10.91	1,023	3.62
II. *Vocational Program*													
All Countries (Average)	14.49	9.19	1,262	18.27	9.89	173	19.02	11.13	521	21.76	12.14	81	7.27
III. *General Program*													
Australia	17.00	11.26	481	19.88	11.37	188	21.27	12.16	281	24.12	13.18	75	7.12
England	13.31	12.20	1,008	13.95	13.50	68	19.75	14.16	257	35.83	15.37	28	22.52
France	18.45	9.27	1,209	20.95	11.91	162	19.00	9.66	518	19.37	7.32	28	0.92
Japan	27.70	16.43	509	26.48	15.55	538	35.02	16.04	653	41.33	15.26	261	13.63
The Netherlands	22.04	9.00	137	23.36	7.12	49	21.27	10.32	75	27.39	9.69	21	5.35
United States	13.98	11.21	1,057	16.03	11.68	218	16.63	11.69	635	21.14	13.41	285	7.16

Population 1 b.

Father's Occupational Status

Country	Lowest (7 and 9)			Second Lowest (5 and 8)			Second Highest (3, 4 and 6)			Highest (1 and 2)			Difference Highest—Lowest
	M.	S.D.	N.	M.	S.D.	N.	M.	S.D.	N.	M.	S.D.	N.	
I. Academic Program													
Australia	18.41	12.04	545	22.39	12.39	161	24.39	11.74	374	25.42	10.87	132	7.01
Belgium	34.15	11.16	432	35.71	9.08	68	34.97	11.56	734	35.53	10.87	182	1.38
England	30.82	17.89	387	39.75	14.09	33	40.95	13.45	382	43.25	11.83	129	12.43
Finland	26.77	8.66	352	28.19	8.48	219	25.85	9.06	150	27.31	11.75	67	0.44
France	17.45	9.80	154	17.62	11.88	67	17.73	10.06	238	19.51	9.56	72	2.06
Germany	25.54	8.51	95	26.10	8.44	25	26.17	8.51	284	26.40	9.60	247	0.86
United States	15.79	12.16	1,371	17.35	13.24	150	22.41	12.38	961	27.01	13.93	539	11.22
All Countries (Average)	26.02	11.19	3,347	26.50	10.57	731	27.96	10.72	3,180	29.51	11.15	1,419	3.49
II. Vocational Program													
All countries (Average)	14.92	9.53	1,501	18.14	10.72	221	19.37	11.03	749	23.07	12.41	108	8.15
III. General Program													
Australia	15.28	11.80	502	18.35	11.06	214	20.77	12.28	294	25.38	12.25	94	10.10
England	16.91	14.01	1,026	14.53	12.37	58	23.33	14.73	298	33.99	13.55	33	17.08
France	19.98	8.97	1,609	29.50	13.07	352	20.30	9.55	707	21.41	7.84	52	1.43
Germany	24.91	11.08	1,951	27.08	12.01	450	25.19	10.83	1,105	24.51	9.32	160	−.40
Japan	27.70	16.43	509	26.48	15.55	538	35.02	16.04	653	41.33	15.26	261	13.63
The Netherlands	19.75	9.96	459	19.34	8.23	137	23.11	10.05	370	23.10	10.92	113	3.53
United States	14.86	11.26	1,438	17.66	11.97	261	20.29	12.43	663	25.03	13.44	296	10.17
All Countries (Average)	19.17	11.29	7,730	21.94	10.74	2,032	22.18	11.96	4,128	26.75	11.80	1,010	7.58

TABLE 5.10. *Total Mathematics Test Scores (Corrected) as Related to Socio-Economic Background (School Program Being Held Constant).*

Population 3 a.

Country	Father's Occupational Status												Difference Highest—Lowest
	Lowest (7 and 9)			Second Lowest (5 and 8)			Second Highest (3, 4 and 6)			Highest (1 and 2)			
	M.	S.D.	N.	M.	S.D.	N.	M.	S.D.	N.	M.	S.D.	N.	
I. Academic Program													
Australia	21.59	8.61	241	21.90	8.53	112	22.09	9.30	348	22.86	9.19	260	1.27
Belgium	34.70	12.70	124	42.41	6.40	19	35.84	12.74	229	37.97	11.53	73	3.27
England	34.61	11.59	172	38.35	6.73	19	35.24	12.44	447	37.45	12.50	251	2.84
Finland	27.02	8.72	77	26.10	8.96	97	25.63	10.76	89	24.12	8.71	105	−2.90
France	34.36	11.13	39	30.29	7.60	25	34.27	10.59	95	33.66	11.36	52	−0.70
Germany	29.73	8.79	37	26.11	9.85	25	28.07	9.50	234	29.47	9.88	342	−0.26
The Netherlands	30.88	5.87	55	28.10	9.61	37	32.12	7.65	202	32.10	8.94	163	1.22
Scotland	25.07	9.35	353	25.06	9.12	53	26.37	9.65	554	25.91	9.96	395	0.84
Sweden	28.24	9.88	168	25.88	9.70	56	27.42	9.60	289	27.55	10.12	220	−0.69
United States	12.30	11.04	477	13.25	7.45	132	16.15	11.88	477	17.04	13.09	286	4.74
All Countries (Average)	30.49	9.44	1,794	29.74	7.83	682	30.28	10.04	3,049	31.19	10.14	2,214	.70
II. General Program													
Japan	30.61	14.65	59	26.47	13.37	106	29.04	14.40	140	31.75	14.42	226	1.14

TABLE 5.11. *Total Mathematics Test Scores (Corrected) as Related to Socio-Economic Background (School Program Being Held Constant).*

Population 3 b.

Country	Father's Occupational Status												Difference Highest—Lowest
	Lowest (7 and 9)			Second Lowest (5 and 8)			Second Highest (3, 4 and 6)			Highest (1 and 2)			
	M.	S.D.	N.	M.	S.D.	N.	M.	S.D.	N.	M.	S.D.	N.	
I. *Academic Program*													
Belgium	26.85	8.41	159	23.92	8.45	54	24.40	8.94	455	23.95	9.64	256	−2.90
England	21.15	9.47	207	24.54	8.61	32	22.81	9.80	595	22.85	9.64	460	1.70
Finland	21.83	7.86	115	24.61	8.07	91	22.71	8.21	95	21.33	7.88	92	−0.50
Germany	27.80	7.59	30	27.79	5.92	29	28.09	7.44	283	27.47	7.54	361	−.33
Scotland	21.04	7.85	600	21.61	7.34	108	20.74	8.02	679	20.59	8.29	459	−.45
United States	6.41	7.73	792	5.55	6.04	146	9.83	9.41	527	13.97	10.20	280	**7.56**
All Countries (Average)	24.26	7.09	2,002	23.91	6.86	492	23.96	8.09	2,779	24.18	8.22	2,070	−.08
II. *Vocational Program*													
Japan	24.92	12.19	337	22.41	11.73	649	22.98	12.12	725	24.37	12.12	240	−.55
III. *General Program*													
Japan	26.03	12.50	231	25.06	13.26	465	26.60	13.22	1,102	27.78	13.73	577	1.75

217

It can be seen in Tables 5.8 and 5.9 that for each occupational group the All Country mathematics scores of students in the academic school programs are highest while they are lowest in the vocational programs. However, our primary interest is in determining the comparisons across socio-economic groups within school programs. It may be seen in Table 5.8 that in Population 1 a in each country, with the exception of Finland, the highest occupational students have higher scores than the students from the lowest occupational groups. The largest differences occur in the general programs while they are on the whole smaller in the academic programs. The differences between the highest and lowest occupational group are relatively large in England, Japan, the United States, and Australia. In Belgium, the Netherlands, and France the differences are considerably smaller.

Population 1 b (see Table 5.9) shows the same pattern. Here again, the differences are least for the academic program. Also, the countries maintain the same order of magnitude of differences between the extreme occupational groups. While it is not clear why the differences for Finland are so small, it is likely that the large differences for the United States, Japan, and Australia reflect the great range of ability and social background to be found in a comprehensive system of education. In Japan, all the students are in the general program, while in the United States and Australia, the different socio-economic groups are well represented in both the general program and the academic program.

At the preuniversity school year (Tables 5.10 and 5.11), the differences are much smaller, and in almost one half of the comparisons the lowest occupational group is equal or superior to the highest occupational group. The major exception to this is the United States, where the majority of students (of all socio-economic groups) are still in school at age 17–18. It is likely that the reason why the hypothesis is not confirmed at the preuniversity school year is because of the selection process in the different countries. Both the initial selection of who goes on to academic secondary education and the attrition during the secondary school program is related to social and economic background. Thus, the students of the lower socio-economic groups who are still in the academic program at the end of secondary school education are a highly select group of students who are able to achieve in mathematics at about the same level as students from the highest socio-economic group.

It is a commonplace observation that school achievement is related to the socio-economic status of the students. The problem taken up in this context is to what extent this holds true when the school program is

218

held constant. For Populations 1 a and 1 b the hypothesis of differences between socio-economic groups within school programs is confirmed, but for Populations 3 a and 3 b it is not confirmed. The explanation for the differences between the 13-year group and the terminal mathematics population is that for the 13-year group the selection or division of students into academic and general school programs still includes almost all members of the age group. In contrast, the 17+group includes only those students who have managed to survive the selection and attrition process in education. The lower socio-economic students who have survived are as able academically as the highest socio-economic students at the preuniversity level. In contrast, countries with a more comprehensive structure and with broad admission procedures to school programs of secondary education will still have major differences in academic achievement of its different socio-economic groups at the preuniversity level.

When speaking of "social bias" in the selection for secondary education (cf. Chapter 3) one should bear in mind that students from higher status homes, where among other things parents have good education, by having been reared in these homes can profit better from higher secondary education than students from mostly working class homes where the parents have only a limited school education. Thus, if one wants to increase the predictive power of admissions procedures an index of social status would add considerably to the validity of these. "Bias" as conceived of in this connection goes into the selection only to the extent that students from upper status homes are more frequently admitted than would be justified on the basis of their probability to perform according to given standards than students from lower status homes.

School Socio-Economic Variability and Mathematics Achievement

As a final hypothesis in our attempt to understand the relationship between socio-economic status and school learning, we wish to consider the effects of school variability in socio-economic status on student achievement. We state the hypothesis (28) as follows: *The difference in mathematics achievement between lower and higher occupational status levels (1) will be least when students are in schools which have the greatest variability in occupational status levels, and (2) will be greatest when students are in schools which are most homogeneous with respect to occupational status levels.*

The present hypothesis is slightly more complex than our previous hypothesis about socio-economic difference in that it is concerned with the relation of both school variability and level of occupational status to student achievement. This hypothesis arises from our interest in the effect of homogeneity of the school population on educational achievement. In some countries there is a tendency to develop schools for different segments of the population, whether the segments be defined in terms of race, economic condition, educational level of parents, or social class. In some communities this segmentation of the population for educational purposes is regarded as imperative for maintaining the quality of education—at least for the upper segments of the population. This is one of the issues in education about which there is much heated discussion on the part of parents as well as educationists. In the United States it is sometimes defined in terms of race, and in some parts of the country individuals are almost prepared to battle "even unto death" over this issue. Undoubtedly, the issue is seen by some in terms of interpersonal relations as well as in educational terms. Here we are concerned only with the cognitive educational outcomes.

Previous research by Svensson (1962) in Sweden demonstrated that academic students did as well in selective academic schools as in comprehensive schools, while non-academic status students did somewhat better academic work in the comprehensive school than they did in a more homogeneous school (homogeneity defined in terms of student level of ability and social status). Katz (1964) has summarized the effects of integrated versus segregated schools in the United States and generally finds that Negro students perform slightly better in the integrated schools, although this may in part be explained in terms of the selectivity of the students. Halsey (1961) also summarized the research on this problem in several European countries and finds improved performance of the less able student under heterogeneous school conditions than under homogeneous school conditions.

In order to test this hypothesis the schools for each country were divided into three categories: schools with little variability, schools with some, and schools with much variability in the occupational status of the fathers. This variability was determined on the basis of the standard deviation of the distribution of responded occupational levels in each school. For each type of school (in terms of occupational variability) the mean mathematics score of students in the four occupational status groups was calculated. We compared the average achievement of pupils from each status group in the three types of schools (Tables 5.12–5.15).[1]

[1] Originally, it was intended to use analysis of variance to test the hypothesis. This, however, proved to be impossible because of the occurrence of zeros in a number of cells.

TABLES 5.12–5.15. *Total Mathematics Test Score as Related to Father's Occupational Status and School Variability.*

School Variability in Father's Occupational Status	Lowest Occupations (7 and 9)			Second Lowest Occupations (5 and 8)			Second Highest Occupations (3, 4 and 6)			Highest Occupations (1 and 2)		
	M.	S.D.	N.	M.	S.D.	N.	M.	S.D.	N.	M.	S.D.	N.
Population 1a (Table 5.12)												
Little	15.53	10.99	2,727	19.27	12.17	981	18.55	11.30	861	21.21	6.96	174
Some	20.75	12.69	4,114	21.70	11.91	1,122	25.41	12.80	2,328	30.26	12.00	586
Much	20.04	12.13	6,533	21.75	12.75	648	25.72	12.72	3,843	30.64	12.98	1,751
Population 1b (Table 5.13)												
Little	18.96	11.16	3,479	20.69	11.55	1,461	21.77	11.10	1,250	23.91	6.10	155
Some	22.44	12.26	6,403	24.62	11.49	1,773	27.17	12.34	3,947	29.64	12.36	911
Much	22.45	12.85	8,996	24.32	12.14	987	27.25	12.53	5,561	31.53	12.26	2,456
Population 3a (Table 5.14)												
Little	24.77	10.65	213	24.57	10.36	205	26.63	9.88	413	26.48	9.26	372
Some	26.38	11.17	474	26.46	8.71	230	26.99	10.23	1,200	28.01	10.75	706
Much	27.66	10.09	1,240	28.03	7.76	299	27.36	10.17	1,714	29.15	10.48	1,358
Population 3b (Table 5.15)												
Little	17.52	7.26	326	18.55	8.19	698	21.06	8.53	768	19.70	8.64	567
Some	24.02	7.21	767	21.73	7.03	562	22.18	8.66	1,909	22.21	8.69	1,193
Much	19.86	8.61	1,743	18.97	7.71	383	20.39	12.11	1,887	21.02	9.02	1,263

The clearest patterns are demonstrated in Tables 5.12 and 5.13 for Populations 1 a and 1 b. In both of these tables there is a clear increase in mean achievement with increase in level of father's occupation. This is true at each level of school variability. Thus, in Population 1 a the mean score for schools with little variability goes from 15.53 for children at the lowest status position to 21.21 for children whose parents are in the highest occupations. A similar pattern is shown in schools with some variability, as well as schools with much variability.

However, our primary interest is in the columns of these tables, rather than in the rows. It is evident that for each economic status group, the students have higher mathematics achievement in the more variable schools (Some and Much) than they do in the least variable schools (Little). This pattern is repeated in both tables and is true for every occupational level. For each of the occupational groups there is little difference between the some and much variability groups.

Of particular importance is the fact brought out by Tables 5.12 and 5.13 that students in both the younger populations who come from low status homes, but who are attending schools which are heterogeneous in social composition, achieve about as well as students from the highest category of homes, when these latter students are studying in socially homogeneous schools. It is no less important that students from the upper-status homes also benefit from being in socially heterogeneous schools. These findings support a comprehensive policy when organizing schools.

Almost without exception increased social heterogeneity of schools is associated with greater variance in achievement among students.

In Population 3 a (Table 5.14) the pattern found in Populations 1 a and 1 b is essentially repeated although the differences are much smaller. This may in part reflect the effect of special selection for specialization in mathematics.

In Population 3 b (Table 5.15) the pattern is very confused between occupational levels since in a number of entries the lower status groups are equal to or superior to the higher status groups. However, the pattern for the columns is clear. Students in schools with some variability are superior in achievement to students in schools with little variability. The schools with much variability characteristically have higher levels of achievement than the more homogeneous schools, but lower than the schools with intermediate levels of variability.

Because of the many cells with zero or near zero entries we have not been able to show this picture at the national level. It is to be hoped that further research within countries can be made to determine whether the international picture is repeated in the various countries.

Viewing this problem only at the international level, it would appear that there is a highly consistent pattern in each of the populations except for Population 3 b, with the clearest pattern and largest differences in the younger age groups. It would seem that pupils from every occupational group profit more from being in schools with some variability than from being in schools with little variability in social class. However, before this conclusion is drawn other explanations should be investigated. Unfortunately this cannot be done with the present data.

With regard to the alternative explanations, it should be said that from the original data it can be seen that schools with little variability tend to recruit their pupils from the lower occupational group. The so-called "elite" school is not well represented in this study. It must be remembered that the lowest social group, which includes all children from manual workers—farm laborers excepted—and from the lower service workers, is a very large group. In many countries it comprises about one half of the population. This group cannot be said to be sociologically homogeneous at the present time, and the possibility that special subgroups are involved in the choice of schools with little variability should be considered. The case of so-called "slum area" schools especially should be borne in mind.

Moreover, in countries where organizational differentiation takes place at the beginning of secondary education, it frequently happens that pupils that are achieving the least are sent to different schools than those to which pupils who achieve at a higher level are sent. In view of the established relation between occupational level of the family and achievement, this would lead to a concentration of low-achieving pupils in schools with little variability. Of course, it is not implied here that these alternatives, or other possible explanations, are more "true" than the one that greater school variability fosters good achievement. Here is a field for future research.

Urban-Rural Differences in Mathematics Achievement

The problem of urban-rural differences has been studied in the past largely from two main points of view. On the one hand, there has been concern with the opportunities for higher education, and on the other, with differences in test scores. The overall consensus from studies of the former type is that the percentage of students going on to higher education from rural homes is considerably below probability expectation (Halsey, Floud, and Anderson (1961), Halsey (1961)), although even here certain qualifying conditions are admitted.

In the long series of studies concerned with differences in test scores, much attention has been given to intelligence. Most of the studies of educational achievement have been carried out with primary school students. The results of these studies are far from conclusive. In a broad review covering the period up to 1959, Barr (1959), discussing research with the 10–12 year group (mainly), reported that in some eight British studies urban children had obtained significantly higher scores than rural children on standardised educational tests. Swedish national surveys of the complete 4th and 6th grade groups in the elementary school arrived at the same results (Ljung, 1958).

A Canadian study of grade 8 pupils by Jackson (1957) reported no significant difference in either arithmetical computation or problem-solving for pupils in two different samples of rural, small city, and large city students, although he observed that differences were greater for problem solving. On the other hand, a National Survey (1961, 1962) taken in Japan in 1961 and repeated in 1962 showed significant differences between the scores of urban and rural students in the second and third grades of junior secondary schools. A national survey in Scotland showed that the differences in attainment in English and in arithmetic among children educated in cities, large towns, small towns, and in other areas were small and not educationally significant (Scottish Council for Research in Education, 1963).

In these studies authors have frequently drawn attention to the fact that differences in test scores may have been in part a function of socio-economic differences (Barr, 1959).

Facilities did not permit the most complex analysis of the data in the present study in terms of all possible variables which might be related to urban-rural differences. However, it was possible to take level of instruction into account, a factor which had not been considered, in other investigations. The major hypothesis was: *When the level of mathematics instruction is held constant, there will be differences in mathematical achievement between students from urban and rural homes* (Hypothesis 29).

In collecting data for this hypothesis, a six-point rural-urban scale was used to classify the place of residence of parents of students in this study:
1. Rural, farm.
2. Rural, village (less than 2,500).
3. Small town (population from 2,500 to 15,000).

4. Medium-sized city (from 15,000 to 100,000).
5. Urban center (more than 100,000).
6. Suburban area adjacent to an urban center.[1]

The mathematics total scores in this section have been treated by regression procedures to remove the effect of level of instruction, since we were interested in urban-rural differences where level of instruction is held constant. Tables 5.16–5.19 present the mean total scores and S.D.'s for each of the four populations for the countries included in the analysis. When an analysis of variance was carried out on each of the four tables, large between-country variation was found but non-significant differences for place of parents' residence and for the interaction between the two main effects.

It seemed likely that the magnitude of the between-country differences might have masked some of the within-country effects, which might help to illuminate the apparently nonsignificant results obtained in the overall tests of significance. Accordingly, the differences between pairs of residence category scores are presented in Table 5.20.

The information in Table 5.20 has been summarized in frequency form in Table 5.21. While there are urban-rural differences in mathematics attainment in each population, in only one country are differences between rural and other groups significant in all populations. In the United States, rural groups have lower scores than town or urban groups at all levels. In Japan, differences similar to those found in the United States are evident in Populations 1 a and 1 b. In Israel, for the only population group for which results were available, 1 b, the differences found are of the same order as those in Japan and United States. In England differences in Population 3 a are significant.

---

[1] Some difficulty was encountered in several countries using this code, the awkward category being the community adjacent to an urban center. In some countries it was interpreted as being adjacent to a large urban center of over 100,000, while in others, such as Australia, the neighborhoods adjacent to a large city (for example, Melbourne) were so coded.

Certain other differences between countries occurred in the way the information for this item was assembled. Either the student was given the agreed scale and asked to place himself on it or he was asked for detailed information about the location of his home, and the staff of the national center then coded the information by reference to official gazetteers or similar sources. In at least one case the students categorized their own information, which was then checked by the national center.

Notwithstanding these factors, it was considered that the data were sound enough for the planned analysis. However, in view of the differences in approach to Category 6 and the relatively small number in some other categories, it was decided to reduce the classification to three categories—rural (1, 2), town (3, 4), urban (5, 6).

TABLE 5.16. *Place of Parents' Residence, Mean Corrected Scores, Standard Deviations, and Sample Size. Population 1a.*

Country	Rural			Town			Urban		
	M.	S.D.	N.	M.	S.D.	N.	M.	S.D.	N.
Australia	19.3	12.0	706	20.4	12.1	675	20.6	12.4	1,535
Belgium	27.6	13.2	741	27.4	13.3	476	27.7	12.3	419
England	21.5	16.5	631	20.1	16.0	963	19.8	16.3	1,215
Finland	23.7	9.1	390	24.0	9.9	230	23.3	8.2	126
France	20.4	10.8	531	17.8	9.3	1,423	17.5	9.4	419
Japan	27.4	15.9	866	32.4	16.5	548	35.8	17.1	626
The Netherlands	25.4	10.9	68	22.6	9.3	273	27.1	10.5	86
Sweden	14.9	9.9	1,332	16.2	10.2	820	17.6	10.7	379
United States	14.1	11.4	1,985	17.2	12.4	2,163	17.6	13.5	1,649
All Countries	21.6	12.2	7,250	22.0	12.1	7,571	23.0	12.3	6,454

TABLE 5.17. *Place of Parents' Residence, Mean Corrected Score, Standard Deviations, and Sample Size. Population 1b.*

Country	Rural			Town			Urban		
	M.	S.D.	N.	M.	S.D.	N.	M.	S.D.	N.
Australia	18.7	11.8	769	18.4	12.4	737	19.2	12.3	1,570
Belgium	30.6	12.5	1,069	29.8	11.8	725	30.6	12.6	770
England	26.1	17.3	666	23.5	17.3	1,046	24.6	17.7	1,321
Finland	25.7	9.5	427	26.9	9.8	275	26.1	9.0	140
France	25.3	12.8	921	19.7	9.3	1,852	18.5	8.6	629
Germany	25.8	12.2	1,757	25.6	10.2	2,044	22.5	7.1	672
Israel	26.0	12.6	540	32.0	14.7	1,687	36.6	12.3	995
Japan	27.4	15.9	866	32.4	16.5	548	35.8	17.1	626
The Netherlands	21.2	10.5	244	21.0	10.7	866	25.8	10.1	333
Sweden	14.5	10.0	1,470	15.7	10.3	901	17.0	10.9	432
United States	16.6	12.0	2,112	18.7	13.0	2,432	18.8	13.7	1,779
All Countries	23.4	12.5	10,821	24.0	12.4	13,113	25.0	11.9	9,267

In several countries a rather unexpected trend has been noted in that the rural scores are higher than either the town or the urban scores; in France in Populations 1 a and 1 b, in Germany at 1 b, and in Finland at 3 a and 3 b. The same trend is observable in the data of England, Belgium, and the Netherlands at Populations 1 a, although in those countries it does not reach significance. Some of the factors discussed below may help to account for this trend.

TABLE 5.18. *Place of Parents' Residence, Mean Corrected Scores, Standard Deviations, and Sample Size. Population 3 a.*

Country	Rural			Town			Urban		
	M.	S.D.	N.	M.	S.D.	N.	M.	S.D.	N.
Australia	21.9	9.0	217	21.6	8.5	235	22.7	8.8	638
Belgium	33.0	11.6	206	34.7	12.3	141	35.8	12.8	401
England	32.8	11.6	223	36.4	12.1	229	35.6	12.6	512
Finland	25.9	9.7	173	26.4	8.8	138	22.1	6.0	60
France	33.4	9.9	63	33.6	10.0	134	34.3	8.6	25
Germany	29.2	10.0	148	29.6	8.5	205	28.0	10.1	297
Japan	31.3	14.4	255	33.2	13.3	276	29.9	15.7	287
The Netherlands	29.1	7.7	64	31.3	7.5	185	32.1	8.2	211
Sweden	26.7	10.3	217	27.5	9.6	391	26.5	8.8	151
United States	10.4	8.4	535	13.0	10.5	497	19.7	13.4	465
All Countries	27.4	10.3	2,101	28.7	10.1	2,431	28.7	10.5	2,807

TABLE 5.19. *Place of Parents' Residence, Mean Corrected Scores, Standard Deviations, and Sample Size. Population 3 b.*

Country	Rural			Town			Urban		
	M.	S.D.	N.	M.	S.D.	N.	M.	S.D.	N.
Belgium	24.5	9.5	380	25.2	9.1	313	23.4	9.5	307
England	20.6	8.8	509	21.4	9.5	515	22.1	9.8	754
Finland	24.7	8.0	183	21.3	7.4	123	21.5	7.3	91
France	24.9	10.6	37	26.1	9.0	99	24.5	10.7	56
Germany	28.2	6.7	128	27.4	7.8	240	28.1	7.7	276
Japan	24.5	13.3	1,695	26.2	13.3	1,659	25.1	13.3	1,011
Sweden	4.8	5.3	55	13.5	6.0	118	10.8	9.0	40
United States	5.3	6.9	661	8.1	8.2	763	12.2	10.4	549
All Countries	20.7	8.7	3,665	21.6	8.7	3,854	21.4	10.8	3,091

No consistent trend is to be found in the urban-town data though the number of significant differences is smaller than for the other comparisons. In some countries, urban scores are higher than town scores although in others they are lower, but in most there is no difference. It is possible that problems of categorizing may be a major factor in this result. It is also possible that urbanization may mean different things in different countries, depending upon the forces that attract people to or away from the large urban centers.

TABLE 5.20. *Differences in Mathematics Scores (Corrected for Level of Instruction) Between Students Grouped by Place of Parents' Residence.*

Country	Population 1a			Population 1b			Population 3a			Population 3b		
	T-R	U-R	T-U	T-R	U-R	T-U	T-R	U-R	T-U	T-R	U-R	T-U
Australia	1.1	1.3	−0.2	−0.3	0.5	−0.8	−0.1	1.0	−1.1	—	—	—
Belgium	−0.2	0.1	−0.3	−0.6	0.2	−0.8	1.7	2.8	−1.1	0.7	−1.1	1.8
England	−1.4	−1.7	0.3	−2.6	−1.5	−1.1	3.6	2.8	0.8	0.8	1.5	−0.7
Finland	0.3	−0.3	0.7	1.2	0.4	0.8	0.5	−3.8	4.3	−3.4	−3.2	−0.2
France	−2.6	−2.9	0.3	−5.6	−6.8	1.2	0.2	1.4	−1.2	1.2	−0.4	1.6
Germany	—	—	—	−0.2	−3.3	3.1	0.4	−1.2	1.6	−0.8	−0.1	−0.7
Israel	—	—	—	6.0	10.6	−4.6	—	—	—	—	—	—
Japan	5.0	8.4	−3.4	5.0	8.4	−3.4	1.9	−1.4	3.3	1.7	0.6	1.1
The Netherlands	−2.8	1.7	−4.5	−0.2	4.6	−4.8	2.2	3.0	−0.8	3.6	2.8	0.8
Sweden	1.3	2.7	−1.4	1.2	2.5	−1.8	0.8	−0.2	1.0	1.7	−1.0	2.7
United States	3.1	3.5	−0.4	2.1	2.2	−0.1	2.6	9.3	−6.9	2.8	6.9	−4.1
All Countries	0.4	1.4	−1.0	0.6	1.6	−1.0	1.3	1.3	0.0	0.9	0.7	0.2

TABLE 5.21. *Summary of Significant and Non-significant Differences Between Urban, Town, and Rural Groups.*

	1 a	1 b	3 a	3 b
R > U	1	2	1	1
U > R	2	4	2	1
N.S.	6	5	7	7
R > T	1	1	0	1
T > R	2	3	2	1
N.S.	6	7	8	7
U > T	2	3	1	1
T > U	0	1	1	1
N.S.	7	7	8	7

In considering the findings there are two aspects to be considered: (1) to account for the support of the hypothesis in three countries, and (2) to account for the refutation of the hypothesis in the remaining countries. The factors that seem relevant fall into two types; those which have been shown by other investigators to be important and for which data exist in the present study and those for which no information was obtained. Neither of these alone is likely to be a major explanatory variable, but the conjoint operation of several may well be important. The following variables will be discussed: range of ability, heterogeneity of samples, organization of school systems, expenditure for and qualifications of teachers, the existence of a national examining system, socio-economic status of parents, sex of students, and types of mathematics scores.

The clearest support for the hypothesis would be expected in the 1 a and 1 b groups, for they contain the full range of pupils in the school system, while the least support might be expected in 3 a, where the percentage of the age group in school is lower and the proportion of those studying mathematics is lower than at any other stage. Moreover, with university preparation being the main aim of this group in most countries, few differences would be expected. In the two countries where significant rural-urban differences were found, these were more marked and consistent at the lower level. In the other countries it is noted that the differences in the 1 a and 1 b groups are larger (although not significantly so) than in the upper population groups. The one country, the United States, where significant differences are consistently found favoring the urban group at the upper level is the one with the highest proportion of the age group in school. It is also the country with the greatest heterogeneity of population from both an ethnic and regional point of view.

A number of studies in the United States, mainly unpublished, provide some support for the heterogeneity of population as a factor. These studies show that the highest average level of achievement is found in certain midwestern states where the population is relatively homogeneous, both in terms of occupational and ethnic origin composition. The other country in the study with wide ethnic range is Israel, and although the data are available for only one group, and the difference between the rural and the urban group is significant (as in the United States), a certain proportion of the population is of relatively recent advent in the country. None of the other countries is considered to have the same range and proportion of occupational and national origin groups.

Several characteristics of the school system may be important in considering the present results. It has been the practice in many countries in the past twenty years to consolidate the secondary school system, either by providing district secondary schools on a geographical basis or by stipulating that a secondary school must be of a certain minimum size. One expected outcome of such a reorganization would be an increase in achievement test scores. Support for this has been provided by Feldt (*personal communication*), who found in a statewide testing program involving over 90 percent of the secondary schools in one state, that during a period of about seven years in which the main consolidation from 3,000 to 400 school districts took place, there was a significant annual improvement in all achievement test scores. Since then, annual increments have been nonsignificant. This type of evidence would lead one to expect to find smaller urban-rural differences in the United States as school consolidation takes place. It is probable that the differences would have been larger had consolidation not taken place. Similar consolidation has occurred in Australia, England, Scotland, and Sweden.

In those countries showing urban-rural differences in achievement, another factor in the school system which could be exerting an influence is the difference in qualification of teachers. It is known that the widest range of teachers' salaries is found in the United States (see Chapter 14 of Volume 1), and that this factor is in part related to level of qualification of teachers. In most of the other countries there is a uniform basic salary scale for all teachers throughout the country. In such countries special allowances and the attractions of larger cities may still lead to the teachers with better qualifications being found in the town and urban areas; but these factors should have less effect in those countries where relatively uniform salary scales are found.

Another national characteristic that seemingly could lead to reduced urban-rural differences in mathematics achievement is that of national or regional examinations, such as are conducted for the General Certificate of Education in England and by the states in Australia and in European countries. In the United States, the absence of any such achievement examination and the use of a variety of testing programs in the secondary school could be an important factor in accounting for rural-urban school differences in achievement.

In addition to the above institutional factors there are several personal and family factors which are likely to enter into the urban-rural dimension. Chief of these is the socio-economic status and the educational level of the family of the child. Some recent work by Sewell (1963) has provided the strongest evidence for establishing the importance of the socio-economic factor in the study of rural-urban differences. At present, the achievement data are being analyzed for this study, but the evidence available on educational and occupational aspirations for a large sample from a single state has implications for the present project. The Sewell study showed that rural youth were likely to be influenced negatively in their educational aspirations by personal characteristics, school environments and communities, and the socio-economic-educational levels of their families. He went on to point out that the relationship of the above variables and rural-urban differences in educational plans is by no means a simple one. He summarizes by saying:

> Separate controls for intelligence and socio-economic status, although reducing the rural-urban differences, did not remove them entirely for boys or girls. However, when both were controlled simultaneously, rural-urban differences for the girls were largely eliminated. For the boys there were still significant differences at all socio-economic status levels, especially in the high ability groups.

Further analyses of the personal variables, controlled both for intelligence and socio-economic status, were not able to account for the original rural-urban differences, leading the author to conclude that the causes of rural-urban differences in youth are by no means simple, and that the differences are real and persistent.

Admittedly, this conclusion refers to aspirations, but in the detailed report Sewell makes it clear that the rural home-school community fosters relatively fewer favorable attitudes to further education. It does not seem too adventurous to hypothesize that the existence of such attitudes could be related to lower achievement in mathematics, a subject often perceived as being primarily college-oriented. Some evidence for this hypothesis may be obtained from further analyses of the present data.

It emerges from the above discussion that the final consideration of the rural-urban differences depends on an analysis which controls not only for level of instruction but also for a number of other variables. Such an analysis is reported in Chapter 6.

When the level of instruction is removed the place of parents' residence (urban, town, rural) is, in general, not significantly related to achievement in mathematics. What seems to happen is that since the level of mathematics instruction is positively correlated with urbanization, the removal of the variance due to level of instruction also removes the effect of parents' residence. The reduction of urban-rural differences in educational criteria is worthy of further study by sociologists and educators. Our study has only pointed up the gross results; further investigations are needed to determine the processes at work in reducing educational differences between urban and rural communities.

Financial Support for Education and Mathematics Achievement

One of our hypotheses states: *Performance on the mathematics test will be related to per-student financial expenditure (1) as a whole and (2) specifically for teachers' salaries* (Hypothesis 30).

The rationale for this hypothesis is quite straightforward: Better teachers will "produce" better achievement, higher salaries will attract better teachers, higher per-student expenditure implies higher salaries (or else smaller classes, which have also been considered to produce more effective teaching), so higher per-student expenditure will be associated with higher achievement.

Correlations were computed for students between the two expenditure variables and total mathematics score, and are shown in Table 5.22. Correlations that are significant at the .05 level are in bold type. For Populations 1 a and 1 b, the coefficients are predominantly positive—14 out of 20 for total expenditure and 13 out of 20 for expenditure for teachers' salaries. However, in Populations 3 a and 3 b there are 13 positive and 5 negative signs for total expenditure and 13 and 5 for salaries.

The variation from country to country indicates that aspects of the administrative and social structure have made per-student expenditure signify quite different things in the different national settings. One such factor may be the degree of centralization of authority for salary sched-

TABLE 5.22. *Correlation of Total Mathematics Score with Total Per-Student Expenditure and Expenditure for Teachers' Salaries.*

	Total Expenditure				Expenditure for Salaries			
try	1a	1b	3a	3b	1a	1b	3a	3b
alia	−.03	−.11	.05	—	−.03	−.11	.05	—
ınd	.25	.24	.02	.05	.20	.21	.04	.02
nd	.16	.11	.12	−.17	.16	.11	.12	−.17
e	.16	.13	−.20	—	.14	.10	−.20	—
any	—	.16	−.24	−.10	—	.23	−.15	−.09
	—	.02	—	—	—	.02	—	—
	.00	.00	.07	.20	−.02	−.02	.12	.23
Netherlands	.39	.25	.11	.03	.41	.25	.06	.07
and	.07	.09	−.02	.11	.07	−.05	.01	.11
en	−.08	−.09	.14	.12	−.13	−.15	.12	.10
d States	.10	.09	.08	.07	.10	.09	.07	.08
{ +	6	8	7	6	6	7	7	6
{ −	2	2	3	2	3	4	3	2
	.08	.07	.02	.04	.09	.06	.03	.04

ules and of resulting uniformity throughout the country. Another may be the policy for staffing small schools. Thus, in certain instances a high per student expenditure may imply a very small school rather than a well-paid and effective teacher. Also, even in countries with uniform salary schedules, there are still relatively high correlations between teacher salary and student achievement for Populations 1 a and 1 b because of the tendency for more qualified teachers to be assigned to the more advanced classes (see Table 2.1). Thus, it appears that the variable that has been studied is a symptom of other conditions in the educational system, rather than a causal factor in its own right, and a symptom of different conditions in different countries.

Sex Differences in Mathematics Achievement and Attitudes Toward Mathematics

Many previous studies have examined sex differences in ability and achievement in mathematics. The present inquiry provides an excellent opportunity to investigate these differences cross-nationally and to seek possible explanations of the differences that emerge. One might even say that sex differences in educational achievement cannot be fruitfully

TABLE 5.23. *Ratio of Male to Female Students.*

Country	Column 1 Ratio M/F at Terminal Secondary School Level	Column 2 Ratio M/F in Population 3 a	Column 3 Ratio M/F in First Year at a University
Australia	1.39	2.32	2.59
Belgium	1.47	7.13	3.35
England	1.29	5.53	3.16
Finland	0.78	1.73	.96
France	1.12	6.00	1.50
Germany	1.24	3.92	3.06
Israel	1.02	3.11	1.77
Japan	1.10	2.11	2.33
The Netherlands	1.80	5.37	4.87
Scotland	1.34	2.02	1.94
Sweden	1.44	2.80	1.51
United States	1.03	2.37	1.38
All Countries (Average)	1.25	3.70	2.37

studied unless cross-cultural differences are considered among the independent variables.

It is, however, necessary to consider first whether there are differences between the sexes in the numbers studying mathematics. From the information collected in this study it is possible to examine these differences in the preuniversity grade levels of secondary schooling. The tendency for a specialist study of mathematics at the preuniversity level to become a predominantly male activity is shown in Table 5.23. Column 1 gives for each country the ratio of male to female students who have reached the terminal level of secondary schooling. Column 2 gives the ratio of male to female students in Population 3 a (mathematics specialists) of this study. Since the study of mathematics at the preuniversity level in some countries is strongly linked with matriculation requirements for entry to a university, column 3 is given, in which the ratio of male to female students in their first year in any course at a university is provided.

From a comparison of the figures given in columns 1 and 2 it is seen that a much higher proportion of male compared with female students who remain at school at the preuniversity level undertake a study of mathematics. Moreover, by comparing the figures given in columns 2 and 3 it is seen that, with exceptions in the cases of Australia and Japan, male students predominate in mathematics classes at the preuniversity year to a greater extent than they do in enrollments at the first-year university level. Even more obvious, however, is the wide range across countries in the ratio of male to female students who are studying mathematics at the preuniversity level. It should be noted that this ratio is greatest in the cases of Belgium, France, the Netherlands, and England. Moreover in these countries, with the exception of France, the universities are predominantly male institutions.

The wide ranges in the ratios discussed are striking, and it is possible that the studying of mathematics at school may be influenced not only by the educational opportunities available to males and females in a community but also by the role of women in society and the freedom for women to enter certain occupations. However, it is important to note that in most of the countries associated with this study the restrictions on women entering certain occupations arise from concepts of the role of women in a society, rather than from a formal lack of freedom.

The differences described above may also stem from the different arrangements made in the countries concerned for educating male and female students.

Table 5.24 gives for the students who are involved in this study in Populations 1 b and 3 a, the ratios of the numbers in single-sex schools to the numbers in coeducational schools.

At the lower school level, Belgium and France have predominantly single-sex schools; in England and the Netherlands the pupils are almost equally divided among single-sex and coeducational schools, and in all other countries the provision of education is largely in coeducational schools. In Australia, however, this study (and the figures here) is limited to government schools, while the nongovernment schools which cater to approximately 25 percent of the pupils are with few exceptions single-sex schools.

At the upper-secondary school level, with the exception of the Netherlands and Belgium, there is a greater tendency to provide single-sex schools. In Australia, Israel, Sweden, and the United States, there are probably no real differences between the levels.

There are three factors involved in allocating students to single-sex schools:

1. A traditional rejection of coeducation (typically for Roman Catholic Schools in France, Belgium, and the Netherlands and for Anglican Schools in England).

TABLE 5.24. *Ratio of Students in Single-Sex Schools to Those in Coeducational Schools.*

Country	Population 1 b	Population 3 a
Australia	.34	.35
Belgium	5.01	4.26
England	.94	3.29
Finland	.08	.21
France	2.08	2.93
Germany	.06	.44
Israel	.26	.30
Japan	.03	.27
The Netherlands	1.28	.30
Scotland	.08	.21
Sweden	.06	.07
United States	.01	.04
All Countries (Average)	.85	1.05

2. The view that girls during adolescence have other needs than are served by the schools that have developed in a male-dominated world.
3. Early differentiation of groups that are given prevocational training: domestic and commercial for girls, technical and agricultural for boys.

In searching for causes of the marked differences between countries in the ratio of male to female students studying mathematics at the preuniversity level, many possible factors can be considered. In this particular study, data were available for an examination of relationships between these ratios and the following:

1. Postsecondary school educational opportunities for women.
2. Occupational freedom for women.
3. Mathematical achievement.
4. Interest and plans to undertake further study.
5. Attitudes toward mathematics.
6. Type of school—either single-sex school or coeducational school.
7. Level of education reached by father and mother.
8. The effectiveness of a male or female teacher with male or female students in mathematics.

The interaction between some of these factors will make their identification and the extent of their contribution difficult to determine. Clearly, the student who is achieving well will show more interest in mathematics and will probably undertake further study in the subject. This and similar cycles will be developed at each stage of secondary schooling and will be repeated at successive stages unless a link is broken, after which the student may cease to study the subject. The examination of these factors and the cycles developed is a complex task. This study can only hope to detect some possible differences between the performance of male and female students, and to reveal lines for further investigation.

Review of Previous Research

In previous studies that have examined sex differences in ability and achievement in mathematics three distinct systems of constructs have been used. The majority of workers in England following Spearman (1927) use a model for the structure of mental abilities that is hierarchical in nature with a central factor of general intelligence. Workers in the United States have frequently preferred to use a second model developed by Thurstone (1938) with a number of primary factors. Both these models make allowance for verbal, numerical, and spatial factors, and, working with one or other of these models, there have been many attempts to detect a mathematical group factor, to determine its nature and the differences in its composition for male and female students.

Early research workers in England tended to report that there was no group factor associated with mathematics, while later workers have been able to detect its existence. Working only with boys, Wrigley (1958) has shown that there is a clearly identifiable mathematical group factor with verbal, spatial, and numerical components.

In the United States, recent investigations have suggested that mathematical ability for boys and girls is built out of a combination of verbal, numerical, and spatial factors, but have suggested that the components of this ability are different between the sexes—the verbal factor being stronger and the numerical and spatial factors being weaker for girls.

A third system of constructs with elements of contents, operations, and products has been more recently proposed by Guilford and Merrifield (1960). Two studies have been undertaken. Hills (1955) was unsuccessful, and Peterson *et al.* (1963) were more successful in showing that there was some evidence of a set of abilities or traits which is associated with success in mathematics, although sex differences were not reported.

It is apparent that no clear pattern has emerged from these studies, but the distinction between verbal and numerical or computational abilities, and their combination with spatial ability to give a general mathematical ability would seem sound. Moreover, any sex differences that have been detected suggest that verbal ability is stronger for girls and that numerical and spatial abilities are stronger for boys. Tyler (1956) and Anastasi (1958) have surveyed the field of sex differences in aptitude and achievement and report that girls usually do better in verbal and linguistic studies, and boys generally have stronger numerical and spatial aptitudes and do better in tests of arithmetical reasoning.

The studies discussed above have not been longitudinal nor have findings been compared between different countries. There might be cultural factors which have a significant influence on school learning in general and mathematical attainment in particular and which produce national differences in attitudes, interests, and achievement between the sexes.

Two significant studies appear to have been undertaken previously to compare differences in achievement in mathematics between students in different countries: Pidgeon (1958) and Foshay *et al.* (1962). The latter study in particular made a substantial contribution in this field. In only one country, the United States, among the 13-year-old students, did the girls surpass the boys in total performance on the five tests (nonverbal intelligence, mathematics, reading comprehension, geography, and science). In addition, girls showed a small but consistent tendency to do better than boys on the mathematical test in eleven out of the twelve countries.

It is likely that sex differences in achievement in mathematics arise from causes other than influences of mental ability or cultural factors. Witkin *et al.* (1962), surveying the evidence for sex differences in problem solving, suggests that in both perceptual and intellectual situations men tend to be more analytic than women and should hence do better in mathematics.

There have been no major published studies concerned with sex differences in attitudes toward mathematics that go beyond the listing of subject preferences and simple expressions of likes and dislikes.

Milton (1957), seeking the cause of sex differences that he found to exist in problem-solving skills, some of which were mathematical in nature, showed that they could be partly accounted for by differences in sex-role identification. A positive relationship existed between masculine identification and problem-solving achievement. Carey (1958) found that sex differences in problem solving are a function of sex differences in attitudes towards problem solving. Lindgren *et al.* (1964) developed this idea to show that attitudes that were favourable to situations involving the solving of problems were positively correlated with achievement in arithmetic. Alpert, Stellwagon, and Becker (1963) showed that student attitudes toward mathematics were linked with parents' conception of the educational goals of a school mathematics course and with the extent of mathematics education desired for the child by the parents.

Although the area of sex differences in achievement in mathematics has been a field for a great deal of previous investigation, conclusive findings as to possible causes of observed differences have not as yet been obtained.

Differences Between the Sexes in Achievement

On the basis of the previous research studies reported above it was hypothesized that: *In all countries, in the 13 year old populations:*

1. *There will be no differences in overall mathematics achievement between boys and girls.*
2. *There will be slight differences favoring the girls on highly verbal problems.*
3. *There will be slight differences favouring the boys on computational problems* (Hypothesis 31).

It should be noted that these hypotheses apply only to the two populations in this study below the compulsory school leaving age. These populations rather than the preuniversity year populations, were chosen to prevent the issue being confused by sex differences in the holding power of both school and subject. For the populations at the upper-secondary school level it was hypothesized that similar sex differences would be found, but that the selection procedures operating would distort the patterns.

Total Mathematics Scores

It is suggested from previous research that there will be no consistent differences in total mathematics score between boys and girls, although slight differences might exist within particular countries.

Table 5.25 gives the results pooled for all countries, showing differences in performance between the sexes. The effects of differences in level of mathematics instruction have been removed from these results by regressing on this factor. These results show clear differences in favor of boys at all levels. In addition, the boys show slightly greater variability.

Differences between the sexes on total mathematics score for each country are shown in Table 5.26. The figure shown in each case is

TABLE 5.25. *Sex Differences in Total Mathematics Scores.*

Population	Boys			Girls			$M_b - M_g$
	M.	S.D.	N.	M.	S.D.	N.	
1 a	22.93	12.65	10,991	20.71	11.73	10,291	**2.22**
1 b	24.99	12.79	17,344	22.93	12.21	15,862	**2.06**
3 a	29.84	10.36	5,636	26.80	9.93	1,853	**3.04**
3 b	23.83	9.14	4,953	19.65	8.51	5,687	**4.18**

TABLE 5.26. *Sex Differences in Total Mathematics Score by Country.*

Country	Population 1 a	Population 1 b	Population 3 a	Population 3 b
Australia	14	11	−2	—
Belgium	38	30	79	80
England	19	25	4	58
Finland	18	19	47	52
France	16	20	51	—
Germany	—	17	86	52
Israel	—	−14	3	—
Japan	25	25	20	39
The Netherlands	22	31	12	—
Scotland[a]	—	—	—	—
Sweden	8	12	4	17
United States	2	5	25	2

[a] The data on level of instruction were incomplete.

obtained by taking the amount by which the mean score of boys is in excess of the mean score for girls for that country, and dividing this difference by the overall standard deviation for the set of scores for that population for all countries and multiplying by 100.

From an analysis of variance of the data it is found that the interaction between countries and sexes is significant at all levels. The variance of interaction between countries and sexes is not significant at any level when level of instruction has been regressed out. There is, however, evidence to suggest that policies of adjusting the level of instruction to suit male and female pupils are already starting to be used in some countries in the lower-secondary school. For Populations 1 a and 1 b, small but significant differences favor the boys in Belgium, Japan, and the Netherlands. For the preuniversity year populations, sex differences in the holding power of both school and subject are

TABLE 5.27. *Comparison of Boys and Girls on Verbal and Computational Problems.*

Population	Boys			Girls			$M_b - M_g$
	M.	S.D.	N.	M.	S.D.	N.	
			Verbal Problem Scores				
1 a	13.50	8.48	13,701	11.55	7.99	13,134	**1.95**
1 b	14.83	8.23	20,236	12.95	8.01	18,657	**1.88**
3 a	13.88	5.04	6,604	12.33	5.16	2,336	**1.55**
3 b	17.68	6.28	5,913	14.02	6.19	6,762	**3.66**
			Computational Problem Scores				
1 a	9.66	6.31	11,164	8.79	5.89	10,223	**.87**
1 b	10.35	6.06	20,154	9.70	5.79	18,750	**.65**
3 a	15.89	6.46	6,580	13.98	6.29	2,322	**1.91**
3 b	6.83	3.48	5,913	5.41	3.35	6,717	**1.42**

operating, but there is a marked male superiority in most countries for the population (3 b) not specializing in mathematics and strong male superiority among mathematics specialists (3 a) in the countries where stringent selection procedures are at work—namely, Belgium, France, and Germany.

Verbal and Computational Problem Scores

It is suggested from the previous research findings that there will be slight differences favoring the girls on highly verbal problems and slight differences favoring the boys on computational problems.

The relevant results obtained from this study are given in Table 5.27. In this table the scores were *not* regressed on the level of mathematics instruction.

There are clear differences in favor of boys in all populations in both verbal and computational scores. Generally boys show greater variability than do girls.

Differences between the sexes on verbal problems scores and computational problems scores are shown for each country in Table 5.28. The figures are obtained by the same procedure as used previously in Table 5.26. In general, there are more significant differences favouring the boys on the verbal problems than on the computational problems, especially at the lower grade level (1 b) and at the preuniversity nonspecialist level (3 b).

From an analysis of variance of the data for this set of results it is

TABLE 5.28. *Sex Differences in Verbal and Computational Problem Scores by Country.*

Country	Verbal Problem Scores				Computational Problem Sco[res]		
	1 a	1 b	3 a	3 b	1 a	1 b	3 a
Australia	11	16	19	—	−2	2	23
Belgium	45	49	59	79	45	48	72
England	25	29	5	5	22	25	−6
Finland	23	28	40	52	4	2	39
France	30	31	56	—	21	22	35
Germany	—	24	57	47	—	10	89
Israel	—	−11	17	—	—	−16	−9
Japan	27	27	27	59	17	18	25
The Netherlands	44	40	7	—	37	28	14
Scotland	15	18	32	86	—	6	24
Sweden	11	14	11	29	−5	−2	29
United States	7	12	34	6	−13	−9	8

found that the variance between countries and between sexes is significant in all cases. However, the variance of interaction between countries and sexes is significant at the 5 percent level or above only for computational problem scores at the lower secondary school level (Population 1 a and Population 1 b).

The similarity in the results for the two lower-secondary school populations is probably a direct consequence of the fact that a considerable proportion of the students is common to both groups. It would seem that there are possibly forces operating differently from country to country to produce differences of response of pupils of the two sexes to mathematical problems. Moreover, these forces would appear to operate differently in the two aspects of mathematical achievement being considered, namely, computational problems and verbal problems. In some countries the difference in performance between the sexes is similar in the two areas, and in other countries—Finland, Sweden and the United States—it is not.

In addition, it should be noted that in the area of computational problems at the lower age level the girls are superior to the boys, although not significantly so, in Israel, the United States, and Sweden, but in the area of verbal problems it is only in Israel that this difference holds.

Sex differences in these two sets of scores are less marked in Population 3 a than 3 b. The differences favor boys in all except four cases which is surprising for these two populations when the extent of the selective processes in some countries is considered.

These results, which are confirmed by the correlation coefficients and regression analysis data given in detail in Chapter 6, lead us to believe that the lower achievement of girls in mathematics at all levels could well influence their decisions to cease studying the subject as they move through secondary school and proceed to university courses. It is suggested that these differences cannot be explained simply by consideration of sex differences in mental abilities or aptitudes for mathematics because the results obtained are contrary to those predicted from previous studies in this field. It emerges in this particular study that boys in most countries perform better than girls on both verbal and computational problems, which is contrary to what might be suggested from consideration of differences in verbal and numerical abilities.

Other factors besides the components of mathematical ability must be sought to explain the differences observed, particularly as these differences are not the same across countries and sometimes show, unexpectedly, slight female superiority in computational problems and total mathematical achievement. Moreover, the significant interactions between countries and sexes in this area suggest that other factors should be considered.

Interest in and Attitude Toward Mathematics

The investigation collected information from students in all populations concerning their interest in mathematics and their plans to study more mathematics and views about the difficulty of learning mathematics. The data obtained are summarized for all countries in Table 5.29.

From these results it is seen that boys are more interested in mathematics while their plans to take more mathematics become significantly

TABLE 5.29. *Correlations Between Sex of Student and (A) Interest in Mathematics, (B) Plans to Take Further Mathematics, and (C) Attitudes Toward the Difficulty of Learning Mathematics.*

Population	A	B	C
1 a	.09	.01	−.04
1 b	.10	.02	.01
3 a	.12	.19	.01
3 b	.12	.10	.01

TABLE 5.30. *Sex Differences in Interest in Mathematics—All Countries.*

Population	Boys			Girls			$M_b - M_g$
	M.	S.D.	N.	M.	S.D.	N.	
1 a	5.95	1.68	13,920	5.65	1.65	13,320	**0.30**
1 b	6.03	1.65	20,096	5.68	1.61	17,912	**0.35**
3 a	6.65	1.88	6,624	6.31	1.89	2,304	**0.34**
3 b	5.16	1.74	5,986	4.84	1.72	6,840	**0.32**

more frequent at the upper-secondary school level. There are only minor differences between the sexes in views concerning the difficulty of learning mathematics.

In all the populations the boys show significantly greater interest in mathematics than do the girls. This was shown in the correlation, and it is also repeated in the comparison of mean scores in Table 5.30.

Another obvious finding, holding across all countries, is the greater interest shown by both boys and girls of the upper-secondary population who are specialists in mathematics when compared with the lower secondary school populations. Also, there is less interest shown by the upper secondary school population who are not specializing in mathematics, which might be expected since these students have ceased to study the subject. The data supporting these findings are given in Table 5.30. Here it may be seen that at all levels the girls have significantly less interest in mathematics than boys.

Differences between the sexes in the various countries in interest in mathematics are shown in Table 5.31 in which the negative signs indicate greater interest on the part of the girls.[1]

It is noted that for Population 3 a only in England and France is the interest shown by girls significantly greater than that of boys. This is not surprising since the girls involved are a highly selected group.

For all other countries and levels (except Sweden at the preuniversity level) the boys have indicated a greater interest in mathematics and this interest must surely be a factor which, in conjunction with their greater achievement, probably leads them to continue with the study of the subject. On the other hand, girls with lesser interest and lesser achievement would tend to cease studying mathematics.

[1] The figure shown in each case is obtained by taking the mean interest score of boys in excess of the mean interest score for girls for each country, dividing this difference by the overall standard deviation for that population for all countries, and multiplying by 100.

TABLE 5.31. *Sex Differences in Interest in Mathematics by Country.*

Country	Population 1 a	Population 1 b	Population 3 a	Population 3 b
Australia	26	12	24	—
Belgium	2	13	20	16
England	20	22	−31	7
Finland	35	26	56	16
France	46	46	−42	—
Germany	—	32	2	1
Israel	—	18	35	—
Japan	23	23	45	30
The Netherlands	7	22	27	—
Scotland	9	13	49	42
Sweden	1	5	−1	−10
United States	20	22	35	24

An analysis of variance of the data establishes that not only is the variance between countries and between sexes significant but that the interaction between countries and sexes is significant particularly for the lower secondary school and populations for the mathematics specialists, as can be seen from the results given in Table 5.32.

Differences between boys and girls in the various countries in their plans to take further mathematics are shown in Table 5.33. A negative value indicates a greater tendency on the part of the girls to plan to continue to study mathematics.

In Populations 1 a and 1 b there are no significant differences, except in the cases of France and the Netherlands, suggesting that at this age level the question is not really meaningful for students in most countries because a real choice does not exist.

In France and the Netherlands the girls tend to indicate their wish to cease studying the subject and the boys tend to indicate their plans

TABLE 5.32. *Interest in Mathematics: Analysis of Variance.*

F-Ratio for interaction between countries and sexes.

Population	F-Ratio
1a	2.71
1b	2.71
3a	2.18
3b	1.60

TABLE 5.33. *Sex Differences in Plans to take Further Mathematics by Country.*

Country	1 a	1 b	3 a	3 b
Australia	3	3	55	—
Belgium	0	11	4	33
England	−3	9	−6	−23
Finland	0	−3	67	46
France	21	20	−11	—
Germany	—	0	40	15
Israel	—	3	26	—
Japan	0	0	68	15
The Netherlands	53	40	−49	—
Scotland	9	11	51	64
Sweden	−11	−9	60	−2
United States	−6	−3	28	31

[a] This table shows the differences in the means divided by the population standard deviation for all countries, and the result multiplied by 100.

to continue. At the preuniversity level the results obtained are significant in most countries and for both populations. The boys more frequently plan to continue and in general, girls more frequently plan to cease the study of mathematics. An analysis of variance of these results indicates a significant interaction at Populations 3 a and 3 b between countries and sexes. The obvious exceptions at 3 a are those of the students in the Netherlands, France, and England, where the girls who are taking mathematics have indicated their plans to continue studying the subject to a greater extent than the boys.

Sex differences relative to the attitude scale concerning the difficulty of learning mathematics were studied and the results are shown in Table 5.34.[1] A negative result in this case indicates that the girls consider the subject less difficult than do the boys. There are few significant differences across the table, which is surprising but indicates that either the scale may not be very sensitive or, alternatively, that the students' perception of the difficulty of learning mathematics is not a factor that influences achievement or interest in mathematics.

There are very few significant differences in Table 5.34, and it is apparent that student attitude toward the difficulty of learning mathematics does not show up in this test (or study) as a major factor in explaining differences in mathematics achievement between boys and

[1] These tables show the differences in the means divided by the population standard deviation for all countries, and the result multiplied by 100.

TABLE 5.34. *Sex Differences in Attitudes Toward the Difficulty of Learning Mathematics by Country.*

Country	1 a	1 b	3 a	3 b
Australia	3	1	15	—
Belgium	−6	−6	21	−12
England	−2	−6	4	7
Finland	26	20	8	−5
France	−6	−3	−3	—
Germany	—	4	−14	−6
Israel	—	0	−39	—
Japan	−6	−6	5	−7
The Netherlands	−1	18	10	—
Scotland	2	6	13	22
Sweden	−3	−2	−10	3
United States	−6	1	3	13

girls. This is further demonstrated in Chapter 1 in terms of the low correlations between mathematics scores and attitudes about the difficulty of learning mathematics.

Single Sex and Coeducational Schools

In Table 5.24 figures were given showing that the countries vary considerably in the relative proportion of students in single-sex and coeducational schools. It was hypothesized that: *Differences between boys and girls in mathematics achievement, interest in mathematics, plans for further mathematics, and attitudes about the difficulty of learning mathematics would be least in coeducational schools, while they would be greatest in single-sex schools* (Hypothesis 32).

Since single-sex schools may be created for religious, vocational, and other special reasons, the differences between the sexes in single-sex schools may reflect selective factors as well as educational differences. On the other hand, in the coeducational schools, the students of both sexes are likely to have more equal opportunities for learning the subject. In Table 5.35 it is evident that the differences in mathematics achievement at all levels are in favor of the boys. However, the differences between the sexes are much less in the coeducational schools, and in the case of Population 3 a, almost disappear. This may, in part, be attributable to the selective factors determining which girls specialize in mathematics.

The countries vary in the patterns of sex differences for single-sex

TABLE 5.35. *Comparison of Mathematics Scores of Boys and Girls in Single-Sex and Coeducational Schools.*

Population	Boys			Girls			$M_b - M_g$
	M.	S.D.	N.	M.	S.D.	N.	
			Single-Sex Schools				
1a	25.69	11.35	3,069	22.73	10.38	3,631	**2.96**
1b	27.74	11.08	4,894	24.36	10.70	3,915	**3.38**
3a	29.71	9.04	1,831	25.12	7.65	460	**4.59**
3b	23.69	8.51	2,043	19.49	8.29	2,566	**4.40**
			Coeducational Schools				
1a	22.70	12.24	7,927	20.72	11.56	7,665	1.98
1b	24.73	12.68	12,449	23.15	12.30	11,951	1.58
3a	28.98	10.12	3,799	28.33	11.94	1,389	.65
3b	23.37	8.98	2,922	20.37	8.36	3,091	**3.00**

TABLE 5.36. *Sex Differences in Total Mathematics Score (Corrected) in Single-Sex and Coeducational Schools by Country.*

	1 a		1 b		3 a		3 b	
	1 Sex	Coed	1 Sex	Coed	1 Sex	Coed	1 Sex	Coed
Australia	−16	12	2	10	10	0	—	—
Belgium	29	15	25	4	95	—	84	50
England	24	4	35	0	7	—	57	26
Finland	—	19	87	21	—	42	—	72
France	13	22	18	21	—	—	—	—
Germany	—	—	58	15	14	18	80	42
Israel	—	—	−19	12	—	—	—	—
Japan	—	22	—	22	—	32	31	5
The Netherlands	5	34	22	42	—	7	—	—
Scotland	—	—	—	—	—	—	—	—
Sweden	—	10	—	10	—	3	—	39
United States	30	10	4	4	—	20	—	31

and coeducational schools, see Table 5.36.[1] However, in only two instances, Australia (1 a) and Israel (1 b), are the girls superior to the boys, and in both instances these are in single-sex schools, and the differences are not significant. In 9 instances the boys are significantly superior to the girls in single-sex schools, while in only 4 instances are the boys significantly superior to the girls in coeducational schools. In

[1] See foot-note p. 246.

only one country, Belgium, is the pattern of male superiority in single-sex schools generally consistent; in Germany and England this pattern emerges, but not as clearly and consistently as in Belgium. In the Netherlands, in Populations 1 a and 1 b, the boys are significantly superior only in the coeducational schools. Over all countries it is evident that the picture is somewhat mixed, and it is likely that special circumstances must account for the extremely varied patterns.

Again, returning to the overall international results in Table 5.35 it is evident that the boys in single-sex schools are generally higher than boys in coeducational schools. Girls in single-sex schools are superior in mathematics achievement at the lower levels of school, but at the preuniversity level, those in coeducational schools are superior. Whether these differences are to be ascribed to selective factors or to the superiority of one type of school over the other is a matter for further research. It would suffice here to point out that countries with an early transfer (before the age of 13) tend to have a higher proportion of single-sex schools than countries with a more comprehensive system. This explains at least part of the superiority of single sex schools at the lower level.

What is clearly evident is that the difference between the sexes in mathematics achievement is considerably smaller in coeducational schools. We would venture the view that the similarity of the two sexes in mathematics achievement in coeducational schools is quite probably attributable to the greater equality of opportunity to learn mathematics in such schools. It is possible that the differences that still remain are likely to be attributable to the role expectations for girls and boys and to the common view in many countries that mathematics is a "male" subject. This is most clearly demonstrated by the consistent pattern in all 12 countries for males to specialize in mathematics in larger proportions than females (see Table 5.23).

We have also summarized the data for interest in mathematics in single-sex and coeducational schools (see Table 5.37). In each comparison, the boys show somewhat more interest in mathematics than do the girls. However, it is in the coeducational schools where the difference between the sexes is the greatest. In each instance, the difference between boys and girls in interest is significant in the coeducational schools while in no instance is the difference significant in the single-sex schools. We find it difficult to explain this reversal of the hypothesis except in terms of the possible greater concern about role and self image where the two sexes are in the same school than in schools where they are isolated.

TABLE 5.37. *Comparison of Boys and Girls in Interest in Mathematics in Single-Sex and Coeducational Schools.*

Population	Boys			Girls			$M_b - M_g$
	M.	S.D.	N.	M.	S.D.	N.	
Single-Sex Schools							
1 a	5.94	1.71	3,240	5.81	1.58	2,892	.13
1 b	6.04	1.65	4,424	5.93	1.55	4,176	.11
3 a	6.50	1.90	2,196	6.31	1.85	472	.19
3 b	4.89	1.62	2,043	4.83	1.75	2,566	.06
Coeducational Schools							
1 a	6.02	1.67	10,680	5.58	1.67	10,402	**.44**
1 b	6.07	1.63	12,124	5.60	1.62	11,420	**.47**
3 a	6.68	1.85	3,799	6.33	2.01	1,389	**.35**
3 b	5.15	1.73	2,922	4.85	1.67	3,091	**.30**

In summary, we have tried to explain the difference in the mathematics achievement in single-sex versus coeducational schools in terms of opportunity to learn mathematics, while the differences in interest in mathematics between the two types of schools may more easily be explained in terms of role and self-image of girls and boys in relation to a subject viewed as a "male" school subject.

Mathematics Achievement and Educational and Vocational Aspirations

We have demonstrated the consistently high relationship between socioeconomic status and the students' plans for additional education. We asked the students to indicate what *plans* they had for further education as well as their *desires* for further education, and in Chapter 1 we have shown the high relationship between student educational and vocational plans and their mathematics scores. It was hypothesized: *Students who (1) plan to go on to higher education or (2) have aspirations for higher education will perform significantly better on the mathematics test than students who do not have such plans or aspirations, even when the level of mathematics instruction is held constant* (Hypothesis 33).

We decided to study this hypothesis by means of partial correlations between mathematics scores and educational plans or educational aspirations, holding level of instruction constant (see Table 5.38).

The results show clearly that educational plans and aspirations are

TABLE 5.38. *Partial Correlation Coefficients Between Mathematics Scores, Educational Plans, and Educational Aspirations—Holding Level of Mathematics Instruction Constant.*

Country[a]	Population 1 b		Population 3 a		Population 3 b	
	Educational Plan	Educational Aspiration	Educational Plan	Educational Aspiration	Educational Plan	Educational Aspiration
Australia	.37	.35	.20	.20	—	—
Belgium	.36	.34	.28	.22	.24	.21
England	.58	.56	.28	.16	.22	.20
Finland	.35	.32	.17	.15	.17	.14
France	.10	.10	.07	.05	—	—
Germany	.06	.06	.19	.13	.14	.04
Israel	.39	.30	.23	.12	—	—
Japan	.52	.50	—	—	—	—
The Netherlands	.41	.36	.25	.08	—	—
Sweden	.36	.31	.26	.23	.12	.10
United States	.35	.34	.31	.31	.35	.31
All Countries	.30	.28	.18	.14	.14	.10

[a] Scotland is not included in this table since its data on level of instruction were insufficient.

related to mathematics score in almost all countries, in Population 1 b as well as in Populations 3 a and 3 b. In Population 1 b the relationships are generally substantial, the typical value being about .35 for either plans or aspiration. The correlations are lowest in France and in Germany. We are not able to offer an explanation for the extremely low correlations in these two countries.

At the higher educational level, although the correlations are typically significant, they tend to be considerably smaller, being best represented by a figure of .15 to .25. This is presumably a reflection of the greater homogeneity of the older students in both plans and aspirations, as a result of the selection that has taken place between age 13 and age 17+.

In general, educational plans are more closely related to achievement than are educational aspirations, although the differences are usually small.

A related hypothesis was: *Students planning or desiring to enter vocations in which mathematics is relevant would have higher mathematics scores than students with other vocational plans or desires, even when level of mathematics instruction is held constant* (Hypothesis 34).

TABLE 5.39. *Partial Correlation Coefficients Between Mathematics Scores, Vocational Plans, and Vocational Aspirations—Holding Level of Mathematics Instruction Constant.*

Country[a]	Population 1 b		Population 3 a		Population 3 b	
	Vocational Plan	Vocational Aspiration	Vocational Plan	Vocational Aspiration	Vocational Plan	Vocational Aspiration
Australia	.15	.14	.08	.09	—	—
Belgium	.25	.31	.18	.13	.20	.20
England	.40	.33	.08	.06	.13	.14
Finland	.12	.17	.02	−.09	.07	.08
France	.01	.01	−.01	.04	—	—
Germany	.07	.08	.07	.08	.07	.05
Israel	.27	.23	−.03	−.04	—	—
Japan	.13	.24	−.06	.01	−.01	.01
The Netherlands	.17	.15	.10	.02	—	—
Sweden	.20	.21	.06	.00	.07	.03
United States	.12	.12	.25	.19	.19	.14

[a] Again, Scotland is missing from this table since its data on level of instruction were insufficient.

Vocational plans or aspirations was a dichotomous variable in which the student's response was categorized as to whether it was or was not a professional or technical type of occupation for which higher mathematics would be a requirement. The partial correlations by countries are shown in Table 5.39.

In Population 1 b, the correlations are once again significant and positive for all countries except France and Germany, although the correlations with vocational plan and aspiration are typically a good deal smaller than those with educational plan and aspiration. In Populations 3 a and 3 b, the correlations are significant for only a few of the countries, again presumably reflecting the greater homogeneity of these more advanced groups.

It seems clear from the data that students who plan or have aspiration for higher education perform better on the mathematics test than do students who do not have such plans or aspirations. And this relation is true even when the level of mathematics instruction is held constant. With a few exceptions Hypothesis 33 is supported. Hypothesis 34, according to which plans or desires to enter scientific and technical occupations are positively related to mathematics scores, is also supported, but the relationships are not very high and there are many exceptions.

Summary and Conclusions

In this chapter we have attempted to investigate some of the relations between social and cultural variables and student achievement in mathematics. While we can and do find relationships which are statistically significant, our special problem in this chapter is to determine on a cross-national basis how social and cultural variables are linked to student learning. We have, with the empirical evidence at hand, made attempts to understand, and in some instances to speculate, about these processes. However, it is clear that future cross-national research will need more powerful conceptual tools as well as more direct types of evidence than have been available in this investigation if the links between school learning and cultural and social forces are to be more clearly understood.

Home Background

The socio-economic status of the student has been determined by his father's occupation and his parents' level of education. In this chapter we have attempted to determine the relationships between socio-economic status and the schooling the student receives, his achievement in mathematics, and his aspirations for the future. In order to understand this set of variables we have also attempted to explore some of the relationships between home background and the interests and attitudes of the student.

We have studied the selectivity of the educational systems in the different countries. Using the home backgrounds (parent's education and father's occupational status) of the 13-year-old sample as an indication of the general base in each country, we have compared this sample with the samples at the preuniversity level. It is evident that all the countries in this study do practice some socio-economic selectivity in determining who reaches the terminal stages of academic public education. In three of the countries (the United States, Japan, and Finland) there appears to be a minimum amount of selectivity at the preuniversity level with regard to socio-economic characteristics. The greatest amount of selectivity is apparent in Germany, the Netherlands, and France. In these countries, preuniversity public education is, to a large extent limited to the children of the better educated or higher occupational status groups.

We have also shown in Chapter 3 that the bias toward the higher socio-economic groups is greatest in the countries which have the smallest percentage of the group completing preuniversity education and which make their selections of academic talent at the earliest ages. Undoubtedly, many countries are increasingly concerned with the development of talent as contrasted with the selection of talent. The finding of this section of our report puts this problem in an international perspective and should help educational policy-makers contrast the situation in their country with that of the other countries in this study. It is clear that many of the countries in this study could do much more than they now do to use education as a means of developing talent.

In all of the countries there is a significant relationship between the parent's characteristics (education or occupational status) and the student's mathematics achievement. This relationship is highest at the 13-year level and lower at the terminal levels, primarily because of the reduced heterogeneity of the parental characteristics at the later levels of public education. In those countries with a minimum of selectivity of students at the later levels, the relationship between parental characteristics and mathematics achievement approaches that found with the 13-year olds.

As we try to understand the ways in which the parental characteristics influence the student's achievement (in the light of the data available in this study), it becomes evident that the home background is associated with the student's educational aspirations (number of years of additional education expected or desired) and his occupational aspirations (expected or desired occupation). In a number of countries the parent's background in part determines the type and level of mathematics instruction the student receives. Thus, there is a complex set of inter-relationships between the home background, the motivations and aspirations of the student, and the curriculum and type of instruction the student receives.

The general consistency of the positive relationship between student's mathematics achievement and parental characteristics is striking. When this finding is seen in the light of the research literature, it appears that parents with higher socio-economic characteristics do a better job of preparing their children for school (no matter what the educational system) than do parents with lower socio-economic characteristics. Furthermore, the motivation and aspirations of the students for further education and for higher occupations are related to the social milieu in which the family lives. Thus, we get a picture of a consistent interplay between the home and the school. However, it should be noted that we did not find strong relationships between the student's interest in mathematics and father's level of education or between the amount of homework the student claimed to do each week and the father's level of education.

In contrast to this consistency within countries is the remarkable difference between countries. If we select the four nations with the highest scores on the mathematics test for Population 1 b and contrast them with the four nations with the lowest scores, we find that students with fathers in the top occupations (professional and managerial) in the four lowest countries achieve in mathematics at a lower level than do the children of the lowest occupations (unskilled and semi-skilled) in the four highest countries. Thus, we find that the differences among countries are so great that those who might be termed "culturally disadvantaged" in one country achieve at a higher level than the most culturally advantaged group in another country. Such differences raise many new questions for further research. In further studies it should become a central area of concern to find out how the schools and the curriculum in one country manage to raise the level of achievement of its least advantaged group to a level comparable with that of the most advantaged group in other countries. For education, these differences represent the equivalent of a quantum leap, and we must find ways of discerning the elements in school life, teaching, and curriculum, as well as the cultural values and motivations which overcome vast differences in home environments. It is evident that "cultural advantage" and "cultural disadvantage" are terms which are primarily relevant to within country variation and that they have little meaning when we cross national boundaries.

As we attempt to understand some of the school characteristics which relate to home background, it is evident that there are selective forces which determine the type of school program to which the student is admitted. In particular countries, the sorting out of students into academic, general, and vocational programs is strongly related to the socioeconomic backgrounds of the students. The higher occupational groups are characteristically found in the academic programs, while the lower occupational groups are more frequently found in the general and vocational programs. It is especially noteworthy that although the mathematics achievement of students still show differences by socio-economic groups when the students are grouped by type of school program, the differences are somewhat reduced. That is, students within a school program become more homogeneous with respect to mathematics achievement—evidently, a combined effect of selectivity and instruction.

Finally, students are found to achieve at a higher level when they are in schools with some or much variability in terms of socio-economic characteristics than they do when placed in schools that are homogeneous with regard to socio-economic background of their fellow students.

Place of Residence

We had expected the historic differences in sophistication and cultural development between urban and rural communities to show up in mathematics achievement. In the results for all countries combined, there are few significant differences between rural, town, and urban communities with respect to mathematics achievement. It should, however, be remembered that we have controlled for the level of instruction in each country, and this could possibly explain why so little difference is found when students are grouped by place of residence.

In only two countries (the United States and Japan) do we find significant differences favoring the urban communities at the 13-year level, and in only one country (the United States) do we find consistent significant differences favoring the urban communities at the preuniversity level. We conclude that whatever the case might have been in the past, there is little evidence in the nations in this study of highly differentiated achievement with regard to mathematics in the rural, town, and urban communities. We conjecture that this may be one of the consequences of economic and social development in the highly developed nations included in this study. The effect of place of residence on quality of education may still be a major factor in less developed countries, but we do not have any evidence on this in the present study.

The clearest example of a country where place of residence does relate to level of achievement is the United States. Here, the strong tradition of decentralized control of education and the urban-rural differences associated with the quality and support of the schools may in part explain the achievement differences found here.

The explanation for the differences in Japan is not so clear. This country is highly centralized with respect to education, and the financial support for schools and teachers does not vary greatly from school to school or from community to community.

Whatever differences we do find between urban and rural communities are very small as compared to between country differences. Even where there are differences of some magnitude, it is possible to find the student under the less favorable rural conditions of education in one country excelling the most favored group of town or urban students in another country.

We are led to the view that in most of the countries participating in this study, the place of residence does not, to any larger extent, influence the learning of mathematics. It is to be hoped that future studies will determine whether quite other findings will be found in other aspects of the curriculum. In mathematics, rural and urban differences in achievement in most of the countries included in this study can largely be dismissed as so small and inconsistent that they no longer pose a major concern for the educator.

There is evidence in our data that the place of residence does in some countries influence motivation and aspiration for further education and that this in turn is related to the parental level of education and occupational status.

Financial Support for Education

Previous studies, especially in the United States, have found positive relationships between level of financial support for the schools and the achievement of the students. These studies are not entirely clear in tracing the relationship between the level of support and school achievement, and there is some indication, at least in the United States, that support for the schools may be regarded as a symptom of the socioeconomic status of the community. It is, however, believed that financial support may be reflected in the quality of teachers especially where the salary of teachers varies from school to school.

In this study we find that at the 13-year level, there is a significant relationship between the mathematics achievement and the level of per student expenditures for both teacher salary and total support for the schools. Significant positive relationships for these variables are found in 6 out of 11 countries. In the other 5 countries, the relationships are either negative or chance. At the preuniversity level, the relationships are less clear. In Sweden, the United States, and Japan the relationships favor higher expenditures, while in other countries the relationships are inconsistent or, in some cases, negative.

It is likely that the inconsistency of relationships may reflect the variation in administrative practice and social structure across countries. Where financial support for the schools and teacher salaries are determined by centralized authorities, it is possible that per-student expenditure may be a reflection of the size of the school (larger teacher-student ratios yield higher per student expenditure) or the effort of the authorities to attract teachers to less favorable communities and school situations.

At least as far as mathematics achievement is concerned, the relationship between financial support and learning outcomes is far from consistent. Although the relationships are significant in a few countries, financial support does not appear to be a major factor associated with student achievement in mathematics.

This is a problem that is far from settled by this study. Further research is needed to determine the effect of variation in financial support on other aspects of the curriculum. More complex procedures will be needed to understand the processes which link financial support of education to quality of instruction to quality of student achievement. In this further research we will have to separate the effects of support for the schools from the effects of the home backgrounds.

Sex Differences

We take the view that educational differences between the sexes are determined more by social and cultural differences than by innate differences in mental ability. From this point of view, we are interested in comparing the achievement of boys and girls in the different countries in order to help us understand something about the influence of social forces on school learning.

All the countries in this study reflect a predilection for mathematics in males. The clearest example is the ratio of males to females in mathematics specialization at the preuniversity level. The ratio of males to females varies from about two to one in one country to seven to one in another. Mathematics specialization, while not restricted to males, is strongly biassed against females.

In each population males achieve at a higher level than do the females, even when the level of mathematics instruction has been held constant. Also, the boys are consistently superior whether the mathematics problems are largely computational or verbal in form.

Although the differences between boys and girls within countries are not always significant, it is rare to find instances in which girls are superior to boys in mathematics achievement.

When we attempt to understand why boys are superior to girls in mathematics, the clearest indication comes in connection with interest in mathematics. At every level, and in most countries, the boys are much more interested in mathematics than are the girls. The exceptions are France and England at the level of mathematics specialization where the girls have higher interest scores than the boys. This may be understood in terms of these girls being so highly selected (almost six boys for every girl) that those girls admitted to mathematics specialization must be among the most highly motivated students.

Another attempt to analyze the differences between girls and boys was to compare the mathematics achievement of boys and girls in coeducational and single-sex schools. With regard to mathematics, the difference between girls and boys is far greater for single-sex schools than it is in coeducational schools. This suggests that when the learning conditions are more similar, the differences in mathematics achievement between boys and girls will be markedly reduced. Thus, in the United States and Sweden, where the majority of students are in coeducational schools, the difference in achievement between the sexes is the lowest.

It is of some importance to note that the sex differences in interest in mathematics is slightly larger in the coeducational schools than in the single-sex schools. It is possible that this may reflect the heightened awareness of sex roles in coeducational schools, even where the opportunities for learning mathematics are more nearly equal.

It is clear that mathematics is generally regarded in the IEA countries as a male subject and that, with few exceptions, boys excel girls in mathematics achievement as well as interest. That girls can learn mathematics almost to the same level as boys is suggested by the reduced differences when boys and girls learn under coeducational school conditions.

We had originally attempted to understand these differences in mathematics as functions of the differential roles of males and females in the different countries. However, we find so few exceptions to the rule of male superiority in mathematics that we are led to believe that variation in the roles of the sexes in these countries will not be helpful in understanding mathematics achievement differences between the sexes.

Once again we must remind the reader of the large between country differences. Although the girls of one country are lower in their mathematics achievement than the boys of that same country, there are a number of countries where the "inferior" girls are superior to the males of other countries. Sex differences are a within-country phenomenon. Across countries, girls may be superior to boys in mathematics achievement—depending on the countries. Here again, we would attribute the between country differences to educational differences (curriculum and instruction), which, in their turn, mirror differences in cultural values. On the whole, educational differences among the countries far outweigh any of the social differences within countries.

Chapter 6

A Regression Analysis

Introduction

The correlation matrix described in Chapter 9 of Volume I served three purposes. First, since it contained 54 of the variables used for the original hypotheses, it provided an alternative means of testing these hypotheses. Second, it was used for the estimate of standard errors, in the way set out in Chapter 9. Third, it was used to explore the relations between the variables by multiple regression. This chapter is concerned with the third purpose.

From the 54 variables in the matrix, the corrected total mathematics score was chosen as the dependent variable or criterion. Of the remainder, 26 were included as independent variables in a reduced matrix. The selection of the 26 was made by excluding the attitude scales, the subscores of the mathematics tests, and one or two other variables that overlapped heavily. In addition, three other variables had to be excluded because their coding was unsuitable for regression analysis. This was regrettable, since the three variables were "Sex of School", "Type of School", and "Type of Professional Training".

The 26 independent variables that were included can be grouped under four heads, as follows:

Parental Variables
Mother's Education.
Father's Education.
Father's Occupation (Status).
Father's Occupation (Scientific or Nonscientific).
School Standard Deviation in Father's Occupational Status.
Place of Parents' Residence.

This chapter was written by G. F. Peaker, C.B.E.

Teacher Variables

Student's Opportunity of Learning the Test Items.
Description of Mathematics Teaching and School Learning.
Length of Training.
Sex of Teacher.
Recent In-Service Mathematical Training.
Degree of Freedom Given to the Teacher.

School Variables

Number of Weekly Hours of Mathematics Instruction.
Number of Hours in the School Week.
Number of Hours of Mathematics Homework in the Week.
Number of Hours of All Homework in the Week.
Total Roll of School.
Percentage of Men Teachers on the School Staff.
Educational Differentiation.
Number of Subjects Taken in Grade 12.
Number of Subjects Taken in Grade 8.
Cost per Student in Teachers' Salaries.

Student Variables

Sex of Student.
Age of Student.
Student's Level of Mathematical Instruction.
Student's Interest in Mathematics.

The choice of variables for a regression analysis is not altogether easy. The object of the analysis is to show first how much information about the criterion each independent variable contributes when it stands alone and second how much it contributes when taken in conjunction with its fellows.[1] The essential point is that the value of any new in-

---

[1] If two variables correlate at r, and are both standardised in the sense of being expressed on a scale such that their variances are unity, then their common variance is r^2, and their residual variance is $1-r^2$. If one is taken as the criterion and the other as the independent, or predictor, variable, then it can be said that the second "accounts for" a proportion r^2 of the variance of the first, provided that there is no other information. If there is other information, in the shape of the correlations of further standardised variables with the first, and with each other, then composite variables can be made up, which are weighted sums of the independent variables, and the correlation of any such composite, or weighted sum, with the criterion can be found. Among such composites there is one which has the highest correlation, R, with the criterion, and this composite "accounts for" a proportion R^2 of the criterion

formation depends heavily on what is already known. Information that would be extremely useful if it stood by itself may be of little interest as a mere supplement to other information. This point is well understood by the subeditors of the daily press and decides what goes into the paper and what goes on the spike. In the present context, it obviously leads to the ruling out of the part scores, not because they are bad predictors of a total score, but because they are too good. They overlap too much. This was an easy decision. But three others were harder. It was plain at the outset that the student's "Interest in Mathematics", his "Opportunity for Learning the Matters Tested by the Test Items", and his "Level of Instruction" were all likely to be good predictors of his total score. The question was whether they would be too good—whether, that is to say, they would account for so much of the total variation as to leave only a small remainder for distribution among the remaining variables. Finally, it was decided to include them. In the upshot they together accounted for rather more than half the total of the assigned variation.

The analysis was rerun excluding them, but since the proportions in which the remaining variation was assigned were much the same as before, a second set of tables has not been printed.

A variable that was omitted from the correlation matrix, because it did not appear among the original hypotheses, was "Number of Staff". This was an unfortunate omission, since it entailed the consequence that the variable could not be included in the regression analysis. If it could have been included, it would have made it possible to see whether it was generally true that schools with better staffing ratios made higher scores. This is because the variable "Total Roll" was included, and

---

variance. The condition that the correlation of the composite with the criterion is to be maximised gives the weights to be attached to each variable.

These weights are called standardised regression coefficients and have the property that when each is multiplied by the corresponding correlation, of the independent variable with the criterion, the sum of these products is equal to R^2, so that each product can be regarded as the contribution, to R^2, of the variable concerned. R^2 is called the assigned variation.

There is a dilemma about the nomenclature. The "independent" variables are not independent in the statistical sense. If they were, the problem would be simpler, since the standardised regression coefficients would merely be the correlations and R^2 would be the sum of their squares. On the other hand, to call them "predictors" carries a suggestion that they predict the future, which is not the case, at any rate in a simple sense.

The analysis was done by the step-wise method, which adds to the composite one variable at a time, and stops when no remaining variable makes a significant addition to R^2. This accounts for some of the blanks in the tables.

when "Total Roll" is held constant "Number of Staff" is equivalent to staffing ratio; a positive regression coefficient for "Number of Staff" means that among schools with the same number of students those with more teachers make higher scores. This was found to be the case in England, for which a national analysis, including the variable "Number of Staff", was subsequently carried out. At first sight this seems to contradict another English result, that on the whole larger classes had higher scores. But the two are in fact compatible. The explanation is merely that in streamed schools the duller students are grouped in smaller classes. This is known to be the case in England, although not perhaps so widely known as it should be. The beneficial effect of improved staffing ratios tends to be masked by the simple correlation with size of class that arises from having smaller classes for duller students. Had it been possible to include the variable in the general analysis it seems likely that similar results might have been obtained for other countries. But time did not permit this.

Owing to various imperfections, some of the samples had to be excluded from the regression analysis. The 36 that remained are shown in the preliminary Table 6.1, which gives the percentage of assignable variation in each case. This ranges from 67 percent for Population 1 a in the Netherlands to 23 percent for Population 3 a in England. The averages for the four populations are 40, 34, 38, and 32 percent. When the analysis is taken over all countries the results are slightly different, being 30, 26, 37, and 34 percent. The main reason for the differences is that the variables are reckoned from the national means within each country, whereas in the "all countries" analysis, they are reckoned from the international means. A minor reasons is that in the "all countries" analysis the weight of each country is given by the number of students in its sample, whereas the entries in the "mean" row are unweighted means, as they should be. To rewrite the program to give each country the same weight in the "all countries" analysis would have been a major enterprise for which time could not be found.

The amount of the total variation that can be assigned would be surprisingly large, were it not for the fact that rather more than half of it is accounted for by "Level", "Interest", and "Opportunity", as shown in the four lowest rows in Table 6.1.

The 20 main tables given in this chapter are all arranged in a uniform pattern. There are five groups. Tables 6.2 to 6.5 cover the Parent Variables, Tables 6.6 to 6.9 cover the Teacher Variables, Tables 6.10 to 6.13 cover the first group of School Variables, and Tables 6.14 to 6.17 cover the second group, while Tables 6.18 to 6.21 cover the

TABLE 6.1. *Percentage of Variance Accounted For.*

Country	Populations			
	1 a	1 b	3 a	3 b
Australia	39	26	41	—
Belgium	37	33	46	30
England	50	50	23	31
Finland	33	27	36	—
France	46	44	—	—
Germany	—	24	—	25
Israel	—	26	—	—
Japan	33	33	42	37
The Netherlands	67	38	38	—
Scotland	44	47	37	40
Sweden	28	28	—	—
United States	28	27	44	28
Mean of Countries	40	34	38	32
All Countries	30	26	37	34
Level	8.6	2.9	10.5	6.2
Interest	5.7	7.1	9.3	7.1
Opportunity	5.2	5.4	4.3	0.3
Total	19.5	15.4	24.1	13.6

Student Variables. Under the name of each variable there are two columns. Column *"r"* gives the simple ocrrelation of that variable with the total corrected mathematics score, and column *"b"* gives the standardized regression coefficient. The product of these two entries gives the contribution to the assigned variation.[1] These products are summed along the rows to give the entry in the right-hand column. Thus, in the English row of Table 6.2, the simple correlation between "Total Score" and "Father's Occupation" is .38, and standardized regression coefficient is .15. The product of these is .057, so that this column contributes 5.7 percent to the total of 7.1 percent shown at the right.

At the foot of each column the numbers of positive and negative coefficients are noted. Below this the column ranges are given, then the column means, and finally the coefficients when the analysis is taken over all countries. Coefficients are printed in bold type if they

---

[1] See footnote 1 on page 261.

differ from zero by more than twice their standard errors or, in other words, if there is a high probability that they have the right sign. The standard errors are given in Chapter 9 of Volume I. They average .04, and this average value is enough to keep in mind in rapidly assessing the value of any coefficient, although the bold type is based on the actual standard errors. In the "signs" row, bold type is used to indicate a heavy preponderance of one kind, while in the "range" rows bold type means that the corresponding column entries are heterogeneous, in the sense that their differences are more than can reasonably be accounted for by sampling fluctuation. Thus, it will be seen that bold type for signs indicates similarities between countries, but bold type for ranges indicates differences. Marked differences between the entries in the "mean" and the "all countries" rows indicate differences between countries in the means for the variables concerned.

The tables in any case are fairly complicated, as is natural in a very complicated situation. To avoid still further complication, the individual rb products have not been given, but only their row sums. It is, however, easy to analyse any row sum into its component parts. There are two exceptions to this rule. In the tables where "Level", "Opportunity", "Interest", "Sex of Student" and "Age of Student" occur, the variance assignable to them has been separately indicated. In Tables 6.18–6.21 the proportions separately assignable to sex and age of the student are shown.

Parental Variables (Tables 6.2–6.5)

In this group we should expect to find that much more of the variation would be accounted for in Tables 6.2 and 6.3, for Populations 1 a and 1 b, than in Tables 6.4 and 6.5, for Populations 3 a and 3 b. We should expect this because in the first two populations everybody is still at school, so that the parents are drawn from the full range of "Education", "Occupation", and "Residence". For the other two populations, the parents are more homogeneous (see also Hypothesis 08, Chapter 3); they do not include the parents of boys and girls who have already left school. Furthermore, we should expect this difference to be least marked where the proportion of early leavers is relatively small. Both these expectations are borne out in the result. If we look at the bottom-right corner of each table, we see that the "Parental Variables" account for about 5 percent of the total variation for Populations 1 a and 1 b, but for only about 2 percent for Populations 3 a and 3 b. If we look at the right-hand column in Tables 6.4 and 6.5, we see that the propor-

TABLE 6.2. *Parental Group of Variables: Population 1a.*

Country	Mother's Education		Father's Education		Father's Occupation (Status)		Father's Occupation (Scientific or Otherwise)		Parent's Place of Residence		Percent of Variance
	r	b	r	b	r	b	r	b	r	b	
Australia	04	02	06	01	22	13	05	00	05	08	3.4
Belgium	05	−02	13	02	24	13	14	−02	−01	−04	3.0
England	24	05	24	01	38	15	27	00	−07	00	7.1
Finland	12	02	13	05	01	04	01	10	05	−17	0.2
France	14	−03	19	01	19	01	09	01	−10	−02	0.2
Japan	32	11	33	11	25	10	27	−02	20	14	11.9
The Netherlands	27	−03	40	04	33	13	16	04	03	−04	5.6
Scotland	12	03	14	01	27	13	12	03	−07	−02	4.5
Sweden	16	05	16	−02	20	07	18	−05	07	05	1.3
United States	29	12	30	11	28	12	18	−02	18	05	10.7
+	10	7	10	9	10	10	10	4	6	4	
−	0	3	0	1	0	0	0	4	4	5	
Range	28	15	34	13	37	12	26	15	30	31	
Mean	18	03	21	04	24	10	15	01	03	00	4.8
All Countries	10	01	15	04	26	13	17	02	04	00	4.4

TABLE 6.3. *Parental Group of Variables: Population 1b.*

Country	Mother's Education		Father's Education		Father's Occupation (Status)		Father's Occupation (Scientific or Otherwise)		Parent's Place of Residence		Percent of Variance
	r	b	r	b	r	b	r	b	r	b	
Australia	06	02	07	01	20	14	05	01	02	05	3.1
Belgium	07	00	14	04	19	10	10	−01	00	−05	2.4
England	26	04	25	02	36	13	25	02	−06	−03	6.9
Finland	10	05	14	07	07	06	04	11	02	−10	2.1
France	13	−01	18	01	22	07	06	03	−20	−12	4.2
Germany	13	02	13	−05	15	00	12	−02	01	01	−0.6
Israel	16	06	18	06	22	12	13	02	29	18	10.2
Japan	32	11	33	11	25	10	27	−02	20	14	11.9
The Netherlands	21	07	26	09	22	08	09	04	08	01	6.0
Scotland	12	03	15	01	26	12	11	03	−04	00	4.0
Sweden	15	04	15	−01	17	06	14	−02	06	06	1.6
United States	28	11	30	11	29	13	17	−03	10	01	9.7
+	12	10	12	10	12	11	12	7	8	7	
−	0	1	0	2	0	0	0	5	3	4	
Range	26	12	26	16	29	14	22	14	49	28	
Mean	17	04	19	04	21	09	13	01	04	01	5.1
All Countries	11	03	16	06	22	12	13	00	06	02	4.0

TABLE 6.4. *Parental Group of Variables: Population 3a.*

Country	Mother's Education		Father's Education		Father's Occupation (Status)		Father's Occupation (Scientific or Otherwise)		Parent's Place of Residence		Percent of Variance
	r	b	r	b	r	b	r	b	r	b	
Australia	03	02	02	00	05	04	02	00	00	−01	0.3
Belgium	09	03	08	01	04	00	08	01	16	04	1.1
England	12	08	09	03	08	04	07	−01	07	03	1.7
Finland	03	05	−04	−02	11	−10	−05	04	−08	−13	0.0
Japan	13	00	13	05	08	01	12	02	01	−11	0.9
The Netherlands	00	−03	05	05	−04	−04	−01	−02	20	10	2.4
Scotland	06	06	04	00	06	02	05	−01	−05	−01	0.5
United States	25	08	32	06	24	04	15	−01	39	09	8.2
+	7	7	7	5	7	5	6	3	5	4	
−	0	1	1	1	1	2	2	4	2	4	
Range	25	11	36	08	28	14	20	06	47	22	
Mean	09	04	09	02	08	00	05	00	08	00	1.9
All Countries	−03	05	07	03	11	01	13	00	13	05	0.8

TABLE 6.5. *Parental Group of Variables: Population 3b.*

Country	Mother's Education		Father's Education		Father's Occupation (Status)		Father's Occupation (Scientific or Otherwise)		Parent's Place of Residence		Percent of Variance
	r	b	r	b	r	b	r	b	r	b	
Belgium	01	01	−02	−03	00	−04	−01	−01	−03	03	0.0
England	02	04	−04	−04	01	04	−01	−05	07	05	0.7
Germany	02	06	−06	−09	−02	06	−08	−08	01	−07	1.1
Japan	11	03	13	06	09	03	11	−00	07	06	1.8
Scotland	06	−04	03	02	03	00	03	−01	−01	−03	−0.2
United States	26	06	31	12	26	10	16	01	30	13	11.9
+	6	5	3	3	4	4	3	1	4	4	
−	0	1	3	3	1	1	3	4	2	2	
Range	25	10	37	21	28	14	24	09	33	20	
Mean	08	03	03	01	06	03	03	−02	07	03	2.6
All Countries	−06	01	04	04	13	03	12	01	05	06	0.9

tion accounted for is very small everywhere except in the United States, which has the lowest proportion of early leavers. The United States has 11, 10, 8, and 12 percent in Populations 1 a, 1 b, 3 a, and 3 b as the proportion of the variance assignable to parents, whereas in all other cases the proportion for 3 a and 3 b does not exceed 3 percent.

If we look now in more detail at Table 6.2, we see that the parental variables make the largest contribution in Japan, where the 12 percent is made up of about $3\,^1/_2$ percent each for "Mother's Education" and for "Father's Education", $2\,^1/_2$ each for "Father's Occupational Status" and "Parental Residence", with a small negative contribution from "Father's Occupation (Scientific or Otherwise)". A negative contribution arises when the simple correlation and the standardized regression coefficient have opposite signs, as in the simple three variable case where the correlations with the criterion are .6 and .4, and the correlation between the independents is .8. Below Japan we have in descending order of contribution the United States, England, the Netherlands, Scotland, Australia, Belgium, Sweden, Finland, and France. In Table 6.3, for Population 1 b, the order is the same, except that Israel comes in high, with 10 percent, and Germany very low, with -0.6 percent, while France moves up to 4.2 percent. Except for the French case, the differences are only trifling, as would be expected from the fact that these two populations have a large common element.

Where the contribution to the assigned variance is very small, the reason is not, for the most part, that the simple correlations are very small, but that the regression coefficients are. It is in the process of combining the simple correlations for the parental variables with other information that the contribution is diminished. This is, for example, the reason why the fourth variable—"Father's Occupation (Scientific or Otherwise)—contributes little. In the simple correlation columns there are many entries in bold type; in the regression coefficient columns there are none because the scientific occupations are those which have high status. The information in the first column is useful for prediction by itself, but not when other things are held equal. Holding other things equal amounts to subclassifying the population on the other variables, and within these subclassifications the correlations for the fourth variable are practically zero, although between them they are considerable.

For the first four variables all the signs in the "r" column are positive for Populations 1 a and 1 b. This is not the case for the fifth variable—"Place of Parent's Residence". For this variable we have four negative signs for Population 1 a and three for Population 1 b, with a long range in each case. The explanation is to be found in the scoring system for this variable, which reads:

1. Rural (farm).
2. Rural (nonfarm or village less than 2,500).
3. Small city or town (population 2,500–15,000).
4. Medium-sized city (population 15,000–100,000).
5. Urban center (population exceeding 100,000) and suburban area (community near or adjacent to an urban center).

This produces correlations ranging from .20 in Japan to $-.10$ in France in Population 1 a, and from .28 in Israel to $-.20$ in France for Population 1 b. It seems unlikely that a classification of parental residence could be produced that would be equally applicable to all countries. In England and Scotland, for example, neighborhood is a much more powerful variable than region or rural versus urban. The difference between adjacent neighborhoods in the same town is far greater than the difference between small and large centers of population. It is this that accounts for the fact that "Place of Parents' Residence" makes no contribution in England or Scotland, although a different classification would have made a large one (as is shown, for example, by the English national sample in the *Newsom Report*). This is in fact the same dilemma that arose over the trichotomies in various hypotheses reported earlier. A consistent classification will be insensitive in some countries and over all countries, while a sensitive classification will be inconsistent. However, in this case it would be easy to make too much of this point, since much of the variation associated with even a sensitive classification of parental residence would overlap with the occupational variance and tend to disappear from the final allocation.

In general, the status of the "Father's Occupation" is the most powerful of this powerful group of variables. In the extreme case of Population 1 a in England, it accounts for over 14 percent of the total variation when taken by itself, and even in conjunction with the other 25 variables, it accounts for nearly 6 percent. At the other extreme, in Finland it accounts for very little. But the remaining countries approximate more closely to England than to Finland. For the two older populations the effects are most powerful in the United States, where the proportion of the age group still at school is largest.

Teacher Variables (Tables 6.6–6.9)

At first sight this group appears to be rather more powerful than the parent group. Thus, for Population 1 a, it accounts for 5.4 percent of

TABLE 6.6. *Teacher Group of Variables: Population 1a.*

Country	Sex of Teacher		Length of Training		Recent in Service Mathematical Training		Degree of Freedom		Description of Teaching		Student's Opportunity of Learning		Percent of Variance
	r	b	r	b	r	b	r	b	r	b	r	b	
Australia	−05	−02	10	−01	04	−02	−09	00	01	−01	—	—	−0.9
Belgium	−17	−06	09	04	09	10	−06	−02	—	—	—	—	2.4
England	−06	01	32	11	02	03	09	02	09	03	55	33	22.1
Finland	17	03	03	06	−15	−21	−04	−02	15	09	04	−05	5.1
France	−11	12	03	−01	10	02	17	04	—	—	26	06	1.1
Japan	02	00	03	−02	−02	−03	−03	02	09	04	09	04	0.8
The Netherlands	−01	−07	−10	−02	12	−09	−06	02	13	05	—	—	−0.3
Scotland	02	−01	08	01	12	05	−02	−01	10	04	56	40	23.5
Sweden	−03	−02	09	05	06	00	−05	−01	14	05	−03	00	1.3
United States	01	−01	08	01	10	04	03	03	02	00	19	11	2.6
+	4	3	9	6	8	5	4	5	8	6	6	5	
−	6	6	1	4	2	4	6	4	0	1	1	1	
Range	34	18	42	13	23	31	26	06	14	10	59	45	
Mean	02	00	08	02	05	−01	00	01	07	03	17	09	5.7
All Countries	−05	02	01	02	03	01	−02	−01	07	03	29	18	5.4

TABLE 6.7. *Teacher Group of Variables: Population 1b.*

Country	Sex of Teacher		Length of Training		Recent in Service Mathematical Training		Degree of Freedom		Description of Teaching		Student's Opportunity of Learning		Percent of Variance
	r	b	r	b	r	b	r	b	r	b	r	b	
Australia	−05	−01	04	02	−07	−08	−01	02	03	03	—	—	0.8
Belgium	−14	−06	09	05	07	08	−08	−02	—	—	—	—	2.0
England	−08	−03	31	12	02	00	−03	−04	05	03	51	29	19.0
Finland	09	01	06	09	−15	−22	−08	−04	18	09	04	−03	5.8
France	−08	19	03	−01	11	02	26	10	—	—	27	09	3.7
Germany	−04	−09	22	03	05	−01	−12	−01	−04	01	03	−04	0.9
Israel	22	14	09	−01	03	02	10	05	−08	−06	10	06	4.6
Japan	02	00	03	−02	−02	−03	03	02	09	04	09	04	0.8
The Netherlands	−12	02	−02	02	15	02	−01	01	07	−03	—	—	−0.2
Scotland	03	00	08	02	14	04	00	−01	06	01	60	48	29.6
Sweden	−04	−01	09	05	07	01	−06	−02	13	05	−04	00	1.3
United States	06	02	04	01	09	03	02	01	02	02	17	10	2.2
+	5	5	11	9	9	7	4	6	8	8	8	6	
−	7	5	1	3	3	4	7	6	2	2	1	2	
Range	36	23	33	14	30	30	38	9	26	15	55	52	
Mean	01	02	09	03	04	−01	00	01	04	02	15	08	5.9
All Countries	−01	03	−01	00	04	03	00	02	05	02	28	19	5.5

TABLE 6.8. *Teacher Group of Variables: Population 3a.*

ry	Sex of Teacher		Length of Training		Recent in Service Mathematical Training		Degree of Freedom		Description of Teaching		Student's Opportunity of Learning		Percent of Variance	Percent of Variance (Excluding Oppurtunity)
	r	b	r	b	r	b	r	b	r	b	r	b		
alia	−10	04	10	−01	−05	−01	10	03	06	00	**40**	**16**	6.2	−0.2
ım	−12	−06	**−17**	**−07**	−01	−06	**08**	−02	—	—	—	—	1.8	1.8
nd	02	00	01	01	−07	−04	04	00	**−11**	**−09**	13	13	3.0	1.3
ıd	02	08	−04	−02	**16**	07	**−16**	**−10**	−03	03	−09	−02	3.0	2.9
	−09	−01	**17**	**11**	04	00	04	**12**	−02	−01	**44**	**30**	15.7	2.5
Jetherlands	−06	−08	08	**22**	−06	05	07	**12**	−10	−09	—	—	3.7	3.7
.nd	**−13**	**−08**	**09**	02	−07	−04	−04	−01	−04	−05	18	08	3.2	1.7
d States	−05	07	**14**	06	**18**	05	03	−04	−02	−01	**29**	**12**	4.8	1.3
	2	3	6	5	3	4	6	3	1	1	5	5		
	6	4	2	3	5	4	2	4	**6**	**5**	1	1		
	15	16	**34**	**29**	25	13	**26**	**22**	17	12	**53**	**32**		
	−06	00	05	03	02	00	02	01	−03	−03	20	13	5.2	1.9
ountries	−13	02	00	05	−13	00	−01	00	−04	−02	30	14	4.0	−0.2

TABLE 6.9. *Teacher Group of Variables: Population 3b.*

ry	Sex of Teacher		Length of Training		Recent in Service Mathematical Training		Degree of Freedom		Description of Teaching		Student's Opportunity of Learning		Percent of Variance	Percent of Variance (Excluding Oppurtunity)
	r	b	r	b	r	b	r	b	r	b	r	b		
m	**−26**	**−28**	**−14**	01	−10	−02	−07	−02	—	—	—	—	7.5	7.5
ıd	—	—	—	—	—	—	—	—	01	01	—	—	—	—
ıny	−10	−06	13	**16**	03	−03	**11**	01	−03	−02	−01	02	2.7	2.8
	−02	03	03	01	**08**	**08**	00	01	00	−01	**31**	**13**	4.6	0.6
ıd	—	—	—	—	—	—	—	—	−08	−01	—	—	—	—
States	01	04	03	00	02	01	00	00	**−09**	−04	04	03	0.5	0.4
	1	2	3	3	3	2	1	2	1	1	2	3		
	3	2	1	0	1	2	1	1	3	**4**	1	0		
	27	**32**	**27**	15	18	11	**18**	3	10	5	**32**	11		
	−09	−04	01	**06**	01	01	01	00	−05	−02	**11**	**06**	3.8	2.8
untries	−08	**08**	−03	**06**	−05	**06**	−02	−02	−04	−02	**16**	02	−0.7	−1.0

the total variation over all countries, compared with 4.4 percent for the parent group. In England, the teacher group accounts for about 22 percent, and in Scotland, for 24 percent, compared with 7 percent and 4 percent for parents. But closer examination shows that this is illusory, since it is the "opportunity" columns that contain nearly all the variance accounted for. This has been shown in the tables for this group by the addition of another column on the right giving the percentage of the variance accounted for when "opportunity" is excluded. Of the 5.4 percent over all countries, only 0.2 percent remains for the other five teacher variables; of the 22 percent for England, only 4 percent; and of the 24 percent for Scotland, only 1 percent. The mean value falls from 5.7 to 1.3 percent, and the highest entry remaining is 5.3 percent for Finland. Similar results can be observed in the tables for the other populations. The highest simple correlation, outside the "Opportunity" columns, is .32 for "Length of Training" in England for Population 1 a. By itself this would account for over 10 percent of the variation, which is considerable, but only a third of the 30 percent accounted for by the simple correlation of .55 for "Opportunity" when taken by itself. In Scotland, "Opportunity" taken in conjunction with the other teacher variables accounts for 22.4 percent out of a total of 23.5 percent. The difference from the English situation (18.2 percent out of 21.8 percent) arises mainly because in Scotland there is less variation in "Length of Training".

Why has "Opportunity" been included among the teacher variables, and why are its effects so much larger in England and Scotland than elsewhere? It has been included among the teacher variables because it is defined as "Teacher Ratings of Student Opportunity to Learn All Mathematics Items". It is larger in the United Kingdom than elsewhere because a longer range was noted in the degree of familiarity between high and low scoring schools. What is involved is not the general suitability of the tests for the students of one country compared with those of another. This can be seen from a comparison of the country means. The comparison of the country correlations and regression coefficients is concerned with differences between schools in the same country. This distinction accounts for the fact that the variation attributable to "Opportunity" is comparatively small for Population 3 a in England and Scotland, although it was for these students that the most serious differences occurred between the mathematics syllabuses and the test content. The whole of applied mathematics, which fills about half the timetable in England, was left untested, and so was much of the work in calculus and coordinate geometry. On the other hand, some of the test items related to work which does not form part of the normal English syllabus. But to a large extent this affected all schools in England alike. Consequently, it does not produce large correlations. For Population 3 a the large entries in the "Opportunity" column are those for Japan, the United States, and Australia.

Let us now take the other variables of the group in order. In the first column, for "Sex of Teacher", positive signs mean that women are more successful than men. If we look at the "All Countries" row, we see that the balance is in favor of men for the simple correlations, but swings the other way for the regression coefficients. This is largely accounted for by the negative correlation between "Sex" and "Level", and suggests that on the whole the men are better mathematicians and the women are better teachers. The differences, however, are slight. In Tables 6.18 to 6.21, the first columns show that boys are more successful mathematical students than girls, even in Population 3 a after the less successful students have left. In the transition from learning mathematics to teaching it there is a further effect of selection, since the proportion of volunteers is likely to be higher among women than among men.

For the next variable, "Length of Training", in Population 1 a the simple correlations range from .32 in England to −.10 in the Netherlands, with a slightly shorter range from .31 in England to −.02 in the Netherlands for Population 1 b. In Population 3 a the range is from .17 in Japan to −.17 in Belgium. The mean simple correlations for the four populations are .08, .09, .05, and .01, with .10, −.01, .00, and −.03 when the simple correlations are taken over all countries. The differences between the mean correlations and the correlations for all countries reflect chiefly the fact that the average "Length of Training" differs from country to country. Over all countries the contribution to the variance is negligibly small for all populations. The differences between countries are more marked than the similarities. This is shown by the long ranges as well as by the differences between the entries for means and the entries for all countries. There are five negative simple correlations, out of 34, and three of these depart from zero by more than twice their standard errors. Since this is contrary both to the natural expectation and to the position elsewhere, the explanation must be sought in special circumstances in these cases, which are the Netherlands (Population 1 a) and Belgium (Populations 3 a and 3 b).

The position is much the same for the next variable, "Recent-In-Service Training in Mathematics". Here again we have long ranges and small contributions to the variation over all countries, so that the differences between countries are more marked than the similarities. We have no fewer than eleven negative simple correlations, out of 34, but only three of them differ from zero by more than two standard errors (Finland, Populations 1 a and 1 b; and Australia, Population 1 b). In Population 1 a the second largest contribution to the variance comes from Belgium, and this may be related to the zeal given to developing new school mathematics in that country. A much larger contribution comes from Finland, but it is the product of two negative coefficients!

For the next variable, "The Degree of Freedom Enjoyed by Teachers", no fewer than 16 of the 34 simple correlations are negative, and 3 of the 16 differ from zero by more than 2 standard errors. In the lowest row, the contribution to the "All Countries" variance is negligible.

For the last variable, "Description of Teaching", we have for Populations 1 a and 1 b two negative signs out of 22, of which one differs from zero by more than twice its standard error (Israel, 1 b). With this exception the effect is in the expected direction. But when we come to Populations 3 a and 3 b, we find nine negative signs, out of 14, of which two differ from zero by more than twice their standard errors.

In short, it may be said that the variables of this group, with the exception of "Student's Opportunity", provide slight and conflicting evidence. The conflict arises not because of the influence of other variables in the matrix, but because of the presence of many simple correlations with negative signs. These negative signs would not be reversed by including other teacher variables, such as "Age" and "Experience", in a group composed of teacher variables alone; but the analysis of such a group might explain, although it would not remove, the conflict. Thus, if "In-Service Training" were given chiefly to young and inexperienced teachers, a negative simple correlation might turn into a positive partial correlation when "Age" and "Experience" were taken into account—similarly, if it were given mainly to teachers who had hitherto been less successful than their fellows. Although these would be sufficient, they are not also necessary conditions. To establish the value of a training course, it is not necessary to show that the teachers who undergo it become better teachers than their fellows, but only that they become better teachers than they would otherwise have been.

School Variables—I (Tables 6.10–6.13)

The four variables in this group are "The Length of the School Week", "The Time Given to All Homework", "The Time Given to Instruction

TABLE 6.10. *First Group of School Variables: Population 1a.*

	All Schooling		All Homework		Time for Instruction in Math.		Math. Homework		Per-cent of Variance
Country	r	b	r	b	r	b	r	b	
Australia	−01	−07	26	20	−09	06	06	−10	4.1
Belgium	−10	−06	24	12	25	07	10	−06	4.6
England	03	−01	32	11	10	05	04	−03	2.9
Finland	06	07	−06	00	−15	−10	−08	−01	2.0
France	−34	−09	26	−02	−41	00	12	04	3.0
Japan	04	02	11	14	03	03	00	−11	1.7
The Netherlands	19	−08	33	03	−20	03	00	−04	−1.1
Scotland	00	−01	20	08	28	13	04	−03	5.1
Sweden	−04	−05	07	03	−02	−06	−09	−09	1.3
United States	−04	−03	14	13	02	04	−01	−06	2.1
+	4	2	9	8	5	7	5	1	
−	5	8	1	1	5	2	3	9	
Range	53	15	39	16	69	23	21	15	
Mean	−02	−03	19	08	−02	01	02	−05	1.4
All Countries	−05	01	19	15	05	04	02	−06	2.9

TABLE 6.11. *First Group of School Variables: Population 1b.*

	All Schooling		All Homework		Time for Instruction in Math.		Math. Homework		Per-cent of Variance
Country	r	b	r	b	r	b	r	b	
Australia	−10	−08	16	21	03	00	−01	−14	2.7
Belgium	−12	−09	18	12	29	15	06	−10	7.0
England	01	−01	31	14	04	07	05	−03	4.5
Finland	17	15	02	00	−16	−09	00	01	4.0
France	−34	−19	25	−01	−41	19	06	00	−1.6
Germany	15	08	11	06	−11	01	−09	−08	2.5
Israel	11	09	14	10	01	04	−07	−11	3.2
Japan	04	02	11	14	03	03	00	−11	1.7
The Netherlands	14	04	22	03	−01	09	01	04	1.2
Scotland	−05	−03	18	07	25	12	02	−02	4.4
Sweden	−02	−05	06	03	01	−04	−09	−08	1.0
United States	−03	−03	12	10	01	02	00	−06	1.3
+	6	5	12	10	8	9	5	2	
−	6	7	0	1	4	2	4	9	
Range	51	34	29	22	70	28	15	15	
Mean	00	−01	16	08	00	05	00	−06	2.7
All Countries	−05	00	20	17	00	03	03	−06	3.2

TABLE 6.12. *First Group of School Variables: Population 3a.*

	All Schooling		All Homework		Time for Instruction in Math.		Math. Homework		Per-cent of Variance
Country	r	b	r	b	r	b	r	b	
Australia	11	07	00	−03	49	08	24	06	6.1
Belgium	01	08	04	09	04	−02	02	01	0.4
England	−08	−08	−08	−07	13	01	04	−01	1.3
Finland	−03	00	−08	−09	—	—	15	09	2.1
Japan	15	13	12	00	21	19	18	07	7.2
The Netherlands	−01	06	−08	−03	00	−10	−04	−02	0.3
Scotland	00	01	−04	03	48	29	19	06	15.0
United States	−16	−03	22	10	−22	−06	09	−04	3.6
+	3	5	3	3	5	4	7	5	
−	4	2	4	4	1	3	1	3	
Range	31	21	30	19	70	39	28	13	
Mean	00	03	01	00	14	05	11	03	4.3
All Countries	−04	03	12	00	13	08	16	05	1.7

TABLE 6.13. *First Group of School Variables: Population 3b.*

	All Schooling		All Homework		Time for Instruction in Math.		Math. Homework		Per-cent of Variance
Country	r	b	r	b	r	b	r	b	
Belgium	−23	−09	−04	−04	00	04	−02	02	2.2
England	−06	−02	02	03	−02	−01	00	00	0.2
Germany	10	06	−04	−01	—	—	06	03	0.8
Japan	09	03	19	07	23	08	19	06	4.8
Scotland	−09	−05	−04	−03	31	09	05	−01	3.3
United States	−14	05	17	11	−08	−01	05	01	1.3
+	2	3	3	3	2	3	4	4	
−	4	3	3	3	2	2	1	1	
Range	33	15	23	15	39	10	21	07	
Mean	−06	00	04	02	07	03	07	02	2.1
All Countries	−16	−09	10	11	−04	04	06	02	2.5

in Mathematics", and "The Time for Mathematics Homework". Together, the four account for about 3 percent of the variation for Populations 1 a and 1 b, and about 2 percent for Populations 3 a and 3 b. For the two younger populations, the important variable is the "Length of Time for All Homework", the contributions of the others being negligible. In the "All Homework" columns of Tables 6.10 and 6.11, the array of positive coefficients in bold type is impressive. In the simple correlation columns for the other variables, the signs are evenly balanced, and the means consequently negligible. The regression coefficients are mostly negative in the columns for "All Schooling", and for Population 1 a the mean is negative and departs from zero by more than two standard errors. This is also the case for "Mathematics Homework" for both 1 a and 1 b, and the "All Countries" regression coefficient is also significantly negative.

The contrast between the solid arrays of positive signs for "All Homework" and negative signs for "Mathematics Homework" is unexpected and needs explanation. It should be noted that it is the regression coefficients, and not the simple correlations, that are negative. Among the simple correlations there is a 2-to-1 preponderance of positive signs; among the regression coefficients, only 2 out of 22 are positive.

Mathematically, the explanation is not difficult. If we consider the "All Countries" row we see that the simple correlations with the dependent variable are .19 and .02, and if we look up the correlation between the two independent variables, we find that it is .43. Consequently, the little matrix for these three variables alone produces a negative numerator for the partial correlation (and the regression coefficient) for "Mathematics Homework". Similarly, for the rows for separate countries. The spherical triangles are in fact obtuse, and this accounts for the negative partials. But why are the triangles obtuse? In other words, why is the simple correlation for "All Homework" so much greater than the simple correlation for "Mathematics Homework"? A possible explanation would be that the wind is tempered to the shorn lamb, although the lamb may not welcome this. If it were the case that on the whole homework is given according to need, so that students who were rather weak in mathematics had rather more homework than their fellows, the correlation between "Mathematics Homework" and "Achievement" would be reduced. If this were the case, we should have an explanation. But is it in fact the case?

For Population 3 a the position is quite different. In the "All Countries" row, "Mathematics Homework" and "Mathematics Instruction" account for all the variation, the contributions of "All Schooling" and

"All Homework" being negligible. In the United States row there is a considerable negative correlation for "Mathematics Instruction", which leads to the large entry in the row for ranges, Scotland being at the other extreme with a very large positive coefficient. Seven out of the eight ranges are significantly large, so that the position varies a good deal from country to country.

For Population 3 b, the position is different again. In the "All Countries" row practically all the variation is accounted for by "All Schooling" and "All Homework". But in the columns for these variables, as for the others, there are large ranges and an even distribution of signs, so that the differences between countries are now more striking than the similarities.

School Variables—II (Tables 6.14–6.17)

For Population 1 a these account for nearly 3 percent of the variation, the most powerful variables being "Total Roll", "Percentage of Men Teachers", and "Number of Subjects Taken in Grade 8". It is to be expected that the fourth variable, "The Number of Subjects Taken in Grade 12" (in the school) should have little effect at this level, since it concerns the older students in a different population.

The fifth variable, "The Cost per Student in Teacher's Salaries", presents a striking contrast between the mean simple correlation and the coefficient for all countries. The explanation lies mainly in the fact that Japan, which is the highest scorer at this level, is also a low-cost country. The correlation within countries is mostly positive, but the correlation between countries is negative.

The sixth variable, "Educational Differentiation", gives relatively high positive simple correlations in England and the Netherlands, and a relatively high negative one in France, the remainder being negligible. For this variable a positive correlation means that scores are higher where there is less differentiation.

There is a long range, from 13.4 percent in the Netherlands to 0.4 percent in the United States, in the proportion of the variation accounted for by this group. Finland has 7.7 percent; England, 4.6 percent; Sweden, 3.6 percent; and Scotland, 3.4 percent; the other proportions being very small.

For "Total Roll" and "Percentage of Men Teachers" the large coefficients are everywhere positive in the simple correlation columns, and for "Cost per Student" they are positive everywhere except in Sweden. For the other variables, the signs are more evenly balanced.

TABLE 6.14. *Second Group of School Variables: Population 1a.*

Country	Total Roll of School		Percentage of Men Teachers		Number of Subjects in Grade 8		Number of Subjects in Grade 12		Cost per Student (Teacher's Salaries)		Educational Differentiation		Percent of Variance
	r	b	r	b	r	b	r	b	r	b	r	b	
Australia	10	−03	08	10	−03	09	00	−06	−03	−12	00	06	0.6
Belgium	15	08	07	−11	−11	−08	−03	−02	—	—	02	04	1.4
England	09	07	08	04	−02	03	05	01	20	04	24	12	4.6
Finland	26	24	04	−03	—	—	—	—	16	10	—	—	7.7
France	12	−09	18	06	08	06	02	01	14	06	−16	01	1.2
Japan	11	−01	−04	06	−13	−11	—	—	−02	00	−01	00	1.1
The Netherlands	41	20	45	−09	—	—	—	—	41	20	15	07	13.4
Scotland	27	09	−07	−02	−10	−04	07	03	06	03	−02	00	3.4
Sweden	−01	−03	21	11	07	−01	—	—	−13	−10	−01	01	3.6
United States	06	−01	01	−05	−03	−02	01	00	10	05	−01	04	0.4
+	9	5	8	5	2	3	4	3	6	6	3	7	
−	1	5	2	5	6	5	1	2	3	2	5	0	
Range	42	33	52	22	21	17	10	05	54	32	40	12	
Mean	16	05	10	01	−02	−01	01	00	09	02	02	04	3.7
All Countries	16	06	13	06	08	09	01	−01	−03	02	11	04	2.8

TABLE 6.15. *Second Group of School Variables: Population 1b.*

Country	Total Roll of School		Percentage of Men Teachers		Number of Subjects in Grade 8		Number of Subjects in Grade 12		Cost per Student (Teacher's Salaries)		Educational Differentiation		Percent of Variance
	r	b	r	b	r	b	r	b	r	b	r	b	
Australia	03	−02	05	−10	05	12	−06	−11	−11	−10	08	09	2.5
Belgium	22	13	08	−09	−03	00	05	05	—	—	06	03	2.6
England	08	05	11	06	−04	03	06	04	21	05	23	14	5.4
Finland	24	24	08	−01	—	—	—	—	11	05	—	—	6.2
France	17	−06	18	04	09	06	04	02	10	04	−09	13	−0.4
Germany	11	−02	18	04	—	—	05	04	23	−02	−33	−08	2.9
Israel	17	09	−18	−03	12	06	—	—	02	03	12	06	3.6
Japan	11	−01	−04	06	−13	−11	—	—	−02	00	−01	00	1.1
The Netherlands	23	04	31	02	—	—	—	—	25	11	03	03	4.4
Scotland	23	02	−12	−08	−11	−05	03	01	05	00	−02	−04	2.1
Sweden	00	−02	25	16	07	−02	—	—	−15	10	−02	02	2.3
United States	−01	−06	−01	−02	−05	−04	01	01	08	06	03	07	1.0
+	10	6	8	6	4	4	6	6	8	7	6	8	
−	1	6	4	6	5	4	1	1	3	2	5	2	
Range	25	30	49	26	25	17	12	16	36	21	56	21	
Mean	16	03	07	00	00	00	02	00	06	01	01	04	2.8
All Countries	10	02	07	03	09	08	02	00	−03	01	12	11	2.4

TABLE 6.16. *Second Group of School Variables: Population 3a.*

Country	Total Roll of School		Percent-age of Men Teachers		Number of Subjects in Grade 8		Number of Subjects in Grade 12		Cost per Student (Teacher's Salaries)		Educa-tional Differ-entiation		P ce V an
	r	b	r	b	r	b	r	b	r	b	r	b	
Australia	−05	04	01	−01	−13	−08	−16	02	05	04	09	08	1
Belgium	48	48	20	02	—	—	—	—	—	—	39	07	26
England	01	02	−02	00	−08	−04	20	21	04	−05	−04	05	4
Finland	−02	07	03	02	—	—	—	—	12	08	—	—	0
Japan	17	19	13	07	—	—	−04	−05	12	12	12	20	8
The Netherlands	30	18	05	−08	—	—	08	00	06	08	08	−08	4
Scotland	00	−01	07	01	−16	−08	−08	−05	01	01	−02	−03	1
United States	32	14	04	06	01	00	10	03	07	06	−18	−04	6
+	5	7	7	5	1	0	3	3	7	6	4	4	
−	2	1	1	2	3	3	3	3	0	1	3	3	
Range	50	49	22	15	17	08	36	26	11	17	57	28	
Mean	15	14	06	01	−09	−05	02	02	07	05	06	04	
All Countries	−01	03	18	07	22	07	30	21	−08	03	16	09	10

TABLE 6.17. *Second Group of School Variables: Population 3b.*

Country	Total Roll of School		Percent-age of Men Teachers		Number of Subjects in Grade 8		Number of Subjects in Grade 12		Cost per Student (Teacher's Salaries)		Educa-tional Differ-entiation		F c V a
	r	b	r	b	r	b	r	b	r	b	r	b	
Belgium	04	−02	24	−27	—	—	—	—	—	—	13	09	−
England	05	−02	28	14	−14	−11	06	08	02	−03	−09	02	
Germany	07	−03	09	08	—	—	11	21	−09	−15	—	—	
Japan	04	07	13	06	—	—	−20	−11	23	20	−11	−04	
Scotland	−03	00	16	04	−08	−05	−02	−02	11	05	−02	−02	
United States	27	08	07	05	01	03	14	12	08	09	−16	−02	
+	5	2	6	5	1	1	3	3	4	3	1	2	
−	1	3	0	1	2	2	2	2	1	2	4	3	
Range	30	11	21	10	15	14	34	32	32	35	29	13	
Mean	07	01	16	02	−07	02	−04	05	06	03	−04	00	
All Countries	02	−06	25	12	15	08	32	34	−16	−04	02	−11	1

The table for Population 1 b is generally similar to that for Population 1 a. For Population 3 a we find that the group accounts for much more of the variation over all countries, mainly because there is now a large contribution from the fourth variable, "Number of Subjects Taken in Grade 12". This, however, is mainly a difference between countries and not within countries, as may be seen by comparing the mean row with the all countries row. There is a striking contrast between England, where both coefficents are relatively large and positive, and Scotland, where both are negative. In both countries both coefficients are negative for "Number of Subjects Taken in Grade 8". All the large coefficients for "Cost per Student" are positive, and consequently so are the means, but the "All Countries" simple correlation is negative, again mainly through the influence of Japan. In general, this table shows marked contrasts between the mean row and the all countries row. The correlations between countries are very different from those within countries.

This is also true in Table 6.17 for Population 3 b. Within countries the signs for simple correlations are mainly positive for "Total Roll", "Percentage of Men Teachers", and "Cost per Student", and are negative for "Educational Differentiation" (for which they are evenly balanced for Population 3 a). The regression coefficients are mainly positive for "Percentage of Men Teachers", but are otherwise evenly divided.

Student Variables (Tables 6.18–6.21)

This group shows much more uniformity. With only trifling exceptions below the level of significance, the signs are uniform in both the simple correlation and the regression columns for "Sex of Student", "Level of Instruction", and "Student Interest". The negative signs in the columns for "Sex of Student" indicate that on the whole the boys do better than the girls, even when the other variables are held constant. The positive signs for "Level of Instruction" are to be expected, and the main interest of these columns is to show the accuracy of this student's rating. Similarly, high and positive correlations were to be expected for the variable "Student Interest", and they have in fact appeared uniformly.

The variable "Age of Student" also behaves as expected. For Population 1 a the correlations are positive; in the other populations they are preponderantly negative. Population 1 a consists of all pupils aged 13, in whatever grade they are to be found. In this population it is the older pupils who have the higher level of achievement. The other

TABLE 6.18. *Student Group of Variables: Population 1a.*

Country	Sex of Student		Age of Student		Level of Instruction		Student's Interest in Mathematics		Percent of Variance			
	r	b	r	b	r	b	r	b	Sex	Age	Level	Int
Australia	−03	−04	16	−04	46	51	30	26	0.1	−.64	23.5	
Belgium	−19	−18	−02	−01	46	35	24	21	3.4	.02	16.1	
England	−10	−04	11	01	27	10	36	24	0.4	.11	2.7	
Finland	−10	−05	02	−02	35	35	27	19	0.5	−.04	12.2	
France	−14	−09	09	07	61	54	27	20	1.3	.63	32.9	
Japan	−10	−06	08	07	—	—	42	36	0.6	.56	0.0	
The Netherlands	−17	−09	24	−01	74	58	31	19	1.5	−.24	42.9	
Scotland*	−04	00	05	04	20	09	30	20	0.0	.20	1.8	
Sweden	−03	−05	03	02	32	20	38	32	0.2	.06	6.4	
United States	−01	01	14	10	31	21	20	18	0.0	1.40	6.5	
+	0	1	9	6	9	9	10	10				
−	10	8	1	4	0	0	0	0				
Range	18	17	26	08	74	58	22	18				
Mean	−09	−06	09	02	37	29	30	24	0.1	.21	14.5	
All Countries	−08	−03	06	03	36	24	27	21	0.2	.18	8.6	

* After this analysis had been run it was discovered that the *Level of Instruction* data for Scotland slightly faulty. This remark also applies to Tables 6.19.–6.21.

TABLE 6.19. *Student Group of Variables: Population 1b.*

Country	Sex of Student		Age of Student		Level of Instruction		Student's Interest in Mathematics		Percent of Variance			
	r	b	r	b	r	b	r	b	Sex	Age	Level	In
Australia	−06	−06	−20	−18	01	00	35	31	0.4	3.60	0.0	
Belgium	−21	−11	−04	−08	39	26	26	23	2.3	.32	10.1	
England	−09	00	04	03	29	12	39	28	0.0	.12	3.5	
Finland	−12	−08	—	—	09	11	32	25	1.0	.00	1.0	
France	−13	−14	06	02	55	55	29	22	1.8	.12	30.2	
Germany	−11	−03	04	00	38	28	28	23	0.3	.04	10.6	
Israel	06	05	−10	−05	—	—	23	20	0.3	.50	—	
Japan	−10	−06	08	08	—	—	42	36	0.6	.64	—	
The Netherlands	−19	−10	00	−14	46	43	27	23	1.9	.00	19.8	
Scotland	−06	−01	−01	−01	18	08	32	23	0.1	.01	1.4	
Sweden	−05	−06	08	−10	30	17	38	32	0.3	−.80	5.1	
United States	−02	−01	−25	−19	21	13	24	31	0.0	4.25	2.7	
+	1	1	5	3	10	9	12	12				
−	11	10	5	7	0	0	0	0				
Range	27	19	33	20	54	55	19	16				
Mean	−09	−05	−04	−05	29	18	31	26	0.8	.73	8.4	
All Countries	−07	−05	−02	−01	20	15	30	24	0.4	.02	3.0	

TABLE 6.20. *Student Group of Variables: Population 3a.*

try	Sex of Student		Age of Student		Level of Instruction		Student's Interest in Mathematics		Percent of Variance			
	r	b	r	b	r	b	r	b	Sex	Age	Level	Interest
alia	−10	00	00	−04	50	23	47	31	0.0	0.0	11.5	14.6
ım	−24	−02	−22	−12	14	14	30	24	0.5	2.6	2.0	7.2
ınd	00	−02	−18	−13	19	14	29	25	0.0	2.3	2.7	7.2
nd	−23	−12	−13	−16	—	—	51	44	2.8	2.1	—	22.4
ı	−11	−02	−07	−05	—	—	37	27	0.2	0.4	—	10.0
Netherlands	−05	03	−19	−08	—	—	52	45	0.2	1.5	—	23.4
and	−14	−05	08	−06	36	16	40	27	0.7	−0.5	5.8	10.8
·d States	−13	−06	−27	−09	39	16	43	27	0.8	2.4	6.2	11.6
	7	1	1	0	5	5	8	8				
	0	6	6	8	0	0	0	0				
e	24	15	35	12	36	7	23	19				
ı	−12	−03	−10	−08	32	17	41	31	0.6	1.3	5.6	13.2
ountries	−16	−08	01	−15	37	28	34	27	1.3	−0.2	10.4	9.2

TABLE 6.21. *Student Group of Variables: Population 3b.*

try	Sex of Student		Age of Student		Level of Instruction		Student's Interest in Mathematics		Percent of Variance			
	r	b	r	b	r	b	r	b	Sex	Age	Level	Interest
ım	−35	−38	−11	−12	04	01	33	31	13.3	1.3	0.0	10.2
nd	−32	−11	−05	−03	34	24	39	33	3.5	0.2	8.2	12.9
any	−30	−25	−04	−06	—	—	29	29	7.5	0.2	—	8.4
ı	−19	−06	03	01	36	20	32	28	1.1	0.0	7.2	9.0
ınd	−39	−20	−07	−08	51	39	29	21	7.8	0.6	19.9	6.1
d States	−01	−05	−22	−13	14	10	16	17	0.0	2.9	1.4	2.7
	0	0	1	1	5	5	6	6				
	6	6	5	5	0	0	0	0				
e	34	33	27	14	51	38	23	16				
,	−26	18	−07	−07	23	19	30	26	5.5	0.8	7.3	8.2
ountries	−22	−13	00	−15	27	23	31	23	2.9	0.0	6.2	7.1

populations are based upon grade, not age, and in these populations the younger pupils may have a higher achievement than the older ones, having been promoted early precisely because they are more forward. Achievement is correlated with both grade and age, and there is also a stronger relation between grade and age. Consequently, if the simple correlation between achievement and grade is smaller than that between achievement and age, the partial correlation between achievement and age may be negative and will in any case be considerably reduced. Although age is not constant within a single grade, it is severely restricted in range, and it is this restriction that produces the small or negative correlations between age and achievement for the grade populations in comparison with the positive correlation for the age population.

Summary and Conclusions

The analysis produced many interesting results, some expected and others not. It was to be expected that "Student's Opportunity", "Student's Interest", and "Level of Instruction" would all make large contributions to the assignable variance, and the fact that they have done so is mainly useful as showing that the questions were well devised and properly interpreted, in many cases, by the teachers and students who have supplied the information.

The general effect of the analysis is to indicate some variables as important, at any rate in some countries, because they have appreciable regression coefficients, and others as unimportant, because they do not. The whole object of the analysis was to determine the standing of the variables in this way. But how far does the standing of each variable, so determined, depend upon the choice of other variables for inclusion in the analysis? What would happen if we added a twenty-seventh variable to the 26 already included? Unless the regression coefficient of the new variable happened to be exactly zero its inclusion would entail some change in the regression coefficients previously determined. Under certain circumstances these changes might be considerable. For example, the students in the sample for the English Population 1 a were given a verbal reasoning test. The scores for this test had a high correlation (.73) with the scores for mathematics. If this variable is included in the regression analysis it makes radical changes in the regression coefficients; for example, it reduces the coefficient for the "Status of Fathers Occupation" from .15 to .03 and completely alters its standing. This example shows two things. In the first place it shows that it is always possible for the inclusion of a new variable to produce large changes in the previously determined coefficients, even when these have been determined on a large number of variables. Second, it shows the importance of a judicious choice of variables. To include the verbal reasoning test in the analysis would clearly be injudicious, because this test covers again much of the ground covered by the mathematics tests, as is shown by the high correlation. Consequently, it absorbs most of the assignable variation in the mathematics scores and leaves little for distribution among the other variables.

A scrutiny of the 54 variables in the large correlation matrix suggests that only one was omitted that might judiciously have been included. This was "Number of Staff". It might have been better to omit "Interest", "Opportunity", and "Level", as has been said above, but fortunately a rerun shows that their inclusion did not materially alter the standing of the rest. On the whole, therefore, it seems reasonable to say that the analysis gives a fair representation of the relative importance of the variables that we have studied.

It was to be expected that the parental group would make a large contribution, and so it did, with "Father's Occupation" supplying the major share, as expected, and "Father's Education" and "Mother's Education" taking the second and third places. The "Parents' Place of Residence" was always a doubtful starter, because of the difficulty of finding a classification which would suit all countries. In the upshot, this variable contributed little except in Japan and the United States.

The teacher group of variables made only small contribution. This does not, of course, cast any doubt on the common sense view that there is a great difference between the effects of good teaching and of bad. It has frequently been shown that simple ratings by experienced observers correlate moderately well with the subsequent results of teaching. It was not to be expected that the variables included in this group would make contributions of this order of magnitude; their contributions in any case were bound to be fairly small compared with that of a simple rating. In fact, the chief contribution in this group came from a form of rating, by the student, in the shape of the "Description of Teaching" variable. "Length of Training" and "Recent In-Service Training" both made rather smaller contributions than might have been hoped, and nearly half the coefficients for "Degree of Freedom" were negative.

The main result from the first group of school variables is the importance of homework at all levels. "All Homework" is more important than "Mathematics Homework" at the lower levels; at the upper level this is reversed. The second group shows large schools are more effective than small, at all levels, and high costs (in teachers' salaries) are more effective than low. The last result holds within countries but is reversed between countries chiefly because Japan is a high scoring and a low cost country. For the other variables in this group the differences between countries are more marked than the similarities.

The parental variables occupy the prime position everywhere. This is apparent in the tables for Populations 1 a and 1 b, and while the coefficients are smaller for the preuniversity populations in countries with low retentivity, this is merely because their range has been shortened by selection. In the United States, with high retentivity, they are as high for the older as for the younger populations.

Chapter 7

Summary of Major Findings

Introduction

The project reported here is the first large-scale attempt to employ empirical methods in comparative education and thereby to arrive at comparable criterion measures. The construction of internationally valid evaluation instruments—achievement tests—however, did not create the most difficult problem for this study. The main obstacle turned out to be the measurement of certain independent variables, operationally feasible indices of basic characteristics of the school systems. Thus, for instance, the concept of "differentiation" was apparently interpreted in quite a variety of ways by the school principals when asked to rate their own schools according to what extent differentiation was provided. When the students were asked about their course or program, confusion existed as to whether a program should be labeled "academic", "vocational", or "general". In order to arrive at a description of the level of mathematics instruction, the English curriculum had to serve as a reference scale which was not easily translated into the situation in other countries. Even measures that *prima facie* would seem to be easily obtained, such as salary budget for the school, expenditures for equipment, or length of teachers' postsecondary training, were not easily ascertainable. In many instances a lack of comparability stemmed from the difficulties of translating terms like "comprehensive" or "postsecondary". It seems that we still have a long way to go before we will have cross-nationally codified independent variables to describe the most important dimensions of school systems.

As shown in Volume I, Chapter 14, even as regards certain basic statistical information, such as per-pupil expenditures or enrollment figures, there is a lack of uniformity in data reporting.

The difficulties indicated above should be kept in mind when interpreting some inconsistencies that appear in the findings.

This chapter has been written by Professor Torsten Husén.

Pattern of Test Results

The IEA study was not designed to compare countries; needless to say, it is not to be conceived of as an "international contest". As was spelled out in the introduction to Volume I, its main objective is to test hypotheses which have been advanced within a framework of comparative thinking in education. Many of the hypotheses cannot be tested unless one takes into consideration cross-national differences related to the various school systems operating within the countries participating in this investigation.

Certain descriptive statistics relating to the main dependent variables (mathematics test and attitude test scores) have been presented in Chapter 1. The international range among means for the 12 participating countries in total mathematics score for the 13-year-old population covers more than one standard deviation of the combined distribution, and is even larger in the terminal mathematics population (3 a). Thus, students who are much above the average in one country might be regarded as mathematically rather backward in another country. At the lower level (1 a), where 100 percent of the age group is still in full-time schooling, Japan has the highest mean and Sweden the lowest. In the preuniversity year students with terminal mathematics in the United States averaged far below the other countries. It is interesting to note the country means as a percentage of the total number of items in the test. The lowest and highest means expressed as a percent of the total number of items is 22 percent in the United States Population 1 b and 46 percent in Israel also for 1 b. The corresponding percentages for Level 3 a range from a low of 20 percent in the United States to a high of 53 percent in Israel. However, neither the "productivity" of an educational system of a country, nor the effect of the instruction given, can be assessed from national means. The age of school entry varies between countries, and the grade placement of mathematics topics varies considerably. In a dual system with early transfer of the more adept students to the academic secondary school there is a tendency to introduce advanced mathematics topics earlier whereas in countries with a comprehensive system one seeks to introduce them later. Furthermore, the high "retentivity" of the comprehensive systems has the effect that within the terminal mathematics group one can, in spite of the low average, identify an elite comparable both in quality (average) and quantity (proportion of age group) with the entire terminal group in a country with a low retentivity and a high average. Neither the national average at a level where schooling is still compulsory nor the average at a major terminal point would suffice to evaluate the "efficiency" of a whole school system.

Variability among student achievements is also marked, especially at the 13-year-old level where all the students are still in school; in the 1 a population the standard deviation is almost twice as large in England as in Finland.

By means of "bridge" tests the increase in mathematical competence from age 13 to the higher (preuniversity) level could be estimated for the individual countries. This increase was highly related to an index of selectivity, that is, the more selective or exclusive the secondary school, the larger the increase in performance for those who reach the terminal grade.

Partly because the part scores tended to be highly intercorrelated, they did not reveal differences of particular interest between countries.

Total mathematics scores were correlated with 45 other, mostly independent, variables characterizing the school, the teacher, and the student in each country and population (see Table 1.12). Since the between-country components tend to average each other out, these correlations are rather low, though some of them were rather high for individual countries. Of the variables characterizing the schools, size tends consistently to be positively related to the total scores at the lower level. At the upper level the picture is inconsistent, due probably to selection mechanisms on which we do not possess complete data. Length of teacher training is positively related to the total score and so are teacher ratings of the students' opportunity to learn the topics represented by the test items. Yet, the latter correlation is surprisingly low, on the average .20 and ranging from below zero in Sweden to .50 in England. Of the students' characteristics, fathers' and mothers' education and status of fathers' occupation tended at the lower level consistently to be positively related (on the level of .20) to mathematics score. In the terminal population these correlations varied quite widely due to variations in degrees of social selection and of retentivity. Students' interest in mathematics correlated .30 to .36 with total scores; students' plans and aspirations for further education correlated with the same scores about .35 to .40.

Single item results (difficulty and discrimination indices, and error distribution on multiple choice items) have been calculated and are presented in Appendix II. These should be helpful in curriculum studies in the different countries.

By means of five scales the students' attitudes toward mathematics, education, and the environment have been assessed. Students in the terminal program tend to a greater extent to view mathematics as a fixed or "frozen" and less open system than students at the 13-year-old level. Differences in this respect between countries are quite marked with a range of about one standard deviation. Mathematics is perceived as more difficult and demanding by the students in the terminal program than by students at the lower level. The between country range is almost the same as for the view of mathematics as an open or closed system. Students in the terminal program, that is, among those with the largest exposure to mathematics, tend to have a more pessimistic appraisal of the role of mathematics in contemporary society, but large international differences were found in this variable also. The "I like-dislike" scale on attitudes toward school and school learning disclosed an international range of about one standard deviation. Attitudes were most positive in Japan and least positive in the United States.

Correlations between country means (on the one side mean total score and on the other the means for the 45 other variables) were computed. These correlations reflect the between-country components of the variables and are thus much higher than within country variables where the international differences tend to level out. The fact that perstudent expenditure or teacher training displays a sizable negative correlation with total mathematics score on the 13-year-old level is explained by the fact that Sweden and the United States have the lowest score and at the same time the highest perstudent expenditure, while in Japan the opposite is the case. The same logic applies to many other factors, like Gross National Product (GNP), which is negatively correlated with the average mathematics performance at both levels. It should, however, once again be emphasized that neither performance at an age before mandatory school attendance expires nor at the preuniversity level can be used as an assessment index of the "efficiency" of a school system (see Chapter 3).

Problems Related to School Organization

Mathematics is regarded as a "strategic" subject in present-day technological society. Until recently it was conceived of as central to general education and an essential component in a liberal arts education. Nineteenth century humanism held mathematics in high esteem since, together with Latin, it was considered to contribute in a major way to mental discipline and to a sharpening of the intellect. Today, mathematics is more typically regarded as a major component of a proper education for science and technology. Its new objectives, however, have in most countries collided with rather strongly institutionalized curriculum content and methods of instruction (Dahllöf, 1963). In all countries a shortage of mathematicians and mathematics teachers is strongly felt.

Most of the problems dealt with in hypotheses related to school organization could be put under the general heading of "selective versus comprehensive education". As was indicated in the introduction to Volume I, the cardinal problem in most West European countries is how to adapt the dual or parallel school structure which is a heritage from a society with strong class distinctions to a changing technological society where class differences are leveling out and the public has become more sensitive to waste of talent. But most striking of all is the need in economically highly developed countries for education of progressively larger proportions of youth not only at the secondary but also at the university level. Apart from all ideological considerations about "democratization of higher education" or providing "equality of opportunity", the increasing need for highly trained man-power is a powerful impetus to the rapidly expanding enrollments at the secondary and postsecondary level. There is also the growing appreciation of the individual's need for general education in a world of change and amidst growing social and technological complexities. Mathematics as an instrument to deal with the quantitative aspect of the environment is an increasingly important part of the spectrum of skills the individual needs to possess.

The IEA countries represent a rather broad spectrum of school organizations. In the United States the schools are by tradition more comprehensive, while in Europe most pupils are differentiated at an early age into separate schools and courses. In Europe, on the one side a small intellectual and social elite has been channeled into an academic, university-preparing program and on the other hand, the broad mass, especially from working class homes, has proceeded with a general or semivocational program until the end of compulsory schooling.[1] The United States has by tradition a school structure whereby the secondary school has been added to the primary, not as a parallel but as a continuation school. Under the impact of broadened enrollment and extended curricula the selectivity of European schools has tended to lessen, and more flexibility in terms of transfer from one type of program or school to another has appeared. There is also a

---

[1] This does not deny that for several decades students from manual workers and similar families have made up small fractions of secondary enrollment in some countries.

growing tendency to postpone selection for academic education and the final choice of program in order to provide a broader general education for all students and to avoid a lengthening of parallelism when compulsory schooling is prolonged and school-leaving age is raised.

"Comprehensive" schools catering for all the students of a given area with all the programs under the same roof have been introduced in some European countries since World War II. Their introduction has in most cases been a political issue since they have been conceived of as a means of "democratizing" secondary education.

The problem of comprehensive education is, however, apart from its social and political implications, regarded by many teachers as a purely pedagogical one, namely that of grouping. How should students be organized into instructional groups so as to provide for optimal learning? Some secondary schoolteachers have conceived of grouping as a fairly simple problem: the academically talented should be grouped together in separate classes (or schools) at an early age in order not to be hampered by their slower-learning non-academic age mates. It is thereby implied that the latter category of students would also profit from the separation because they would not be discouraged by the presence of brighter classmates. "Undifferentiated" classes, embracing the whole range of ability, are supposed to be impractical and unproductive, since the standards of the university-bound students would be lowered and the slow-learning students would lag more behind than they do in separated classes.

Grouping can be executed in two main ways. One can either try to homogenize the students throughout all the academic subjects according to some criterion of scholastic ability, such as marks, teachers' rating, achievement or aptitude test scores. Thereby, one obtains so-called homogeneous grouping. "Streaming", which takes place in most schools in England on the primary level and in the comprehensive schools on the secondary level, is another example of homogeneous grouping. The other type of grouping takes into account the *intra*individual differences and is a way of grouping students according to their varying proficiency in separate subjects; this is usual in English grammar schools and is called "setting". Thus, parallel instructional groups in mathematics on a certain grade level are set up according to the performance in that particular subject. This within-subject grouping does not imply separation of the students over all the subjects even if there might be a quite sizable correlation between subjects (cf. Yates, 1966).

As is pointed out in Chapter 3, the setting up of a comprehensive or selective system and grouping practices are affected not only by the society, which the school is supposed to serve, but also by the teacher's philosophy of his role in that society. The beliefs they hold about what can be accomplished with their students have a strong effect on the outcomes of school instruction over and above the effects of school structure or grouping procedure as decided upon by the educational policy-makers.

When designing the IEA study it was decided to collect data which would enable us to test certain basic assumptions related to the issue of selective versus comprehensive school structure. These problems might be grouped in the following categories.

1. What effects does comprehensive and selective organization have upon the performances in mathematics of high- and low-ability pupils, respectively? What effects do the two school structures have upon the students' interest in school subjects, in this case in mathematics?
2. Does a comprehensive system with a higher retentivity of pupils at the secondary level produce an elite of the same standard as a selective system? Is this elite smaller or larger in the comprehensive system?
3. What is the "yield" of the comprehensive system as compared with the selective in terms of "how many are brought how far"?
4. What are the socio-economic implications of the two systems in terms of equality of opportunity?

As might be expected, the average level of mathematics performance was inversely related to the proportion of the age group in school at the terminal level. When the national scores at this level were adjusted for age at testing, performance at age 13, and retentivity (that is, the proportion of a yeargroup still at school at the preuniversity stage), they became closely concordant as may be seen from Table 3.8.

According to the assumptions many teachers hold about the effects of selection and grouping, students in a system with specialized courses and schools should perform better than those in an undifferentiated, more comprehensive system. Furthermore, it was expected that the variability would be lower in the specialized schools. Apart from the difficulties of classifying schools as comprehensive, academic-selective, etc. and of obtaining relevant information from the students about the program they were following, we were not in a good position to make sensible comparisons between systems within countries because of the differences between comprehensive and selective schools in social class composition and facilities. Thus, in England the students in grammar schools performed much better on average than students in the academic stream of the comprehensive schools. Of course this does not say anything decisive about the relative "productivity" or "efficiency" of the two systems, since the selection mechanisms determining the input to the two types of schools are not under control.

The opportunity to make cross-national comparisons enabled us, however, to shed light on the related important problem, namely, to what extent it is possible to educate an elite within a comprehensive and retentive system. This required that equal proportions of the age groups be analyzed. When the top four percent of the age group in each country were extracted, countries with a more comprehensive and thus a more retentive system then showed the greatest upward shift. The most striking example was provided by the United States. There the total group of mathematics students in the senior high school grade averaged far behind their age-mates in other countries but an elite existed with, on the whole, the same average score as in most of the countries with a much more selective system and with a lower degree of retentivity. This kind of analysis was extended by setting up international percentiles on the basis of the composite distribution of all the countries and determining the proportion of the age group at the terminal level in each country with attainments above given percentile scores. The question "does more mean worse" cannot be answered unequivocally. As can be seen more in detail on pp. 128 et seq., the increase of the intake into the preuniversity school in general and the mathematics program in particular increases the size of the elite, defined, for instance, by the number of students reaching the 95th international percentile. At the same time, increased intake means an increase in variability, due mainly to increased intake of lower ability students. This, however, is achieved not specifically by the pedagogical qualities of the system, but by its retentivity. In selective systems the students, who do not keep up with the fairly uniform requirements, either drop out or are held back as grade-repeaters.

The opening up of opportunities for preuniversity education for more students and retaining students at school thus produced an elite group comparable and in some cases superior in size and quality to the one accomplished by a selective system. What, then, is happening to the major part of the students, those who by statistical or other criteria are not referred to as the elite? The "productivity" or the "yield" of a system cannot be assessed solely by the size and quality of the elite. The more a society needs highly trained manpower, the more the "yield" of its schools should be measured by its capacity to promote optimum achievement among *all types* and at *all levels* of aptitude. A selective system with the dual-track school structure succeeds in bringing the few who survive to graduation up to outstanding accomplishments, whereas those who were not selected are left far behind. A system with a high degree of retentivity can bring a larger proportion of the students of average ability up to a higher level of performance than the selective system. An assessment of the "yield" in mathematics would have required a much more extensive testing program than was possible in the present project. Apart from testing representative samples of students at all the terminal points in the primary and secondary school, we would have had to test students who for various reasons dropped out of the secondary level.

Everybody in educational circles pays lip service to the principle of "equal opportunity". But the *conditions* for this principle to operate vary considerably from system to system. Previous research has furnished ample evidence that the criteria employed in evaluating and selecting students for secondary school admission are often loaded with social class factors and so is grade repeating and dropping out. It could therefore be hypothesized that the more marked the selectivity in terms of entrance examinations, grade repeating and "flunking-out", the stronger the social bias in favor of middle-class students. We found, when using the proportion of an age group of students who reached the preuniversity year as a composite index of selectivity, that social bias in favor of upper- or middle-class students was more pronounced in countries with a low retentivity. Furthermore, the earlier the selection, the stronger the social bias at the preuniversity level.

On the basis of previous research it could be hypothesized that students in the selective academic programs would be more interested in mathematics than those in comprehensive and "remaining" programs. Students in selective programs on the average have a more "school-minded" social background, and the selective academic schools have better educational provisions; for instance, they have more qualified teachers. We found that students in the selective academic program showed the greatest interest and the "remainder" the lowest interest, whereas those in the comprehensive courses occupied an intermediate position. This fits into findings from other research, according to which the average or below-average students tend to be better motivated in undifferentiated classes or schools than if they are allocated to separate classes or schools.

Studies of Certain Features of School Organization

Not only the age of mandatory entry to school but also actual number of years of formal schooling vary considerably between countries. The effect of schooling is often considered as being proportional to the number of years and hours per week students have been exposed to teaching. It was found that the age of school entry bore little relationship to outcomes of instruction in mathematics at the age of 13; students entering school at 6 tended to be somewhat superior to those entering at either 5 or 7, respectively. When the enrollment was broken down according to social status, with most of the culturally deprived children having fathers who were unskilled manual workers, it was found that contrary to expectations the middle-class children seemed to gain more from earlier school entry than did working-class children.

The age of graduation from the academic secondary school varied from 17–18 in some countries to 19–20 in others. There was also diversity in the length of the course to the preuniversity year, in the degree of selection of students, and in the number of subjects studied to the end of the course. When the progress made during the course was measured with due regard for these variations we found that the rate of progress from 13 years to preuniversity was much the same in all the participating countries. Where many subjects were studied, the rate of progress in mathematics alone was slower; where there was a high degree of selection, the small proportion of the students went further than did the large proportion where there was greater retentivity, but the corresponding top group of students from the latter advanced at much the same rate. Thus, the amount of talent available seems to be pretty much the same in all countries but the policies in developing it vary considerably from country to country.

In Population 1 b, the differences in scores could be accounted for only partly by the differences between countries in the mean ages of students in these grades. For the terminal mathematics population, the score was positively related to age, but for the terminal nonmathematics population, the relation was negative. These results are not surprising; they imply that further time devoted to study can increase scores, but further time, without study of the subject, can have the opposite effect.

The appropriate size of class has been a pervasive issue among educationalists for many decades. A considerable number of national studies of the relationship between class size and student achievement have been carried out, especially in the United States. The appropriate size of the school, particularly on the secondary level, has been another issue, since school size is related to the provision of adequate teaching staff to meet students' choices and to disciplinary problems. Comparatively few studies have been devoted to this problem, none of them with an international scope.

The average size of class in the present study ranged from 24 in Belgium to 41 in Japan on the 13-year-old level and among terminal mathematics students from 12 in England to 41 in Japan. At the 13-year-old level larger classes tended to achieve higher scores. The rank-order correlations among countries between average mathematics score and average class size were .29 and .43, respectively, in the lower age populations. This is in accordance with previous research, when studies have been conducted employing students covering the whole ability range. At the preuniversity level the findings were rather conflicting, which could be anticipated since the size of classes was determined by complicated and varying selection mechanisms affecting both the ability distribution and the social class composition of the classes. The rank-order correlations among countries were $-.41$ and $-.23$ for the mathematics and nonmathematics preuniversity populations. On the whole, the results suggest that the relationship between class size and student performance is not a simple one.

The correlation between size of school and mathematics performance varied strongly between countries and programs. In general, schools with an enrollment exceeding 800 performed better with the younger students (the 13-year-olds). The evidence was more conflicting among the preuniversity students.

In accordance with the general belief among educators it was hypothesized that schools' achievement will be better when students are allocated to special schools or programs instead of being kept together in a comprehensive system. The hypothesis was supported by the present study as far as the younger students were concerned. Those in selective academic schools tended to perform better than those following an academic program in a comprehensive secondary school. By the preuniversity year, however, the difference had disappeared.

Interest in mathematics tended to be strongest in selective schools, less strong in comprehensive schools, and weakest in schools for the "remainder". However, as between countries, emphasis on comprehensive education is accompanied by greater interest in mathematics.

As indicated earlier, the average level of mathematics performance is lower where larger percentages of the age group are still in school at the terminal level. So far, more means worse. But when equal proportions of the age groups in various countries are compared at preuniversity age, countries differ much less. The results suggest that in countries with a high retentivity a high standard is preserved among the best students coupled with a dipping into lower levels of mathematical ability for the rest of the group. Countries with high retentivity, such as the United States, Sweden, and Japan, display a higher "yield" at this age level than countries with low retentivity, such as Germany and France. Thus, the proportions of an age group reaching various international levels is positively related to the size of the population retained in school up to the preuniversity level.

There is a social class bias (social selectivity) in the selection for or the drop-out before the preuniversity year in all the participating countries. Social selectivity is very small in the United States, where about three fourths of the entire age group is kept in school up to secondary school graduation age. It is related to retentivity in that the lower the number of students reaching the graduation age in the secondary school, the higher the bias. Furthermore, the earlier the transfer and/or selection takes place for academic secondary education, the stronger the bias.

The rank-order correlation between retentivity (proportion of an age group in full-time schooling at a given age) in the preuniversity year and per capita income is .32, with percentage of students in comprehensive education .89 and with the index of industrialization .17.

Problems Related to Instruction and Curriculum

Student Achievements and Noncognitive Outcomes[1]

It was hypothesized that learning centered or "discovery" approaches, as contrasted to drill or rote methods, would be positively correlated with mathematics performance. The results of a scale describing mathematics teaching and school learning derived from the opinionnaire supported this hypothesis at the lower level. In the preuniversity populations, however, the mean correlation tended to be negative. When either the level of mathematics instruction or the opportunity to learn items was held constant the above findings remained true. A positive relationship between "discovery" approaches and interest in mathematics also existed at the 13-year-old level.

In countries with high scores in mathematics, students tend to consider mathematics an important subject for society and at the same time as a "closed" system and a difficult subject, whereas the opposite is the case in countries with low achievement. The two patterns are

---

[1] Student perception of mathematics as a subject and of mathematics instruction was assessed by a student opinion booklet. An index giving the interest in mathematics was derived from the student questionnaire. Measures from these two instruments were related to the achievement test scores and to each other (cf. Chapters 6 and 12, Volume I).

associated with a selective and a more comprehensive school structure, respectively.

As could be anticipated, students' achievement and interest in mathematics were positively correlated; the highest correlation (.34) was in the preuniversity group specializing in mathematics and sciences. This category of students showed an interest in mathematics which exceeded that among students in other programs by .75 of a standard deviation. The correlations for "wishes to take more mathematics" and scores were somewhat lower for all the populations. Students' interest and their views of mathematics as an open or closed system were negatively correlated at both levels, and especially for the older students; that is, students who viewed mathematics as a closed system were more interested in it than those who viewed it as an open system.

We consistently found at the 13-year-old level that the student's description of mathematics teaching as promoting inquiry was positively correlated with interest in mathematics, whereas no consistent correlation existed at the preuniversity level. The latter finding could be interpreted as a result of more conservative methods of instruction. It could also be seen as a repercussion of the examination system.

Interest in mathematics and father's occupational status showed a weak positive correlation; this was true also for interest and father having a scientific occupation. The mean difference in interest between students whose parents held scientific and non-scientific occupations was highly significant in Japan, significant in England and Sweden, but non-significant in the other countries.

Achievements as Related to Opportunity to Learn and to Teacher Competence

Teachers were asked to rate each test item as to their students' opportunities to learn the topic represented by the item. The correlation of this rating with the student achievement over all countries was less than .3; in some countries the correlations were not even significant. The highest correlations were found for England and Scotland (.5 to .6 at the 13-year-old level) and for Japan (a correlation of over .4 for the science-mathematics students at the preuniversity level). These findings are consistent with the conception that teaching and learning are not identical. They indicate the international character of the tests, which did not markedly lean upon the syllabus of any particularly country. The lack of correlation in countries with a centralized curriculum (for example, Sweden) might also reflect to some extent homogeneity of program.

The emphasis on a given topic in the respective national curricula as rated by the teacher was related to students' achievement. "Profiles" were drawn up indicating for each topic the emphasis (in percentage) and the average achievement. Except for demonstrative geometry emphasis and achievement followed each other closely. Thus, national differences (but not necessarily individual differences) can in part be explained by differences in emphasis in curriculum.

The length of the teachers' postsecondary education was related to the total mathematics score; in Populations 1 a and 3 a, the longer the training, the better is the students' achievements. In the 13-year-old population, students with university-trained teachers performed better than those with teachers trained at other institutions, provided the teachers had four years of more of training, whereas no difference occurred between teachers from the two types of institutions who had three years or less of training.

Within countries, the correlation of in-service training with mathematics score and with various student attitudes and interest scores had 50 signs supporting the hypothesis in favor of useful in-service training and 26 signs non-confirming, though the relationship was weak. The exception was a consistent positive correlation between in-service training and student interest in mathematics in the preuniversity population. The correlation using countries as units, however, showed in-service training going with low scores, an open view of mathematics as a process, a high estimate of the place of mathematics in society and high student interest in the subject.

Mathematics Achievement as Related to the Opportunities Provided by the School

The traditional ideal behind organizing schools is to expose students to teaching that is structured according to a time schedule and divided between teacher-guided and independent work. Most, if not all, learning is supposed to be a direct effect of teaching. Four measures related to these variables were available: namely, the total hours per week in school, hours per week allocated to mathematics teaching, hours per week of homework, and hours per week devoted to mathematics homework. At the 13-year-old level, only three percent of the assignable variation in total mathematics score was accounted for by these four variables, of which total homework made the largest contribution. The other variables, especially mathematics homework and hours of mathematics instruction, were relatively more important in the mathematics-science program in the preuniversity year. On the whole, the number of hours per week of schooling seemed to bear little or no relationship to mathematics achievement, while total homework at the lower level and mathematics homework at the preuniversity level seemed to be of greater importance. On the average, more than one third of all homework time is spent on mathematics (38 percent in Population 1 a and 34 percent in Population 3 a). Population 3 a students spent about twice as much time (5.2 hours as compared to 2.7 hours) on mathematics homework as do students from Population 1 a.

Special opportunities, such as extra courses and extracurricular activities in mathematics, showed a positive relation to achievement in all countries, which may be an effect of self-selection, for example, that better students tend to take extra courses.

Certain test items could be considered to be crucial to the "New Mathematics". If, according to teachers' ratings, most students had an opportunity to deal with items of this category, this was taken as an indication of familiarity with the content of "New Mathematics". At all levels, the students having had "New Mathematics" had a significantly higher mean score on the total tests, even though the results at the preuniversity level were not entirely conclusive.

Mathematics Achievement and Societal Factors

Performance on the mathematics tests was consistently related to father's and mother's education and father's occupation. From a wealth of previous studies it is well known that school achievements are related to the student's socio-economic background, but so far very little is known about *how* consistently the pattern recurs in many *countries.* It could be anticipated that both the social status structure and the school structure of a country would determine the strength of the relationship.

All the participating countries are in the process of rapid urbanization. Therefore, it was considered to be of great interest to analyze to what extent the quality of education, as measured by the outcomes of instruction in mathematics, was related to urban-rural differences. It was hypothesized that the student's expectations and aspirations, with respect to both staying in school and to taking more mathematics, as well as his attitudes toward mathematics, would be related to his social background and that this relationship would vary between countries due to differences in school structure.

As it could be expected that the average level of performance, especially at the 13-year-old level, would be related to the financial support, certain measures were taken of operational costs and teachers' salaries.

Students' Social Background and Mathematics Performance

Although there is a wealth of research in individual countries, especially in the United States, which elucidates the relationship between parents' social status and children's scholastic successes, no comparative studies among countries have been conducted.

The mean level of the number of years of fathers' education ranged from 7.2 years in Finland to 11.4 in the United States in Population 1 b and from 8.7 years in Finland to 11.7 years in the United States for Population 3 a. On the average, parents of children at the preuniversity level have 1.7 more years of formal schooling, which indicates that social selection takes place. The differences varied from .5 years in the United States to 3.6 years in the Federal Republic of Germany. The same picture was obtained by comparing the representation of the three highest occupational groups (on the 9 category scale) at the 13-year-old and the preuniversity levels. In the United States the difference was only 10 percent as compared with 35 percent in England and 51 percent in the Federal Republic of Germany. Thus, graduates from the academic secondary school, that is, students eligible for the university, are in most European countries, and especially in those with competitive selection to the secondary academic school, to a large extent a social elite.

The total mathematics score at the 13-year-old level correlated .16 with fathers' education and .12 with mothers' education. The correlation between fathers' occupational status and the mathematics scores at the same level was .25 and .22, in Population 1 b. At the preuniversity level the correlation was much lower, as could be expected since the selection and elimination processes operating at the secondary level tend to homogenize the enrollment for social status. The correlations between parents' education and mathematics score vary considerably between countries and may be regarded as an indication of the relationship between social status and educational opportunity characteristic of the respective countries or of the degree to which cultural advantages at home are associated with parental occupational and/or educational status. The grade where the majority of the 13-year-olds are to be found (Population 1 b) is of special interest here since within each country we are dealing with pupils who have a relatively homogeneous school environment. Japan and England display the highest correlations between achievement and parents' education and father's occupational status, respectively. At the preuniversity level only one country, the United States, comes out with significant correlations (.32 and .24); part of the explanation is the wider range of social background and ability represented among United States high school students. But considering the size of the correlation at the 13-year level (.29), the explanation also might be that social background has a closer relationship to educational achievement in the United States than in several European countries (see Table 5.3). This interpretation is supported by the multiple regression analysis. The same picture was obtained by holding the school program (academic, general, vocational) constant. Over all countries the difference between the achievements of high-status and low-status students was less than one half of a standard deviation in the academic but as large about one standard deviation in the general program at the 13-year-old level. In England, the difference between highest and lowest status groups in the general program amounted to 1.5 standard deviations. In the academic program, only the United States and England showed significant differences in mathematics scores between the highest and lowest groups. In the general program, Japan and Australia came up with significant differences.

When the school program (academic, general, vocational) was held constant, the mean level of mathematics score was lower among students from lower socio-economic status. A large proportion of high occupational level students were found in the academic program, whereas low status students tended to be in the general or vocational program. The differences in achievement scores between occupational groups were larger in the general and vocational program than in the academic.

Differences in total scores between high status and low status students were related to the social class variability of the school enrollment. A consistent pattern emerged at both the 13-year-old and the preuniversity level, whereby social class differences in score were smaller in schools with a larger spread of the social background of the students. The data suggest that low-status students profit more from instruction in schools with larger spread in the socio-economic background of the students than they do in more homogeneous schools. Further research focusing on alternative explanations would have to be carried out in order to elucidate the enrollment mechanisms.

The motivational effect of the social background (measured by fathers' and mothers' education and fathers' occupational status, respectively) were assessed by correlating these measures with the number of years of additional schooling that the students expected. Fathers' education correlated .32 with the expectations in both the 13-year-old populations, whereas mothers' education correlated .21 and .29, respectively. The expectations correlated .46 with fathers' and mothers' education in Japan at the 13-year-old level and only .24 in Finland. Fathers' occupational status correlated .28. The interesting thing is that the coefficients in this case run as high as from .15 to .31 at the preuniversity level in spite of the much greater homogeneity of this group. The students were also asked about how many years of additional schooling they desired; this was used as a measure of their aspirations. On the whole, these correlations turned out to be of the same order of magnitude as for the expectations, but slightly lower than those between expected additional schooling and social background. The desire to study mathematics in the future also correlated relatively highly with both parents' education and fathers' occupation (scientific or nonscientific) in Japan.

The differences in total mathematics score with regard to place of parents' residence were analyzed for all four populations. Only in two countries, the United States and Japan, did the students from urban and town areas perform better than the rural students in all four populations. The all-countries average for rural students did not differ significantly from that of the town and urban students. In some countries the rural students tended to be superior to the other groups.

From previous research it was hypothesized that there would be a positive relationship between achievement and per student expenditure. However, although a positive relationship was found in some countries, particularly at the 13-year-old level, there appeared to be many other factors operating in the various countries which resulted in inconsistencies.

Sex Differences in Performance and Attitudes

A special aspect of the societal background was provided in connection with sex differences in achievement and interest. The problem of the extent to which sex differences in mathematics are determined by environmental factors and by inherited differences, respectively, cannot be studied unless one takes into account cross-national or cross-cultural variations.

One could expect that sex differences in the cognitive and noncognitive outcomes of mathematics instruction might be attributable to differences in role expectations between various cultural patterns as mediated through the school organization. It was found that the ratio of male and female students varied considerably between the 12 countries in both the preuniversity populations and in university enrollment. Striking differences appeared especially in the science-mathematics program. Whereas in Belgium, the Netherlands, France and England five to seven times as many male as female students are taking mathematics, the ratio is close to two to one in the United States, Finland, Japan, and Scotland. In the first university year there are on the average somewhat more than twice as many male as female students; the Netherlands leads with five times as many, whereas Finland has about equal numbers. Figures like these indicate disparities in educational opportunities, and since higher education is the gateway to professional life, one could expect these inequalities to be related to differentiation of the sexes in professional opportunities, or at least in their utilization.

Sex differences could be expected to be molded by the school organization. Therefore, the proportion of single-sex to coeducational schools was computed. In Population 1 b, where still one hundred percent of the age group is in full-time schooling, in most countries there were few single-sex schools. However, Belgium had five times as many single-sex as coeducational schools, France had twice as many, and the Netherlands and England had about equal proportions.

So far no large-scale cross-national studies of sex differences in mathematics achievements have been carried out, and even the findings from within-nation studies are often inconclusive. The general outcome of previous research, however, is a superiority of boys on mathematics tasks. It was hypothesized in the present study that there would be no total score differences but that the girls would slightly excell on verbal problems and the boys slightly on computational ones. We could make 42 comparisons and all of these except two, on total test score, came out in favor of the boys. In the two younger populations the difference was about one sixth of a standard deviation, and in the two preuniversity populations more than one third. The variability tended throughout to be greater among the boys. The mean sex difference in Population 1 a was highest in Belgium, Japan, and the Netherlands, and in Population 1 b, in the three countries mentioned plus England. The smallest differences throughout were found in the United States and Sweden; in these two countries the boys only slightly excelled the girls at the 13-year-old level.

In all the populations boys displayed more interest in mathematics than girls. This difference was statistically significant in Population 1 b for all the countries except Sweden; at this level, France had the largest difference. With the exception of England and France in the 3 a population the boys showed greater interest. There is an indication of a very marked selection in the two countries mentioned, since the girls represent only a very small proportion of the mathematics-science students in these countries.

One might ask to what extent these differences are due to role expectation as effected by the school structure. It should be noted that the differences are greater in countries with a high proportion of single-sex schools as compared to countries where the majority of students are taught in coeducational schools. On the whole, sex differences in scores are much larger in single-sex than in coeducational systems. It was found, however, that the sex differences in interest were not significant in the single-sex schools, whereas they were strongly significant in the coeducational systems. The most sensible interpretation of this at first sight surprising result is that in single-sex systems the girls are given the opportunity to take the kind of mathematics they want and that their self-image as "nonmathematicians" is not affected to the same extent as in coeducational schools. It should also be noted that more female teachers teach mathematics in single-sex than in coeducational schools in Population 1 b.

It was planned that both cognitive and noncognitive outcomes of mathematics instruction should be related to a variety of measures of the role of women in professional life in various countries. Such measures were obtained in the case study questionnaire but were found to be not sufficient for our purposes. Since sex differences both in the cognitive and noncognitive domain are cross-nationally established, and since it has been shown that these differences are related to the degree of coeducation that characterizes a system, it would seem to be fruitful to carry out further research analyzing more deeply the relationships between the sex roles in various societies and the place of mathematics in school education.

As already mentioned in the preface, this chapter has presented only some of the major results. It has been impossible to avoid certain overgeneralizations since the outcomes of hypothesis testing sometimes vary from country to country. For details, the reader is referred not only to Chapters 3 to 5, where the hypotheses analyses are reported, but also to Chapters 2 and 6, where the between-country correlational analysis and the multiple regression analyses are presented.

It will be realized from Volume I that many more data have been collected than have been analysed. Therefore, a data bank has been established and a manual for the use of the data bank has been prepared. Copies of this manual may be obtained by writing to the International Project for the Evaluation of Educational Achievement, UNESCO Institute for Education, 2000 Hamburg 13, Feldbrunnenstrasse 70, Federal Republic of Germany.

Further Research

The International Project for the Evaluation of Educational Achievement (IEA) emerged from the need to establish international criteria of evaluation when dealing with problems such as school failure and examinations on a comparative basis. It also was recognized from the outset that there were certain problems that could not be elucidated unless advantage was taken of international variation. Thus, the effects of, for example, selectivity and age of school entry, concomitants of sex differences, the relationship between factors such as teacher competence, curriculum, and per pupil expenditure on the one hand and "productivity" of the schools in terms of learning outcomes on the other could be tackled on a survey basis by utilizing data from different countries. The set of hypotheses advanced by the team of IEA researchers and tested in the first stage of the project reflects the conception the group had of the objectives of the project.

Objectives for the Next IEA Stage

But, as could be expected in a pioneering study, more problems were raised than solved—or fruitfully elucidated—by the mathematics stage of the study. It would seem from the experiences so far gained that two major improvements would have to be aimed at for the next stage.

(1) A conceptual model would have to be developed for cross-national research of this kind in comparative education. The major problems would be to identify the independent strategic variables which account for most of the cross-national variation and to bring these variables into a hierarchical relationship. From the mathematics study it would seem that the socio-economic background of the students has a much more powerful impact than most of the "pure" school variables such as number of hours or number of years of study, teacher competence, curriculum sequencing, etc. The educational system, therefore, would have to be seen much more closely in its social context than was the case when the mathematics study was designed. Education can be regarded both as an investment and as a consumption. It prepares the manpower needed in a given or anticipated economy, but it also prepares the growing human being to share the cultural heritage and to participate in developing it further.

Furthermore, the various school subjects could be expected to vary with regard to their relationships both to the society at large and the vocational requirements put upon its citizens. Among other things, a challenging task would be to analyze the curriculum and teaching of a given subject in relation to the varying structure of the economy in a group of countries.

Some members of the IEA group have therefore prepared a document spelling out the possibilities of developing a model for cross-national research dealing with educational problems related to national policies in general and vocational education. This conceptual model will be developed during two conferences in 1966 and 1967.

(2) It would seem that some of the problems dealt with in the mathematics phase will have to be tackled anew using data from subject areas other than mathematics. The outcomes of school learning in mathematics cannot be regarded as the sole criterion of the "effectiveness" of a school system.

Preliminary planning for the next IEA stage with a simultaneous study of four subject matter areas has already been undertaken. Eight small committees in 1964–65 prepared working papers dealing with the feasibility of evaluating outcomes (cognitive and noncognitive) in the following areas: science (physics, chemistry, biology), mother tongue (reading comprehension and literature), foreign languages (English and French as second languages), and civics. It would, in accordance with what was stated above, be necessary to develop for the next stage of the project more sophisticated techniques when dealing with social factors underlying educational achievements in the different countries.

Both types of objectives spelled out above would require a closer cooperation with comparative educationists, economists, and sociologists. The international "model building" conference is envisaged to be cross-disciplinary in order to enable the researchers to encompass the variables which would have to be taken into account when trying to explain cross-national educational differences and to elucidate the interaction between societal and educational factors.

Particular Problems to Be Considered

In the mathematics study, the concept of "educational yield" was developed when dealing with problems related to the "productivity" of the selective and comprehensive school system, respectively. The "yield" can be regarded as a function of the various factors constituting an educational system and therefore can be mathematically stated. In a very simple way, it can be expressed as "how many are brought how far" at the various levels (age or grade) in the school system. This would then in principle imply that samples of students at each *major terminal point* in the system were tested. Furthermore, the "yield" would have to be related to the time, money, and staff, etc. by means of which the students are brought to a certain level of competence. But in order to achieve what was envisaged above, that is, analyzing the relation between learning outcomes and the economy of a given country, one would have to be rather specific. The "yield" would have to be related to the manpower need in the present society, and also, for example, ten years hence. Research procedures would have to be devised which would relate manpower estimates and analyses of requirements to very detailed studies of the different parts (topics) of the curriculum.

Developing New Subject Matter
Attempts have been made in several countries to develop a new mathematics curriculum with the purpose of teaching mathematics more as a coherent discipline than has hitherto been the case. These attempts have been guided by the conviction that mathematics should be taught more by "insight" or "discovery" methods than by "drill" methods. Neither "drill" nor pure pragmatism in terms of teaching only the concepts and skills which the majority of the citizens in present-day society apply would suffice, since the requirements are rapidly changing and new applications would have to be made by a flexible utilization of the basic concepts and principles.

But in order to arrive at a curriculum with adequate topics and sequencing, analyses would have to be made of the requirements made in present-day society and projections of future developments.

If the outcomes of such analyses are related to the occupational structure of societies representing various stages of economic development, certain leads as to the construction of adequate curricula could be derived. So far very little has been done, even at the national level, to assess the long-range, out-of-school relevancy of school learning. If representative groups of adults from various occupations in several countries are tested, inferences about the relevancy could be drawn.

Case Studies of Particular Countries
In the mathematics stage of the IEA study, no detailed case studies of particular countries have been carried out. Such studies would have to imply thorough analyses of curricula, textbooks, and other learning materials. In some instances, special attention would have to be directed to incentives used in instruction, the relation between learning at school and at home, to what extent the parents are supporting the school, etc. Considering the high "yield" in mathematics in Japan, for example, it would seem to be especially fruitful to apply the case study method to this country.

Cognitive Styles

It is reasonable to expect that differences in the "cognitive styles", closely related to linguistic differences characterizing the various cultural patterns would affect school learning. It could be anticipated that, for example, the *explication des textes,* being by tradition a powerful method in teaching literature in France, would influence the learning outcomes in secondary school students in France. In one of the projected IEA studies dealing with the teaching of literature, selections of literature (for example, short stories) are given to the students with the request that they write a short essay expressing their immediate reactions. These essays can be subjected to content analysis according to a categorization scheme which has been tried out in an international pilot study. The variables in the content analyses could then be related to, among other things, the scores obtained by administering reasoning tests to the same students.

Appendix I

Stability Estimates

a) Factors by which the corresponding simple random sampling estimates should be multiplied to give the complex standard errors.
b) Complex standard errors for correlations.

	Populations							
	1 a		1 b		3 a		3 b	
Country	(a)	(b)	(a)	(b)	(a)	(b)	(a)	(b)
Australia	1.7	.03	1.7	.03	**2.0**	.06	—	—
Belgium	1.7	.04	2.0	.04	1.6	.07	1.9	.06
England	1.7	.03	1.7	.03	1.3	.04	1.3	.03
Finland	1.7	.05	1.8	.05	1.3	.06	1.3	.06
France	2.1	.04	3.1	.05	1.1	.06	—	—
Germany	—	—	**3.3**	.05	1.3	.05	1.0	0.4
Israel	—	—	1.8	.03	**0.9**	.07	—	—
Japan	**1.4**	.03	**1.4**	.03	1.4	.05	**2.0**	.03
The Netherlands	1.7	.08	1.9	.05	1.6	.07	—	—
Scotland	**2.9**	.04	3.1	.04	1.5	.04	1.8	.04
Sweden	2.3	.04	2.5	.04	1.6	.05	**0.9**	.05
United States	1.7	.02	1.7	.02	1.6	.04	1.8	.04
Mean	1.9	.04	2.1	.04	1.4	.06	1.5	.04

[1] In each of the factor columns (a) the highest and the lowest factor are in bold type.

Appendix II

Test Item	Target Population	Average Difficulty	Average Discrimination	Order of error frequency	Easier in	Harder in
Test no. A						
1. 43.0 − 17.6 is equal to ——. *Ans.* 25.4	1a	.84	.26		Finland, France, Sweden	England, Japan
	1b	.85	.25		Finland, France, Sweden	England, Japan
2. How many seven-man teams can you make out of 7 nine-man teams? A. 7 B. 8 *C. 9 D. 16 E. 63	1a	.77	.45	E, A	—	—
	1b	.77	.44	E, A	—	Israel
3. (22 × 18) − (47 + 59) is equal to *A. 290 B. 300 C. 384 D. 408 E. 502	1a	.85	.29	E, D	Finland	England
	1b	.87	.29	E, D	Finland	England
4. In the figure on the right the little squares are all the same size and the area of the whole rectangle is equal to 1. The area of the shaded part is equal to	1a	.64	.47	B, A	—	Finland
	1b	.66	.46	B, A	—	Finland

the right, rainfall in inches is plotted for 13 weeks. The average weekly rainfall during the period is approximately

A. 1 inch   D. 4 inches
*B. 2 inches E. 5 inches
C. 3 inches

	1b	.49	.40	C, A	Japan	Israel

6. The value of $2^3 \times 3^2$ is

A. 30 B. 36 C. 64 *D. 72 E. none of these

| | 1a | .43 | .46 | E, B | Belgium, the Netherlands, Japan | Finland, Sweden |
| | 1b | .46 | .43 | E, B | Belgium, the Netherlands, Japan | Finland, Sweden |

7. A box has a volume of 100 cc. Another box is twice as long, twice as wide and twice as high. How many cc is the volume of the second box?   *Ans. 800*

| | 1a | .21 | .39 | | — | Australia, Sweden, United States |
| | 1b | .24 | .40 | | Belgium, Israel | Australia, Sweden, United States |

8. There is a brass plate of the shape and dimensions shown in the adjoining figure. What is its area in square inches?

A. 16 *B. 24 C. 32 D. 64 E. 96

| | 1a | .42 | .37 | C, D | France, Japan | Scotland United States |
| | 1b | .44 | .37 | C, D | France, Israel, Japan | The Netherlands, United States |

Appendix II (cont.)

Test Item	Target Population	Average Difficulty	Average Discrimination	Order of error frequency	Easier in	Harder in
9. What is the square root of 12 × 75?						
A. 6.25 *B. 30 C. 87 D. 625 E. 900	1a	.27	.54	E, A	Australia, Belgium, England	Finland, Japan, Sweden
	1b	.30	.51	E, A	Australia, Belgium, England, Israel, Scotland	Finland, Germany, Japan, the Netherlands, Sweden
10. Three straight lines intersect as shown in the figure on the right. What is x equal to in degrees?	1a	.20	.34	E [A, B, C]	England	—
A. 30 B. 50 C. 60 *D. 110 E. 150	1b	.22	.31	E, C	England, Japan	United States
11. A shopkeeper has x lb of tea in stock. He sells 15 lb. and then receives a new lot weighing $2y$ lb. What weight of tea does he now have?	1a	.70	.49	E, B	Japan	—
A. $x - 15 - 2y$ D. $x + 15 - 2y$	1b	.72	.47	E, B	Australia, Japan	Germany

314

A. $x < \frac{7}{2}$ B. $x < 5$ *C. $x < 14$ D. $x > 5$ E. $x > 14$ | 1b | .34 | .28 | A, E | Sweden | Germany

13. A piece of tin with dimensions as shown is to be folded along the dotted lines to make a box. What is the volume, in cubic centimeters, enclosed in the box? *Ans.* 60 | 1a | .38 | .52 | | — | Australia, Scotland, United States

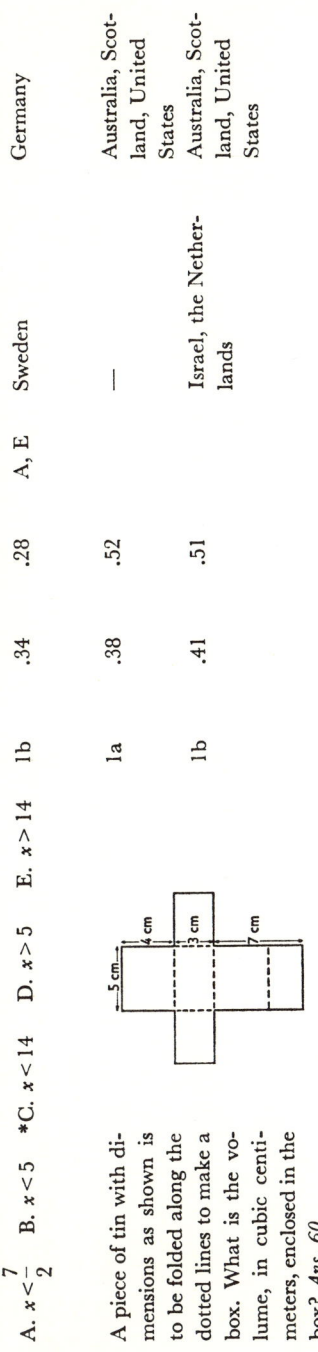

| 1b | .41 | .51 | | Israel, the Netherlands | Australia, Scotland, United States

14. If $\frac{4x}{12} = 0$, then x is equal to

*A. 0 B. 3 C. 8 D. 12 E. 16 | 1a | .31 | .32 | B, C | Australia, United States | —

| 1b | .30 | .29 | B, C | United States | Germany

15. The floor of a room is covered with wooden rectangular blocks. When blocks measuring a inches by b inches are used, M blocks are needed. If blocks fit exactly, how many blocks will be needed if each block measures x inches by y inches?

*A. $\frac{Mab}{xy}$ B. $\frac{ab}{Mxy}$ C. $\frac{(a+b)M}{x+y}$ D. $\frac{ab \cdot xy}{M}$ E. $\frac{Mxy}{ab}$ | 1a | .27 | .37 | C, D | — | —

| 1b | .29 | .37 | C, D | Belgium | Germany, the Netherlands

315

Appendix II (cont.)

Test Item	Target Population	Average Difficulty	Average Discrimination	Order of error frequency	Easier in	Harder in
16. Which of the following sets of conditions is *not* sufficient for the congruence of $\triangle FGH$ and $\triangle PQR$ on the right when f is less than g?	1a	.19	.30	D, B	Belgium, France	—
	1b	.20	.30	D, B	Belgium, France	—
	3b	.58	.40	D, B	Belgium	Germany

*A. $\angle F = \angle P$
  $g = q$
  $f = p$
B. $\angle F = \angle P$
  $h = r$
  $\angle G = \angle Q$
C. $g = q$
  $\angle F = \angle P$
  $h = r$
D. $h = r$
  $g = q$
  $f = p$
E. $f = p$
  $\angle G = \angle Q$
  $h = r$

316

I. $(53 \times 73) \times 17 = 53 \times (73 \times 17)$ II. $133 \times (78 + 89) = (133 \times 78) + 89$ III. $133 \times (78 + 89) = (133 \times 78) + (133 \times 89)$ A. I only D. I and II only B. II only *E. I and III only C. III only	1b 3b	.23 .57	.34 .34	A, C A, C	— France	United States

18. There are 227 boys in a school. Every boy in the school belongs to either the music club or the sports club, and some boys belong to both clubs. The music club has 120 members, and 36 of these are also members of the sports club. What is the total membership of the sports club? *Ans. 143*

	1a 1b 3b	.66 .67 .86	.38 .37 .23		— — Sweden	Japan Japan France, Japan

19. The lengths of the sides of a triangle XYZ are 4, 7, and 10. If a similar triangle has a perimeter of 147, what is the length of its shortest side? *Ans. 28*

	1a 1b 3b	.17 .18 .69	.51 .46 .48		Israel, Japan Israel, Japan Sweden	Finland, Sweden Finland, Sweden United States

20. In the solution of the following system of equations,
$$\left.\begin{array}{l}2x+y=7 \\ x-4y=4\end{array}\right\}$$
the value of y is equal to

A. $-\dfrac{5}{3}$ B. -9 C. $\dfrac{1}{9}$ *D. $-\dfrac{1}{9}$ E. $\dfrac{5}{3}$

	1a 1b 3b	.10 .10 .64	.22 .17 .40	B, E B, E E, C	England, Scotland England, Scotland —	Finland, Germany Finland, Germany Sweden

21. Which of the following is true for any parallelogram $ABCD$ which has an acute angle at B and diagonals AC and BD?

A. $AB < BC$ *D. $AC < BD$
B. $AB = BC$ E. None of them
C. $AB > BC$

	1a 1b 3b	.22 .19 .31	.08 .08 .27	E, B E, B E, C	Israel Sweden —	— Germany Belgium

317

Appendix II (cont.)

Test Item	Target Population	Average Difficulty	Average Discrimination	Order of error frequency	Easier in	Harder in
22. The distance between two towns, A and B, is 150 kilometers. This distance is represented on a certain map by a length of 30 centimeters. The scale of this map is	1a	.26	.33	B, D	France	Australia, England, Scotland, United States
*A. 1/500,000 D. 1/5,000 B. 30/150 E. 1/200,000 C. 1/20,000						
23. Which of the following equals $7 \times (3+9)$?	1b	.26	.32	B, D	France, Israel, the Netherlands	Australia, England, Scotland, United States
*A. $(7\times3)+(7\times9)$ D. 7×27 B. $(7\times9)+(3\times9)$ E. $21+9$ C. $(7\times3)+(3\times9)$	1a	.63	.44	D, E	—	—
	1b	.66	.40	D, E	—	—

Test no. B

Test Item	Target Population	Average Difficulty	Average Discrimination	Order of error frequency	Easier in	Harder in
1. $\frac{2}{5}+\frac{3}{8}$ is equal to	1a	.72	.44	A, B	—	—
A. $\frac{5}{13}$ B. $\frac{5}{40}$ C. $\frac{6}{40}$ D. $\frac{16}{15}$ *E. $\frac{31}{40}$	1b	.74	.42	A, B	The Netherlands	England
2. Peter and Paul decided to start saving money. Peter can save 3 dollars each month and Paul can save 5 dollars. At this rate, after how many months will Paul have exactly 10 dollars more than Peter?	1a	.81	.42	A, C	—	—
	1b	.81	.42	A, C	The Netherlands	—

3. In the division on the right, the correct answer is $.004\overline{)24.56}$ A. .614 B. 6.14 C. 61.4 D. 614 *E. 6140	1a	.58	.43	B, D	Finland, France, the Netherlands	Australia, England, Scotland
	1b	.59	.42	B, D	Finland, France, the Netherlands, United States	Australia, England, Scotland
4. The arithmetic mean (average) of: 1.50, 2.40, 3.75 is equal to A. 2.40 *B. 2.55 C. 3.75 D. 7.65 E. none of these	1a	.60	.46	D, E	Finland	—
	1b	.62	.44	D, E	Finland	—

5. Which of the following operations with whole numbers will *always* give a whole number?
  I. Addition
  II. Multiplication
  III. Division
A. I only *D. I and II only
B. II only E. II and III only
C. III only
	1a	.61	.48	A, B	Finland	Japan

	1b	.62	.46	A, B	Belgium, Finland, Germany	Japan
6. If the Selling Price of an article was $55 and a profit of 10% was made on the Cost Price, what was the cost price in dollars? *Ans. 50*	1a	.19	.38		The Netherlands	Sweden, United States
	1b	.20	.35		Israel, the Netherlands	United States
7. The value of 0.2131×0.02958 is approximately A. 0.6 B. 0.06 *C. 0.006 D. 0.0006 E. 0.00006	1a	.38	.36	E, B	—	Australia, Scotland
	1b	.40	.35	E, B	Germany	Australia, Scotland

319

Appendix II (cont.)

Test Item	Target Population	Average Difficulty	Average Discrimination	Order of error frequency	Easier in	Harder in
8. Joe had three test scores of 78, 76, and 74, while Mary had scores of 72, 82, and 74. How did Joe's average compare with Mary's? A. Joe's was 1 point higher. B. Joe's was 1 point lower. *C. Both averages were the same. D. Joe's was 2 points higher. E. Joe's was 2 points lower.	1a 1b	.88 .88	.25 .25	E, D E, D	— —	— —
9. Which of the following is *false* when a and b are different real numbers: A. $(a+b)+c = a+(b+c)$ B. $ab = ba$ C. $a+b = b+a$ D. $(ab)c = a(bc)$ *E. $a-b = b-a$	1a 1b	.38 .40	.51 .50	D, B D, B	Sweden —	— Germany
10. If $P=LW$ and if $P=12$ and $L=3$, then W is equal to A. $\frac{3}{—}$ B. 3 *C. 4 D. 12 E. 36	1a 1b	.60 .63	.52 .50	E, A E, A	— —	France France, Germany

A. $7x+7y$ *D. $7x-y$
B. $8x-2y$ E. $7x+y$
C. $6xy$

1b	.41	.50	A, C	England, Scotland	Sweden
				Australia, Belgium	Finland, Japan,
				England, Scotland	Sweden

12. What is the value of $(-6)-(-8)$?
Ans. $+2$

1a	.37	.36		Finland, Japan	—
1b	.39	.36		Finland, Japan	Israel

13. If AB is a straight line, what is the measure in degrees of angle BCD in the figure on the right?

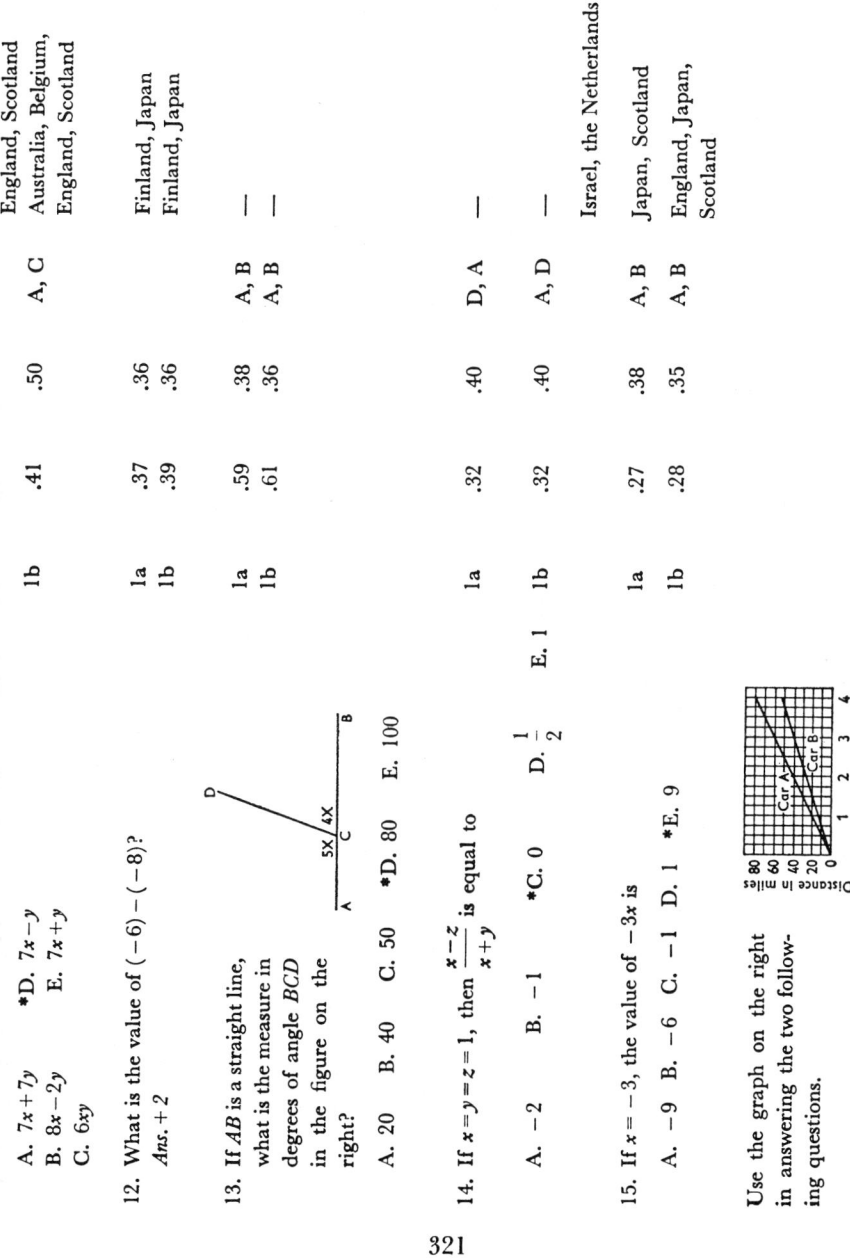

A. 20   B. 40   C. 50   *D. 80   E. 100

1a	.59	.38	A, B	—	The Netherlands
1b	.61	.36	A, B	—	The Netherlands

14. If $x=y=z=1$, then $\dfrac{x-z}{x+y}$ is equal to

A. -2 B. -1 *C. 0   D. $\dfrac{1}{2}$ E. 1

1a	.32	.40	D, A	—	—
1b	.32	.40	A, D	—	—
				Israel, the Netherlands	United States

15. If $x=-3$, the value of $-3x$ is

A. -9 B. -6 C. -1 D. 1   *E. 9

1a	.27	.38	A, B	Japan, Scotland	—
1b	.28	.35	A, B	England, Japan,	Israel
				Scotland	

Use the graph on the right in answering the two following questions.

321

Appendix II (cont.)

Test Item	Target Population	Average Difficulty	Average Discrimination	Order of error frequency	Easier in	Harder in
16. Three hours after starting, car A is how many miles ahead of car B? A. 2 B. 10 C. 15 *D. 20 E. 25	1a 1b 3b	.55 .57 .90	.49 .47 .25	E, A E, A E, C	Japan Japan Sweden	— — United States
17. How much longer does it take car B to go 50 miles than it does for car A to go 50 miles? *A. 1 hour 15 minutes D. 2 hours 30 minutes B. 1 hour 30 minutes E. 2 hours 35 minutes C. 2 hours	1a 1b 3b	.47 .48 .86	.47 .45 .30	B, C B, C B, D	England England Sweden	Belgium, France Belgium, France Belgium, France, United States
18. In $\triangle KLM$ on the right, $KL = KM$, $PO \perp LM$, and LP is a straight line. Then $\triangle NKP$ is isosceles because *A. $\angle P = \angle KJNP$, since both are complements of the equal angles L and M. B. $NK = PK$, since $\angle P = \angle M$. C. its sides are parallel to the	1a 1b 3b	.18 .18 .51	.23 .21 .32	B, E B, E E, B	— — —	Finland Finland, Israel, the Netherlands —

E. ∠P = ∠KNP, since both are half the supplement of angle M.

19. The distance between two schools on a map with a scale of 1 : 10,000, is 20 cm. What is the actual distance in kilometers between the two schools? *Ans.* 2

1a	.26	.36	Belgium, Finland, France, the Netherlands	Australia, England, Scotland, United States
1b	.27	.38	Belgium, Finland, France, Germany, Israel, the Netherlands	Australia, England, Scotland, United States
3b	.56	.15	Belgium, Finland, France, Sweden	The Netherlands, Scotland, United States

20. The equation of the line shown in the graph is

A. $x + 4y = 4$
B. $2x - y = 4$
C. $2x = y - 2$
*D. $x - 4y + 2 = 0$
E. $4x - y = 2$

1a	.10	.04	C, B	United States	Germany
1b	.11	.03	C, B	Sweden, United States	Belgium, Germany
3b	.46	.43	C, E	France	Sweden

21. Which of the following numbers in base two is (are) even?

I. 110011
II. 110010
III. 110101
IV. 100100

A. I only
B. III only
C. I and III only
*D. II and IV only
E. I, III, and IV

1a	.62	.38	C, E	Finland	—
1b	.63	.36	C, E	Finland, Germany	Israel
3b	.70	.16	C, E	Germany	France, the Netherlands

323

Appendix II (cont.)

Test Item	Target Population	Average Difficulty	Average Discrimination	Order of error frequency	Easier in	Harder in
22. The expression $\dfrac{a}{b-c} + \dfrac{a}{c-b}$ where $a \neq 0$, is equal to *A. 0 D. $\dfrac{a}{2b}$ B. $\dfrac{2a}{b-c}$ E. $2a$ C. $\dfrac{a}{b^2 - c^2}$	1a 1b 3b	.19 .19 .60	.20 .19 .36	B, C B, C E, C	Australia, Finland Australia, England, Finland —	Belgium Belgium, Germany, Israel, the Netherlands —
23. Soda costs a cents for each bottle but there is a refund of b cents on each empty bottle. How much will Henry have to pay for x bottles if he brings back y empties? A. $ax + by$ D. $(a+x) - (b+y)$ *B. $ax - by$ E. none of these C. $(a-b)x$	1a 1b	.38 .39	.49 .47	E, D E, D	— —	— —
24. From a long stick of wood a man cut 6 short sticks each 2 feet long. He then found he had 1 foot left over. Which of the following would tell him the length of the original stick of wood? A. $6 \times (2+1)$ D. $(6 \times 2) - 1$ *B. $(6 \times 2) + 1$ E. $(6 \div 2) + 1$	1a 1b	.75 .76	.44 .42	A, D A, D	Finland —	Japan Israel, Japan

Question						
1. Which of the following is the same as a quarter of a million? A. 25,250 B. 40,000 C. $\frac{1}{4,000,000}$ *D. 250,000 E. 2,500,000	1a	.80	.38	E, C	—	Japan, Sweden
	1b	.81	.39	E, C	Germany	Japan, Sweden, United States
2. 0.40 × 6.38 is equal to A. .2552 B. 2.452 *C. 2.552 D. 24.52 E. 25.52	1a	.70	.29	A, E	Finland	Australia, England
	1b	.71	.26	[E, A]	The Netherlands	Australia, England
3. The sum of $9\frac{4}{5}$ and $13\frac{1}{4}$ is equal to A. $22\frac{5}{9}$ B. $22\frac{9}{20}$ C. 23 *D. $23\frac{1}{20}$ E. $23\frac{1}{5}$	1a	.67	.46	A, B	The Netherlands, United States	France
	1b	.69	.46	A, B	The Netherlands, United States	France
4. The ratio of 2 to 5 equals the ratio of what number to 100? *Ans. 40*	1a	.39	.55		—	France
	1b	.42	.53		Germany, United States	France, Sweden
5. In a given triangle the measures of two angles in degrees are 60 and 70. What is the measure of the third angle in degrees? *Ans. 50*	1a	.64	.50		—	Finland, the Netherlands, United States
	1b	.66	.44		Australia, Germany, Israel	Finland, the Netherlands
6. On level ground, a boy 5 feet tall cast a shadow 3 feet long. At the same time a nearby telephone pole 45 feet high casts a shadow the length of which, in feet, is A. 24 *B. 27 C. 30 D. 60 E. 75	1a	.68	.45	C, A	The Netherlands	Finland
	1b	.69	.44	C, A	Germany, the Netherlands	Finland

Appendix II (cont.)

Test Item	Target Population	Average Difficulty	Average Discrimination	Order of error frequency	Easier in	Harder in
7. A runner ran 3,000 meters in exactly 8 minutes. What was his average speed, in meters per second? A. 3.75 *B. 6.25 C. 16.0 D. 37.5 E. 62.5	1a 1b	.50 .51	.49 .48	A, D A, D	— —	— —
8. On the scale to the right, the reading indicated by the arrow is between A. 51 and 52 D. 62 and 64 B. 57 and 58 *E. 64 and 66 C. 60 and 62	1a 1b	.42 .44	.45 .44	B, D B, D	England, Japan England, Japan	Belgium, France Belgium, France, Israel
9. If $x+y=4$ and $x-y=2$, then x is equal to A. 0 B. 1 C. 2 *D. 3 E. 6	1a 1b	.41 .42	.51 .50	C, E C, E	— —	France —
10. One bell rings every 8 minutes, while another bell rings every 12 minutes. They have rung together once at the same moment. After how many minutes will they ring together again A. for the first time? *Ans.* 24	1a 1b	.19 .20	.43 .41		— Germany	France France

hand and the hour hand of a clock in, degrees, is	1b	.58	.44	C, B		Germany, Japan		The Netherlands
A. 30 B. 45 C. 60 D. 90 *E. 120								
12. Any two regular polygons with the same number of sides are	1a	.33	.16	E, A	Sweden	Germany, Japan		Japan
	1b	.33	.15	E, A	Sweden			Germany, Japan
A. congruent D. not similar								
B. noncongruent E. equal in area								
*C. similar								

Questions 13–15.

Imagine that the geometrical figures K, L, M, N and O have been drawn on a rubber sheet. The lines are assumed to have no width. The rubber sheet is stretched parallel to the X-axis while leaving all the distances measured parallel to the Y-axis unchanged. The stretching is uniform, that is, the same for every part of the sheet.

13. For which of the segments K, L, M will the length remain unchanged?	1a	.24	.23	[E, B]	Sweden			Japan
	1b	.24	.24	E, B	Sweden			Japan
*A. only K B. only L C. only M D. K and L								
E. K and M								

327

Appendix II (cont.)

Test Item	Target Population	Average Difficulty	Average Discrimination	Order of error frequency	Easier in	Harder in
14. What will happen to the measure of angle θ of triangle N?						

A. It will remain the same.
 *B. It will become larger.
   C. It will become smaller.
   D. One cannot tell from the data whether A, B, or C is correct.
	1a	.38	.23	A, D	Sweden, United States	Japan

	1b	.38	.22	A, D	Sweden, United States	Japan
15. What will happen to circle O?						
A. It will still be a circle. *B. It will no longer be a circle. C. One cannot tell from the data whether A or B is correct.	1a	.55	.40	A, C	—	Japan
	1b	.55	.40	A, C	—	Japan
16. A factory produces m units per week. How many units per week will it produce after production is increased p percent? A. $100p + m$ C. $\dfrac{m + mp}{100}$ E. $\dfrac{p}{100} + m$ B. $100m + mp$ *D. $m + \dfrac{mp}{100}$	1a	.17	.27	E, A	—	England
	1b	.17	.25	E, A	Finland	—
	3b	.51	.32	E, C	Finland, Sweden	United States

and b. For example, $\overline{3,7}$ cosists of the integers 4, 5, and 6. Which of the following pairs of sets has a larger number of integers in common than any of the other pairs?

A. $\overline{0,15}$ and $\overline{7,20}$ *D. $\overline{4,18}$ and $\overline{8,20}$

B. $\overline{5,15}$ and $\overline{16,30}$ E. $\overline{0,12}$ and $\overline{6,12}$

C. $\overline{5,14}$ and $\overline{5,17}$

18. What are all values of x for which the inequality

$$5x + \frac{5}{3} \leq -2x - \frac{2}{3}$$

is true?

A. $x \leq -\frac{7}{9}$ C. $x > 0$ E. $x \geq \frac{9}{3}$

*B. $x \leq -\frac{1}{3}$ D. $x \geq \frac{7}{3}$

19. The symbol $P \cap Q$ represents the intersection of sets P and Q and the symbol $P \cup Q$ represents the union of sets P and Q. Which of the following represents the shaded portion of the diagram below?

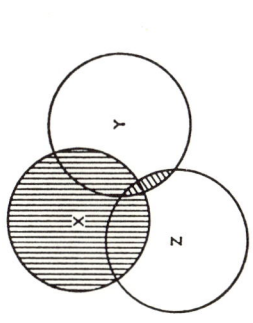

A. $(X \cap Y) \cup Z$ C. $X \cap (Y \cup Z)$ E. $(X \cup Y) \cap Z$

*B. $X \cup (Y \cap Z)$ D. $(X \cap Y) \cap Z$

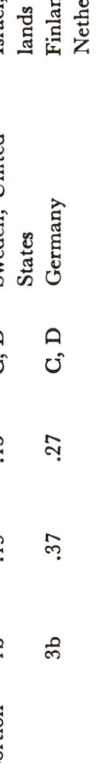

1b	.13	.08	B, A	Finland, Japan, Sweden	Germany, Israel
3b	.24	.25	A, B	England, Finland, Japan, Scotland, United States	France, Germany, the Netherlands
1a	.18	.17	D, C	United States	—
1b	.18	.13	C, D	United States	Germany, the Netherlands
3b	.62	.31	A, C	France, Japan, the Netherlands	England, Germany, Sweden
1a	.18	.17	C, D	Sweden, United States	—
1b	.19	.15	C, D	Sweden, United States	Israel, the Netherlands
3b	.37	.27	C, D	Germany	Finland, the Netherlands

Appendix II (cont.)

Test Item	Target Population	Average Difficulty	Average Discrimination	Order of error frequency	Easier in	Harder in
20. If, in the given figure, PQ and RS are intersecting straight lines, then $x+y$ is equal to A. 15 B. 30 *C. 60 D. 180 E. 300	1a 1b 3b	.41 .43 .83	.42 .38 .22	B, D B, D B, D	— — The Netherlands	Israel United States —
21. Each of 9 boys had t marbles. In order to play a game, they divided the marbles among 12 boys in such a way that each had the same number. How many marbles did each of the 12 have? *A. $\frac{3t}{4}$ B. $t-3$ C. $\frac{4t}{3}$ D. $9t-12$ E. $12t-9$	1a 1b 3b	.31 .33 .80	.53 .47 .38	D, E D, E C, D	— — Scotland	— — Belgium, United States
22. The length of the circumference of the circle on the right with center at O is 24 and the length of arc RS is 4. What is the measure in degrees of the central angle ROS? A. 24 B. 30 C. 45 *D. 60 E. 90	1a 1b 3b	.48 .49 .84	.47 .45 .33	C, B C, B B, C	Japan Germany, Japan Sweden	— The Netherlands United States
23. Given any fraction whose numerator is less than the de-	1a	.37	.28	A, D	Australia	Belgium, Japan

*B. larger than the original fraction
C. twice the original fraction
D. smaller than the original fraction
E. 1 more than the original fraction

Test no. 5

1. If $a = 20$, $b = 0$, $c = 10$, $x = 8$, $y = 12$, then the value of $2aby + 2cx$ is
   A. 100 *B. 160 C. 400 D. 640 E. none of these

	3a	.94	.19	E, D	—	United States
	3b	.71	.40	D, E	Germany, the Netherlands	Scotland, Sweden, United States

2–5. For each of the following equations or pairs of equations, concerned with real numbers, mark on the answer sheet

A. if there is *no* solution
B. if there is *one* solution
C. if there are *two* solutions
D. if there are *three* solutions
E. if there are *more than three* solutions

2. $x + y = 12$, $x - y = 4$
   Ans. B

	3a	.71	.20	C, A	Finland, the Netherlands, Sweden	England, Japan
	3b	.58	.28	C, A	The Netherlands	Japan

3. $m + n = 2$, $3m + 3n = 9$
   Ans. A

	3a	.76	.21	B, C	Finland, France, the Netherlands	Germany, Israel, Japan
	3b	.62	.35	B, C	The Netherlands	Japan

4. $x^2 - 5x + 6 = 0$
   Ans. C

	3a	.94	.13	B, A	Scotland	United States
	3b	.71	.39	B, A	Germany, the Netherlands	Sweden, United States

Appendix II (cont.)

Test Item	Target Population	Average Difficulty	Average Discrimination	Order of error frequency	Easier in	Harder in
5. $3p + q = 16$	3a	.61	.36	A, B	Belgium, France, the Netherlands	Germany, Scotland, Sweden
Ans. E	3b	.32	.29	A, B	Belgium, France, the Netherlands	Germany, Scotland, Sweden
6. If $xy = 1$ and x is greater than 0, which of the following statements is true? A. When x is greater than 1, y is negative. B. When x is greater than 1, y is greater than 1. C. When x is less than one, y is less than 1. D. As x increases, y increases. *E. As x increases, y decreases.	3a 3b	.93 .69	.24 .43	C, D C, A	The Netherlands The Netherlands	Japan Japan, Sweden
7. In the figure on the right, $KX = \frac{1}{3} KL$ and $KY = \frac{1}{3} KM$. Which of the following statements are true? I. $XY = \frac{1}{3} LM$ II. Line XY is parallel to	3a 3b	.74 .58	.31 .35	A, D A, D	Finland, Japan, Scotland —	Israel —

IV. Area $KXY = \frac{1}{9}$ area KLM

A. I and II only
B. II and III only
C. I and III only
D. I, II and III only
*E. I, II, and IV only

8. In the figure on the right, m represents a plane, and PQ is a line segment which is perpendicular to the plane at the point Q. Points A, B and C lie on the plane. If $QA = QB = QC$, then the triangles PQA, PQB, and PQC are

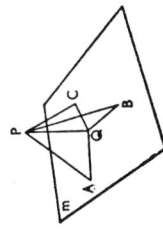

*A. congruent by SAS
B. congruent by SSA
C. congruent by ASA
D. similar but not congruent
E. neither similar nor congruent

3a	.88	.21	B, C	Australia, Finland, the Netherlands, Scotland	France, Germany, Sweden
3b	.74	.41	B, C	Finland	Germany

333

Appendix II (cont.)

Test Item	Target Population	Average Difficulty	Average Discrimination	Order of error frequency	Easier in	Harder in
9. In the figure below, $PQ \perp OQ$, and $RS \perp OQ$. If the measure of OQ and of OR equal 1 and θ is the measure of $\angle POQ$, then the measure of segment PQ is equal to A. $\sin \theta$ B. $\cos \theta$ *C. $\tan \theta$ D. $2 \sin \theta$ E. $1 - \cos \theta$	3a 3b	.76 .43	.33 .34	A, E A, B	France Belgium, Germany	United States The Netherlands, Sweden
Questions 10 and 11 are based upon the graph of a quadratic equation which is shown in the figure on the right						
10. For what value of x is the quadratic function a minimum? A. -1 B. $-1/2$ *C. $1/2$ D. 1 E. $1\frac{1}{2}$	3a 3b	.84 .46	.27 .38	A, B A, B	Finland, Germany, the Netherlands France, Germany, the Netherlands	Japan, United States England, Scotland, Sweden, United

						United States
*A. $-1 < x < 1$ D. $x > 0$ B. $x < -1$ and $x > 1$ E. $x > y$ C. $-3/4 < x < 1\frac{1}{4}$	3b	.29	.34	B, C	The Netherlands	England, Sweden
12. A square plate of the largest possible size is cut from a circular plate of 16 cm diameter. The area of the square plate (in sq cm) will be A. 64 B. 96 *C. 128 D. 192 E. 256	3a 3b	.69 .40	.29 .37	A, E A, E	Sweden Sweden	— —
13. The locus of all mid points of chords drawn from a point of a circle is A. a semicircle D. an oblong *B. a circle E. none of these C. a straight line	3a 3b	.59 .39	.25 .26	E, D A, E	Germany, Finland —	Israel —
14. A piece of wire 52 inches long is cut into two parts and each part is bent to form a square. The total area of the two squares is 97 square inches. What is the length in inches of the side of the smaller square? *Ans. 4*	3a 3b	.67 .43	.32 .37		Israel Belgium, Germany	— Finland
15. The complex number $(1+i)^2$ is equal to	3a	.58	.37	E, A	Belgium, Finland, France, Japan, United States	Australia, Israel, the Netherlands, Scotland
A. 0 B. 2 *C. $2i$ D. $1+i$ E. $2+2i$	3b	.20	.19	E, D	Finland, Japan, United States	England, France, the Netherlands, Scotland

Appendix II (cont.)

Test Item	Target Population	Average Difficulty	Average Discrimination	Order of error frequency	Easier in	Harder in
16. Given $\log_b 2 = \frac{1}{3}$, $\log_b 32$ is equal to A. 2 B. 5 C. $-\frac{3}{5}$ *D. $\frac{5}{3}$ E. $\frac{3}{\log_2 32}$	3a	.71	.41	E, B	Israel, the Netherlands, Scotland	France, Germany, Sweden
	3b	.27	.28	E, B	Japan, the Netherlands	Germany
17. Below there are several definitions of new operations named * in terms of the usual operations on real numbers. For which of the definitions is the property $y * x = x * y$ valid for all positive real numbers x and y? A. $x * y = \frac{x}{y}$ *D. $x * y = \frac{xy}{x+y}$ B. $x * y = x - y$ C. $x * y = x(x+y)$ E. $x * y = x^2 + xy^2 + y^4$	3a	.33	.43	A, C	France, United States	Israel, the Netherlands, Sweden
	3b	.12	.18	A, C	United States	Finland
18. Solve the equation $\sqrt{x+5} - \sqrt{x-3} = \sqrt{x}$ Ans. 4	3a	.38	.43		—	Germany
	3b	.10	.22		United States, France	Finland
19. The graph on the right is the representation of one of the following equations. Which one does it represent? A. $y = (1-x)(x-2)$ D. $y = (1-x)^2(x-2)$ B. $y = (1-x)(2-x)$ *E. $y = (1-x)^2(2-x)$ C. $y = (1-x)^2(2-x)^2$	3a	.48	.42	B, D	—	—
	3b	.11	.23	B, C	Germany, Japan	Finland

336

20. The expression $\dfrac{2}{\sqrt{5}} + \dfrac{\sqrt{45}}{5} + \dfrac{1}{\sqrt{5}-2}$ is equal to *A. $2\sqrt{5}+2$ D. $2\sqrt{5}$ B. $2\sqrt{5}-2$ E. $2-2\sqrt{5}$ C. 2	3a 3b	.60 .18	.41 .21	B, E B, D	Australia, Scotland Japan, United States	The Netherlands England, Germany, the Netherlands
21. Chords of the same length are drawn in two circles of unequal radii. Which of the following is true? *A. The chord in the larger circle could be equal to the radius of the smaller circle. B. The chord in the smaller circle could not be a diameter. C. The distance from the center to the chord is less in the larger circle. D. The minor arc intercepted on the larger circle is longer. E. The minor arc intercepted on the larger circle subtends the greater angle at the center.	3a 3b	.69 .44	.33 .30	B, D B, D	— —	— —

Test no. 6

1. On the outside of two sides of a rectangular plot 25 by 20 meters there is a path of uniform width. The area covered by the path equals one half the area of the plot. What is the width of the path in meters? *Ans.* 5	3b	.33	.31		Germany, Sweden	United States

337

Appendix II (cont.)

Test Item	Target Population	Average Difficulty	Average Discrimination	Order of error frequency	Easier in	Harder in
2. According to one plan for travelling to Mars, the round trip would take almost exactly three years, including a stay on Mars of 449 earth days. If one must go 34,000,000 miles each way on the trip, which of the following can be used to give an estimate of the average speed of travel in miles per hour? A. $\dfrac{(3 \times 365 - 449) \times 24}{34,000,000}$ B. $\dfrac{(3 \times 365 - 449) \times 24}{34,000,000 \times 2}$ C. $\dfrac{34,000,000}{(3 \times 365 - 449) \times 24}$ D. $\dfrac{34,000,000 \times 24}{2 \times (3 \times 365 - 449)}$ *E. $\dfrac{2 \times 34,000,000}{(3 \times 365 - 449) \times 24}$	3b	.50	.28	C, B	—	—
3. Of three wires, each 36 in. long, one is bent into a square, another into a rectangle with length and width in the ratio of 2:1 and the third into an equilateral triangle. Which one of the following statements describes the correct relationship between the enclosed areas? *A. The area of the square is the greatest and that of the triangle is the least. B. The area of the rectangle is the greatest and that of	3b	.48	.37	B, E	Sweden	France

338

square is the least.
D. The area of the triangle is the greatest and that of the rectangle is the least.
E. The areas of the square and the rectangle are the same, but the area of the triangle is less than that of the square or the rectangle.

4. If $3^{x+y}=81$ and $25^{x/2}=5$, then y is

A. 0   B. 2   *C. 3   D. $\frac{7}{2}$ E. $\frac{15}{4}$

3b   .40   .36   D, B   The Netherlands, United States   France

5. A certain number of students are to be accommodated in a hostel. If 2 students share each room, then 2 students will be left without any room. If 3 students share each room, then 2 rooms will be left unoccupied. How many rooms are there in the hostel?   *Ans. 8*

3b   .37   .36   Belgium, Germany   Finland, the Netherlands

6. Four persons whose names begin with different letters are placed in a row, side by side. What is the probability that they will be placed in alphabetical order from left to right?

A. 1/120   *B. 1/24   C. 1/12   D. 1/6   E. 1/4

3b   .37   .28   E, C   Sweden   Japan

7. A number is the multiplicative inverse of another number if the product of the two numbers is 1. Which of the following sets of numbers is identical to the set of its multiplicative inverses?

A. {1, 2, 3}   D. {2, 3, 5, ½, ⅓}
B. {1, 1½}   E. {2, 3, ⅔}
*C. {1, 2, ½}

3b   .42   .32   B, A   Belgium, Germany, United States   The Netherlands, Sweden

Appendix II (cont.)

Test Item	Target Population	Average Difficulty	Average Discrimination	Order of error frequency	Easier in	Harder in
8. In the figure to the right the circle with center C is internally tangent at point T to the circle with center O. P is a point on the larger circle such that TP is not a diameter. If TP intersects the smaller circle at A, then what additional information is needed to prove that AC and PO are parallel? *A. None D. $CO = 2TC$ B. $PO = 2AC$ E. $TA = AC$ C. $TA = AP$	3b	.49	.24	B, D	Sweden, United States	Japan

9. Which of the following is (are) true for all values of θ for which the functions are defined?
 	
   I. $\sin(-\theta) = -\sin\theta$
   II. $\cos(-\theta) = -\cos\theta$
   III. $\tan(-\theta) = -\tan\theta$
	3b	.24	.21	A, B	Belgium, France, United States	Finland, the Netherlands

340

B. II only   E. II and III only
C. III only

10. A radio-active element decomposes according to the formula
$$y = y_0 \cdot e^{-k \cdot t}$$
where y is the mass of the element remaining after t days and y_0 is the value of y for $t = 0$. Find the value of the constant k for an element whose half-life (i.e., time to decompose half of the material) is 4 days.

*A. $\frac{1}{4} \log_e 2$ D. $(\log_e 2)^{1/4}$

B. $\log_e \frac{1}{2}$ E. $2e^4$

C. $\log_2 e$

3b   .10   .15   D, E   Japan, United States   Finland, Germany, the Netherlands

11. A stationer wants to make a card 8 cm long and of such a width that when it is cut into halves, the original width becomes the length and the shape of each half is similar to the original card. What width, in cm, should he make the original card?

A. 4   *B. $4\sqrt{2}$ C. $5\sqrt{2}$ D. $5\sqrt{3}$ E. 6

3b   .26   .26   E. A   Sweden, United States   Belgium, France

12. The arithmetic mean or average of one group of 100 pupils is exactly 80 and the mean of another group of 50 pupils is exactly 65. What is the mean of the combined group of 150 pupils?

A. 79   B. 72.5   *C. 75   D. 77.5

E. It is impossible to determine exactly.

3b   .22   .23   B, E   Japan   France, Germany, the Netherlands

341

Appendix II (cont.)

Test Item	Target Population	Average Difficulty	Average Discrimination	Order of error frequency	Easier in	Harder in
13. In the diagram on the right, the numbers represent regions. The circle X represents the set of regular polygons The circle Y represents the set of quadrilaterals The circle Z represents the set of equilateral triangles Which are the parts of the schema that are empty (have no elements)? A. 1, 3 and 5 *D. 1, 3, and 7 B. 2, 3, and 4 E. 3, 6, and 7 C. 1, 6, and 7	3b	.07	.06	B, C	Japan, Sweden, United States	Finland, the Netherlands
14. In the figure on the right, $FGHJ$ is a parallelogram. Which of the following statements is a condition which implies that	3b	.42	.37	E, A	—	—

342

B. ∠HJG = ∠JGF
*C. ∠HJF = ∠JHG
D. ∠HJF and ∠JGH are supplementary
E. HF and JG are perpendicular bisectors of each other

15. What is the sum of the infinite geometric series

$$1 - \frac{1}{2} + \frac{1}{4} - \frac{1}{8} + \ldots?$$

A. $\frac{5}{8}$ *B. $\frac{2}{3}$ C. $\frac{3}{5}$ D. $\frac{3}{2}$ E. ∞

3b .06 .16 E, A Germany, United States Finland, France

16. A freight train traveling at 50 miles per hour leaves a station 3 hours before an express train which travels in the same direction at 90 miles an hour. How many hours will it take the express train to overtake the freight train?

A. $\frac{5}{9}$ B. $\frac{9}{5}$ C. $\frac{12}{5}$ *D. $\frac{15}{4}$ E. $\frac{18}{4}$

3b .43 .37 B, C — —

17. In right triangle PQR (at the right) the measure of PQ is 4 and θ can be any angle between 30° and 45°. What are all possible values for x, the length of RQ?

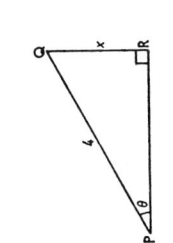

A. $0 < x < 4$ *D. $2 < x < 2\sqrt{2}$
B. $\frac{1}{2} < x < \frac{\sqrt{2}}{2}$ E. $2 < x < 2\sqrt{3}$
C. $\frac{1}{2} < x < \frac{\sqrt{3}}{2}$

3b .21 .33 A, C Japan, United States England

343

Appendix II (cont.)

Test Item	Target Population	Average Difficulty	Average Discrimination	Order of error frequency	Easier in	Harder in
Test no. 7						
1. The expression $|x-1|=1$ implies that A. x is between 0 and 2 D. x is 0 *B. x is either 0 or 2 E. x is 2 C. x is less than 2	3a	.60	.36	E, A	Finland, Japan, the Netherlands	England, Scotland
2. When $(1+p)^6$ is expanded, the coefficient of p^4 is A. 6 B. 10 *C. 15 D. 20 E. 30	3a	.63	.32	A, D	Australia, England	Israel
3. What is the converse of the statement, "If two angles are vertical, then they are equal"? A. If two angles are vertical, then they are not equal. *B. If two angles are equal, then they are vertical. C. If $\angle x$ and angle $\angle y$ are vertical angles, then $\angle x = \angle y$. D. If two angles are not vertical, then they are not equal. E. If two angles are not equal, then they are not vertical.	3a	.37	.06	E, C	Japan, United States	England, France, Israel
4. Suppose you have proved the two theorems: I. If p then q.	3a	.62	.20	B, E	Australia, United States	Israel, Japan, the Netherlands

344

A. If p then s.   *D. If s then not p.
B. If not p then not q.   E. If not s then q.
C. If p or q then s.

5. A train travelled a certain distance at a constant speed. Had the speed been 8 mph greater, the trip would have taken one hour less. Had the speed been 12 mph less the trip would have taken two hours more. How many miles did the train go?   *Ans.* 720

3a   .33   .39   Belgium, Israel, Sweden   Germany, Japan, the Netherlands

6. A wholesale merchant bought a television set at a certain price and then sold it to a retail merchant at an increase of P per cent of this price. The retail merchant sold the set to a consumer for P per cent more than he paid for it. If the customer paid 65 per cent more than the price originally paid by the wholesale merchant, then P satisfies the equation:

3a   .47   .36   A, C   Japan, Sweden   France, United States

A. $1 + \dfrac{2P}{100} = 1.65$ D. $1 + P^2 = 1.65$

*B. $\left(1 + \dfrac{P}{100}\right)^2 = 1.65$ E. $1 + 2P = 1.65$

C. $1 + \left(\dfrac{P}{100}\right)^2 = 1.65$

Appendix II (cont.)

Test Item	Target Population	Average Difficulty	Average Discrimination	Order of error frequency	Easier in	Harder in

7. If a relation R is such that xRy and yRz implies xRz for each $x, y,$ and z of a given set, the relation R is said to be transitive on that set. Which of the following relations are transitive?
 	
    I. "is father of"
   II. "is contemporary of"
  III. "is admirer of"
  IV. "is multiple of"
   V. "is perpendicular to"
 	
A. II, IV, and V   *D. II and IV
B. I and II   E. V only
C. II, III, and IV
	3a	.30	.28	E, B	France, United States	Finland, Israel

8. In the figure shown to the right, which vector is a graphical representation of the complex number $4-2i$? *Ans. *D*	3a	.38	.29	A, C	Belgium, France, Germany, United States	Israel, the Netherlands
9. Solve $0 < x^2 - 3x + 3 < 7$ *Ans.* $-1 < x < +1$ or $2 < x < 4$	3a	.38	.40		Finland, France, Israel, the Nether-	Australia, England, Germany, Scot-

346

only if given an $x \in S$ there exists at most one $y \in T$ such that xRy.

Which of the following relations are functions?

I. x divides y
II. x has y for mother
III. x is parallel to y
IV. x has y for double
V. x has y as majorant (i.e., $x < y$)
VI. $x^2 = y$

A. I, II, and III
B. II, IV, and V
*C. II, IV, and VI
D. IV, V, and VI
E. I, IV, and V

			Japan, United States	Israel, the Netherlands

11. What is the equation whose roots are the squares of the roots of

$$x^2 - 5x + 3 = 0?$$

*A. $x^2 - 19x + 9 = 0$
B. $x^2 + 19x + 9 = 0$
C. $x^2 - 20x + 9 = 0$
D. $x^2 + 19x - 9 = 0$
E. $x^2 - 9x + 19 = 0$

3a .48 .34 C, B Israel Germany, the Netherlands, United States

Questions 12 and 13. Six operations are defined as follows:

$$A = \begin{pmatrix} 1 & 2 & 3 \\ 2 & 3 & 1 \end{pmatrix} \quad B = \begin{pmatrix} 1 & 2 & 3 \\ 3 & 1 & 2 \end{pmatrix} \quad C = \begin{pmatrix} 1 & 2 & 3 \\ 1 & 3 & 2 \end{pmatrix}$$

$$D = \begin{pmatrix} 1 & 2 & 3 \\ 3 & 2 & 1 \end{pmatrix} \quad E = \begin{pmatrix} 1 & 2 & 3 \\ 2 & 1 & 3 \end{pmatrix} \quad F = \begin{pmatrix} 1 & 2 & 3 \\ 1 & 2 & 3 \end{pmatrix}$$

Appendix II (cont.)

Test Item	Target Population	Average Difficulty	Average Discrimination	Order of error frequency	Easier in	Harder in
The operation $A = \begin{pmatrix} 1 & 2 & 3 \\ 2 & 3 & 1 \end{pmatrix}$, for example, means that the numbers in the upper row are transformed into the digits in the lower row, so that $1 \rightarrow 2$ (1 becomes 2), $2 \rightarrow 3$ (2 becomes 3), and $3 \rightarrow 1$ (3 becomes 1). $A \cdot B$ shows that operation B is to be performed after operation A; that is, according to A, $1 \rightarrow 2$, $2 \rightarrow 3$, $3 \rightarrow 1$, and then, according to B, $2 \rightarrow 1$, $3 \rightarrow 2$, $1 \rightarrow 3$. Therefore $A \cdot B$ will be $1 \rightarrow 2 \rightarrow 1$, $2 \rightarrow 3 \rightarrow 2$, and $3 \rightarrow 1 \rightarrow 3$. This produces the same outcome as $F = \begin{pmatrix} 1 & 2 & 3 \\ 1 & 2 & 3 \end{pmatrix}$; let us write this $A \cdot B = F$. In like manner, $A \cdot C$ is $1 \rightarrow 2 \rightarrow 3$, $2 \rightarrow 3 \rightarrow 2$, $3 \rightarrow 1 \rightarrow 1$, and is the same as D; that is to say, $A \cdot C = D$.						
12. Which one operation is equal to $C \cdot D$? *Ans. B*	3a	.53	.39		England, Israel	Finland, France
13. What operation must be performed after operation B so	3a	.47	.40		England, United	France

Q, and R are defined as follows,

$$P = \{(x,y) \mid x^2 + y^2 = 4\}$$
$$Q = \{(x,y) \mid x - y = 2\}$$
$$R = \{(x,y) \mid (x^2 + y^2 - 4)(x - y - 2) = 0\},$$

which of the following is true?

A. $R = P \cap Q$
*B. $R = P \cup Q$
C. $R = \{(2,0)(0,2)(-2,0)(0,-2)\}$
D. $R = \{\}$ (the empty set)
E. $R = \{(2,0)(0,-2)\}$

15. The value $\begin{vmatrix} 4 & 2 & 1 \\ 0 & 0 & 1 \\ 1 & 1 & 0 \end{vmatrix}$ is

*A. -2 B. 0 C. 2 D. 7 E. 9

3a	.23	.25	B, C	Belgium, Germany, Israel, United States	Finland, France, Japan, Sweden

16. Each root of $x^2 - 2x + 5 = 0$ differs from the cube of the other by a positive constant c. What is the value of c?

    Ans. 12

3a	.08	.30		Belgium, England, Finland, Sweden	France, Israel, the Netherlands

17. Two of the roots of the equation $x^4 - 27x^2 - 14x + 120 = 0$ are 2 and 5. Find the two other roots of the equation.

    Ans. -3, -4

3a	.46	.40		Sweden	Israel, the Netherlands

Test no. 8

1. If $\log_a 8 = \frac{9}{2}$, what is the value of a?

A. 2/3 B. 2 *C. 4 D. 5 E. 6

3a	.75	.43	B, A	Finland, Israel, the Netherlands, Scotland	Belgium, France

Appendix II (cont.)

Test Item	Target Population	Average Difficulty	Average Discrimination	Order of error frequency	Easier in	Harder in
2. If x and y are real numbers, for which x can you define y by $$y = \frac{x}{\sqrt{9-x^2}}?$$ A. All x except $x=3$ *D. $-3 < x < 3$ B. All x except $x=3$ and $x=-3$ E. $x<3$ C. $x<-3$ and $x>3$	3a	.57	.34	B, C	Germany, the Netherlands	Australia
3. A set of 24 cards is numbered with the positive integers from 1 to 24. If the cards are shuffled and if only one is selected at random, what is the probability that the number on the card is divisible by 4 or 6? A. 1/6 B. 5/24 C. 1/4 *D. 1/3 E. 5/12	3a	.59	.29	E, C	Finland	France, Israel
4. An angle θ is known to be between 180° and 270° and $\cos^2\theta = 16/25$. The value of $\sin 2\theta$ is then A. $-\frac{24}{25}$ B. $-\frac{15}{25}$ C. $-\frac{7}{25}$ D. $\frac{7}{25}$ *E. $\frac{24}{25}$	3a	.49	.37	A, D	Israel	Germany, the Netherlands
5. For some functions the relationship holds that $f(x+y) = f(x) + f(y)$ for all numbers x and y. For example, when	3a	.51	.24	C, A	United States	—

tive. Which of the following functions is additive by this definition?

A. $f(x) = x^2$ D. $f(x) = 2^x$
B. $f(x) = \sin x$ *E. None of them are additive
C. $f(x) = \log_{10} x$

6. If determinants are used to solve the system of equations $\begin{cases} 2x+y=3 \\ x+4y=7 \end{cases}$, then y is equal to

A. $\begin{vmatrix} 2 & 1 \\ 1 & 4 \end{vmatrix}$ *C. $\begin{vmatrix} 2 & 3 \\ 1 & 7 \end{vmatrix}$ E. $\begin{vmatrix} 3 & 7 \\ 2 & 1 \end{vmatrix}$
   $\begin{vmatrix} 2 & 3 \\ 1 & 7 \end{vmatrix}$ $\begin{vmatrix} 2 & 1 \\ 1 & 4 \end{vmatrix}$ $\begin{vmatrix} 2 & 1 \\ 1 & 4 \end{vmatrix}$

B. $\begin{vmatrix} 2 & 1 \\ 1 & 4 \end{vmatrix}$ D. $\begin{vmatrix} 3 & 1 \\ 7 & 4 \end{vmatrix}$
   $\begin{vmatrix} 3 & 1 \\ 7 & 4 \end{vmatrix}$ $\begin{vmatrix} 2 & 1 \\ 1 & 4 \end{vmatrix}$

| | 3a | .33 | .22 | A, B | Belgium, United States |

Items 7 to 9 refer to the information below.

Consider the following abstract mathematical system:

Undefined terms: elements $a, b, c \ldots$ of class C; operations λ and $*$ relation $=$, having the conventional meaning of "equals".

Postulates: If a, b, and c are any elements of C, then

(1) $a\lambda b$ and $a*b$ are elements of C.
(2) $a\lambda b = b\lambda a$.
(3) $a*(b*c) = (a*b)*c$.
(4) $a*b \ne b*a$, provided $a \ne b$.
(5) $a\lambda(b*c) = (a\lambda b)*(a\lambda c)$.

The Netherlands

Appendix II (cont.)

Test Item	Target Population	Average Difficulty	Average Discrimination	Order of error frequency	Easier in	Harder in
DIRECTIONS: Answer each item, using the code A – if the proposition follows logically from the postulates. B – if the proposition is inconsistent with the postulates (i.e., contradicts the postulates). C – neither A nor B (i.e., the proposition neither follows from the postulates nor is contradicted by them).						
7. $(a*b)\lambda c = (c\lambda a)*(c\lambda b)$ Ans. A	3a	.43	.35	B, C	Scotland, United States	Finland, Israel
8. $(a*b)\lambda c = (a\lambda c)*(b\lambda c)$ Ans. A	3a	.39	.35	B, C	United States	Finland, Israel
9. $a\lambda(b*c) = (a\lambda c)*(a\lambda b)$ Ans. B	3a	.33	.34	A, C	United States	Finland, Israel
10. The graph of $y = f(x)$ is a parabola with axis parallel to the Y-axis. If the maximum value of y is 2, and if the parabola crosses the X-axis at $x = -\frac{1}{2}$ and at $x = \frac{3}{2}$, then its equation is *A. $y = -2x^2 + 2x + \frac{3}{2}$ D. $y = 4x^2 - 4x - 3$ B. $y = 4x^2 - 4x + 3$ E. $y = 4x^2 + 4x - 2$	3a	.50	.27	A, C	Germany, Sweden	—

352

11. For what values of the real number x is $y = -\dfrac{1}{x}$ a decreasing function? A. No x B. $x<0$ *C. $x \neq 0$ D. $x>0$ E. All x	3a	.22	.16	D, E	United States	England, Israel
12. Solve: $2 \cdot 7^{2-x} + 3 \cdot 7^{3-x} = 161$ Ans. $x = -1$	3a	.57	.44		Israel, the Netherlands	Australia, England, France, Scotland
13. Given two arbitrary sets X and Y. Which of the following sets is equivalent to the set $(X \cup Y) \cap (X \cap Y)$? A. X *D. $X \cap Y$ B. Y E. $(X \cup Y) \cup (X \cap Y)$ C. $X \cup Y$	3a	.18	.14	E, C	Belgium, France, Scotland, United States	England, Germany, Israel, the Netherlands
14. Consider the matrices $A = \begin{pmatrix} 1 & x \\ 0 & 1 \end{pmatrix}$ and $B = \begin{pmatrix} 1 & 0 \\ y & 1 \end{pmatrix}$ where x and y are real numbers and $x^2 + y^2 \neq 0$. For which values of x and of y is the product of the matrices commutative? I. $x = 0$ II. $y = 0$ III. $x = y$ A. Only I *D. Both I and II B. Only II E. I, II, and III C. Only III	3a	.12	.11	C, E	Australia, Japan, Scotland, United States	Israel, the Netherlands

Appendix II (cont.)

Test Item	Target Population	Average Difficulty	Average Discrimination	Order of error frequency	Easier in	Harder in
15. Calculate arc $\sin\frac{1}{2}$ + arc $\sin\frac{1}{\sqrt{2}}$ [Arc sin means "angle between $-\frac{\pi}{2}$ and $\frac{\pi}{2}$ whose sin is"] *A. $\frac{5\pi}{12}$ D. arc $\sin\left(\frac{1+\sqrt{2}}{2}\right)$ B. $\frac{7\pi}{18}$ E. arc $\sin\frac{\sqrt{3}}{2}$ C. $\frac{\pi}{3}$	3a	.45	.44	D, C	Australia, England, Israel	Finland, Japan, the Netherlands
16. For what values of x is the function $$\frac{(1-x)(1+3x)}{(2x-1)(x-2)}$$ positive? *Ans.* $-\frac{1}{3}<x<\frac{1}{2}$ and $1<x<2$	3a	.23	.38		Belgium, Finland, France, Israel, the Netherlands	Australia, England, Germany, Scotland, United States

Test no. 9

Test Item	Target Population	Average Difficulty	Average Discrimination	Order of error frequency	Easier in	Harder in
1. In a Cartesian coordinate system, what is the equation of the straight line passing through the point $(0, -5)$ and parallel to the straight line whose equation is	3a	.87	.28	E, C	Germany, the Netherlands	Finland, United States

*B. $2x - y - 5 = 0$ E. $2x + y + 5 = 0$
C. $2x + 3 = -5$

2. An open cylindrical vessel of capacity $9{,}000\pi$ cc is to be made with the curved surface of sheet metal and a wooden base. If the weight of 1 sq cm of the metal is three times the weight of 1 sq cm of the wood, each being of uniform small thickness, what will be the radius of the vessel (in cms) when its total weight is a minimum?
Ans. 30

| | 3a | .16 | .29 | | England, Israel, Sweden | Belgien, Finland, France, the Netherlands |

3. The derivative with respect to x of $\dfrac{4}{\sqrt{3x-4}}$ is

A. $12\sqrt{3x-4}$ *D. $\dfrac{-6}{(3x-4)^{3/2}}$

B. $4\sqrt{3}$ E. $6\sqrt{3x-4}$

C. $\dfrac{-2}{(3x-4)^{3/2}}$

| | 3a | .61 | .39 | C, E | Germany, Israel, Scotland | Japan, Sweden, United States |

4. The value of $\displaystyle\int_0^1 \dfrac{dx}{x^2 - 5x + 6}$ is:

A. $\dfrac{1}{2}\log_e 2$ *C. $\log_e \dfrac{4}{3}$ E. $\dfrac{1}{2}$

B. $\dfrac{1}{3}$ D. $\tan \dfrac{-11}{4}$

| | 3a | .19 | .19 | B, E | England, Germany, Israel, United States | Belgium, France, the Netherlands |

355

Appendix II (cont.)

Test Item	Target Population	Average Difficulty	Average Discrimination	Order of error frequency	Easier in	Harder in
5. $\int (x-1)^2 \, dx$ is equal to A. $2(x-1)+c$ D. $\frac{1}{3}(x^3-x)+c$ B. $\frac{1}{2}(x-1)^2+c$ E. $\frac{(x-1)^3}{x}+c$ *C. $\frac{1}{3}(x-1)^3+c$	3a	.71	.40	A, D	England, Finland, Germany, Israel, the Netherlands, Scotland	Belgium, France, Japan, United States
6. Determine k so that the graph of the function $y = 3x^3 + 6x^2 + kx + 9$ has a point of inflection and a horizontal tangent for the same value of x. *Ans. 4*	3a	.37	.44		Finland, Germany, Israel	Belgium, Scotland, United States
7. What is the equation in x and y of the curve with parametric equations $x = t + \frac{1}{t}, \quad y = t - \frac{1}{t}$? A. $x+y=1$ *D. $x^2-y^2=4$ B. $x+y=2$ E. $2x^2-y^2=4$	3a	.47	.42	B, C	England	Israel

356

nomial function of x is shown in the diagram to the right, the equation of the curve being $y=f(x)$. Which of the following statements is (are) true for that part of the curve for which $a \leq x \leq b$?

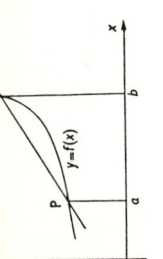

I. $f''(c)=0$ for some value c between a and b.

II. $\dfrac{f(b)-f(a)}{b-a}=f'(c)$ for some value c between a and b.

III. If there is a point of inflexion at Q, $f''(b)$ can have no value but 0.

IV. $\displaystyle\int_a^b f(x)\,dx < \dfrac{1}{2}(b-a)\,[f(a)+f(b)]$

A. All four   D. I and III
*B. II, III, and IV   E. II and IV
C. I and II

9. Given that $3\dfrac{dy}{dx}=x^2-5$, and $y=1$ when $x=2$.

What is the value of y when $x=0$?

A. $-5/3$ B. $-2/3$ C. $1/3$ D. $25/9$ *E. $31/9$

Problems 10 and 11 are based on the figure on the right, which shows a graph of $y=f(x)$, a being less than b. S_1 is the area enclosed by the x-axis, $x=a$, and $y=f(x)$. S_2 is the area enclosed by the x-axis, $x=b$, and $y=f(x)$. S_1 and S_2 are to be considered positive.

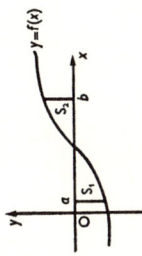

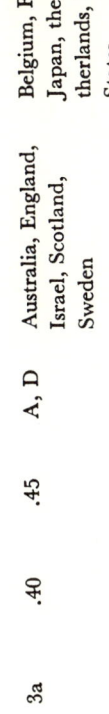

3a   .40   .45   A, D   Australia, England, Israel, Scotland, Sweden   Belgium, Finland, Japan, the Netherlands, United States

Appendix II (cont.)

Test Item		Target Population	Average Difficulty	Average Discrimination	Order of error frequency	Easier in	Harder in
10. The value of $\int_a^b f(x)dx$ is		3a	.34	.29	A, D	Germany	Belgium
A. $S_1 + S_2$	D. $\|S_1 - S_2\|$						
B. $S_1 - S_2$	E. $\dfrac{1}{2}(S_1 + S_2)$						
*C. $S_2 - S_1$							
11. The value of $\int_a^b \|f(x)\| dx$ is		3a	.28	.29	C, D	Germany, the Netherlands	Belgium, France
*A. $S_1 + S_2$	D. $\|S_2\| - \|-S_1\|$						
B. $S_1 - S_2$	E. $\dfrac{1}{2}(S_1 + S_2)$						
C. $\|S_2 - S_1\|$							
12. The function $f(x) = \dfrac{x^2 - 1}{x - 1}$ is defined and continuous for all x except $x = 1$. What value must be assigned to $f(x)$ for $x = 1$ in order that the function be continuous there?		3a	.28	.38		Germany, Israel	Belgium, England, Scotland, United States

13. Find the difference $\vec{b}-\vec{a}$ of the vectors $\vec{a}=(4,2)$ and $\vec{b}=(0,3)$.

  A. $(-4,-2)$ B. $(-4,1)$ *C. $(4,-1)$ D. $(4,2)$
  E. $(4,5)$

  3a .40 .23 C, E Australia, United States Finland, Sweden

14. In a triangle with area a, the midpoints of the three sides are joined so as to form a new triangle. In the triangle thus constructed, another new triangle is inscribed in the same way. This process is continued indefinitely. What is the sum of all of the areas of this sequence of triangles, including the original one?

  A. $\dfrac{9a}{7}$ *B. $\dfrac{4a}{3}$ C. $\dfrac{7a}{5}$ D. $\dfrac{3a}{2}$ E. $\dfrac{5a}{3}$

  3a .50 .34 D, E England, Israel Belgium, France, United States

15. The value of $\displaystyle\lim_{h\to 0}\dfrac{\sqrt{2+h}-\sqrt{2}}{h}$ is

  A. 0   *B. $\dfrac{1}{2\sqrt{2}}$ C. $\dfrac{1}{2}$ D. $\dfrac{1}{\sqrt{2}}$ E. ∞

  3a .22 .32 E, A Belgium, France, Israel, Japan England, Finland, the Netherlands, Scotland

Appendix III

Finland Unweighted Statistics

I. *General Background of the Pupils*

		Population 1a	Population 1b
Number of pupils		1,156	1,325
Age in months	Mean	163	165
	SD	3.4	6.7
Male[a]		49	48
Father's education in years	Mean	6.2	6.2
	SD	2.6	2.5
Mother's education in years	Mean	6.3	6.3
	SD	2.2	2.1
Father's occupation[a]			
Higher professional and technical		2	2
Executive, managerial		2	2
Subprofessional		8	8
Working proprietor		3	3
Proprietor in agriculture		28	28
Clerical, sales		1	1
Skilled, semiskilled		42	43
Laborer, agriculture		5	4
Unskilled manual		9	9
Place of residence[a]			
Rural farm		24	22
Nonfarm under 2,500		37	37
Small town (2,500–15,000)		19	19
Medium city (15,000–100,000)		11	12
Large city (over 100,000) or suburb		9	10

[a] Reported as percent of total.

Appendix III (cont.)

II. *Educational Experiences and Reactions of Pupils*

		Population 1a	Population 1b
Size of class	Mean	28	28
	SD	10	10
School course			
Academic		34	34
Vocational		0	0
General		66	66
Level of mathematics	Mean	1.9	2.0
	SD	.3	.1
Hours mathematics instruction	Mean	3.4	3.4
	SD	.8	.8
Hours mathematics homework	Mean	2.4	2.5
	SD	1.6	2.0
Expect to take more mathematics		65	66
Desire to take more mathematics		49	49
Interest score	Mean	5.8	5.8
	SD	1.6	1.6
Years education expected	Mean	3.4	3.2
	SD	2.3	2.2
Years education desired	Mean	3.4	3.3
	SD	2.4	2.3

III. *Total Mathematics Test Score Distributions*[a]

Score	Population 1a	Population 1b	Score	Population 1a	Population 1b
0	1	1	41–45	2	2
1–5	17	18	46–50	1	1
6–10	19	17	51–55	0	0
11–15	18	16	56–60	0	0
16–20	14	12	61–65	0	0
21–25	11	13	66–70	0	0
26–30	9	10	Mean	15.4	16.1
31–35	5	6	SD	10.8	11.6
36–40	3	4	Number of cases	1,156	1,325

[a] All scores have been corrected for guessing. Entries are percentages of the total group.

References

Alpert, R., Stellwagen, G., and Becker, D. (1963). "Psychological Factors in Mathematics Education", *School Mathematics Study Group Newsletter*, No. 15, 17–24.
Anastasi, A. (1958). *Differential Psychology*, 3rd Edition, New York, Macmillan.
Anderson, C. (1940). "The Development of a Level of Aspiration in Young Children", unpublished Ph.D. dissertation, State University of Iowa.
Anderson, I. H. (1964). *Comparisons of the Reading and Spelling Achievement and Quality of Handwriting of Groups of English, Scottish, and American Children*, University of Michigan.
Banghart, F. W., et al. (1963). "An Experimental Study of Programmed Versus Traditional Elementary School Mathematics", *The Arithmetic Teacher*, April 199–204.
Barr, F. (1959). "Urban and Rural Differences in Ability and Attainment", *Educational Research 1*, No. 2, 49–60.
Bernstein, B. (1961). "Social Class and Linguistic Development: A Theory of Social Learning", in A. H. Halsey, J. Floud, and C. A. Anderson (eds.), *Education, Economy, and Society*, New York, The Free Press of Glencoe, 288–314.
Bloom, B. S. (ed.), Engelhart, M. D., Furst, E. J., Hill, W. H., and Krathwohl, D. R. (1956). *Taxonomy of Educational Objectives, Handbook I: Cognitive Domain*, New York, David McKay.
Boalt, G., and Husén, T. (1964). *Skolans Sociologi*, Stockholm, Almqvist and Wiksell.
Brownell, W. A. (1963). "Arithmetical Abstractions: Progress Towards Maturity of Concepts Under Differing Programs of Instruction", *The Arithmetic Teacher*, October, 322–329.
Carey, G. L. (1958). "Sex Differences in Problem Solving Performance as a Function of Attitude Differences", *Journal of Abnormal and Social Psychology*, 56, 256–260.
Chausow, H. M. (1955). "The Organization of Learning Experiences to Achieve More Effectively the Objective of Critical Thinking in the General Social Science Course at the Junior College Level", unpublished Ph.D. dissertation, Department of Education, University of Chicago.
Dahllöf, U. (1960). *Kursplaneundersökningar i matematik och modersmålet. Empiriska studier över kursinnehållet i den grundläggande skolan, 1957 års skolberedning III (Curriculum Investigations related to Mathematics and Swedish, Empirical Studies in the Curriculum of the Basic School)*, Stockholm, Government Printing Office. (Statens Offentliga Utredningar 1960: 15).
— — (1963). *Kraven på gymnasiet — Undersökningar vid universitet och högskolor, i förvaltning och näringsliv (Demands on the Gymnasium—Investigations Conducted at Universities, Other Institutions of Higher Learning, Civil Service, Industry and Commerce)*, Stockholm, Government Printing Office (Statens Offentliga Utredningar 1963: 22).
Dave, R. H. (1963). "The Identification and Measurement of Environmental Process Variables that are Related to Educational Achievement", unpublished Ph.D. dissertation, Department of Education, University of Chicago.
Davie, J. (1963). "Social Class Factors and School Attendance", *Harvard Educational Review*, 23, 175–185.

Deutsch, M. (1964). "Facilitating Development in the Pre-School Child: Social and Psychological Perspectives", *Merrill–Palmer Quarterly 10*, 249–263.
—— (1965). "The Role of Social Class in Language Development and Cognition", *American Journal of Orthopsychiatry 25*, 78–88.
Douglas, J. W. B. (1964). The Home and the School. London, MacGibbon and Kee.
Dressel, P. L., and Mayhew, L. B. (1954) *General Education: Explorations in Evaluation; Final Report of the Cooperative Study of Evaluation in General Education of the American Council on Education*, Washington, American Council on Education.
Floud, J. (1961). "Social Class Factors in Educational Achievement", in A. H. Halsey (ed.), *Ability and Educational Opportunity*, Paris: OECD.
Floud, J., and Halsey, A. H. (1961). "English Secondary Schools and the Supply of Labor", in A. H. Halsey, J. Floud, and C. A. Anderson (eds.), *Education, Economy, and Society*, New York, The Free Press, 80–92.
Foshay, A. W. (ed.). (1962). *Educational Achievements of 13-Year-Olds in Twelve Countries*, Hamburg, UNESCO Institute for Education.
Ginther, J. R. (1964). *Conceptual Model for Analyzing Instruction—Programmed Instruction in Medical Education: Proceedings of the first Rochester Conference, June, 25–27*. Rochester, The University of Rochester.
Gray, S., and Klaus, R. A. (1963). "Interim Report: Early Training Project", George Peabody College and Murfeesboro, Tennessee, City Schools, mimeographed.
Guilford, J. P., and Merrifield, P. R. (1960). "The Structure of Intellect Model: its Uses and Implications", report from the Psychological Laboratory No. 24., University of Southern California.
Halsey, A. H. (ed.), (1961 a). *Ability and Educational Opportunity*, Paris: OECD.
Halsey, A. H., Floud, J., Anderson, C. A. (eds.), (1961 b). *Education, Economy, and Society*, New York, The Free Press of Glencoe.
Härnqvist, K. J. (1960). *Individuella Differenser och Skoldifferenser, (Individual Differences and School Differences) 1557 års skolberedning II*, Stockholm, Government Printing Office (Statens Offentliga Utredningar 1960: 13).
Havighurst, R. J., and Neugarten, B. L. (1962). *Society and Education*, Boston, Allyn and Bacon.
Hess, R. D., and Shipman, V. C. (1965). "Early Experience and the Socialization of Cognitive Modes in Children", *Child Development, 36*, 869–886.
Hieronymous, A. N. (1951). "A Study of Social Class Motivation: Relationships Between Anxiety for Education and Certain Socio-Economic and Intellectual Variables", *Journal of Educational Psychology, 42*, 193–205.
Hills, J. R. (1955). "The Relations between Certain Factor-Analyzed Abilities and Success in College Mathematics", report from the Psychological Laboratory, No. 15, University of Southern California.
Husén, T. (1961). "Educational Structure and the Development of Ability", in A. H. Halsey (ed.), *Ability and Educational Opportunity*, OECD.
—— (1965 a). "An International Perspective on the Academic Secondary School", in *The High School in an Interdependent World*, Edmonton, Alberta.
—— (1965 b). "Curriculum Research in Sweden", *International Review of Education, 11*, 189–208.
—— (1966). "The Relation between Selectivity and Social Class in Secondary School", *International Journal of Educational Sciences*, Vol. I. No. 1.
Jackson, R. W. B. (1957). *Achievement in the Skill Subjects in Public School in Four*

Areas of Ontario: 1955–56 Rural Schools Survey, Toronto University, Ontario College of Education.

James, H. T., Thomas, J. A., and Dyck, H. J. (1963). *Wealth, Expenditure, and Decision Making for Education*, U.S. Office of Education, Cooperative Research Project, No. 1241, Stanford University School of Education.

Kahl, J. A. (1953). "Educational and Occupational Aspirations of Common Man Boys", *Harvard Educ. Review, 23*, 186–203.

Katz, I. (1964). "Review of Evidence Relating to Effects of Desegregation on the Intellectual Performance of Negroes", *American Psychologist, 19*, 381–399.

Koshe, G. S. (1957). "A Comparative Study of the Attainment and Intelligence of Children in Certain Comprehensive, Modern, and Grammar Schools", unpublished M.A. thesis, University of London.

Krathwohl, D. R., Bloom, B. S., and Masia, B. B. (1964). *Taxonomy of Educational Objectives—Handbook 2: Affective Domain*, New York, David McKay.

Landers, J. (1963). "The Higher Horizons Program in New York City", in *Programs for the Educationally Disadvantaged*, USOE, Washington, Government Printing Office.

Lindgren, H. C., et al. (1964). "Attitudes Towards Problem Solving as a Function of Success in Arithmetic in Brazilian Elementary Schools", *Journal of Educational Research 58*, 44–45.

Ljung, B. O. (1958). *Konstruktion och användning av standardprov (Construction and use of standardized achievement tests)*, Stockholm, School of Education.

Lucow, W. H. (1963). "Testing the Cuisenaire Method", *The Arithmetic Teacher*, November, 435–438.

Marklund, S. (1962). *Skolklassens storlek och struktur, (Size and Homogeneity of Class as Related to Scholastic Achievement)*, Stockholm, Almqvist and Wiksell.

McIntosh, D. M. (1962). "Educational and Social Obstacles to the Emergence of Talent with Special Reference to the United Kingdom", *Year Book of Education*.

Miller, T. W. G. (1957). "A Critical and Empirical Study of the Emergence, Development, and Significance of the Comprehensive School in England", unpublished Ph.D. thesis, University of Birmingham.

—— (1961). Values in the Comprehensive School. Educational Monographs, Birmingham: University of Birmingham.

Milton, G. A. (1957). "The Effects of Sex Role Indentification upon Problem Solving Skill", *Journal of Abnormal and Social Psychology, 55*, 208–212.

Ministry of Education (1954). *Early Leaving*, HMSO, London.

Mogstad, L. P. (1958). "Et tilskud til klarlegging av centraliseringsspörsmålet i landsfolkeskolen", in *Forskning og Danning*, Oslo: Oslo University Press.

National Surveys of Scholastic Achievement (1961). Japan, Ministry of Education.

—— (1962). Japan, Ministry of Education.

OECD (1961). *New Thinking in School Mathematics*.

—— (1961). *Policy Conference on Economic Growth and Investment in Education*.

—— (1962). *Planning of Education in Relation to Economic Growth*.

Orring, J. (1959). *Flyttning, kvarsittning och utkuggning i högre skolor, (Promotion, Grade-Repitition and Drop-out in Academic Secondary Schools)*, Stockholm, Government Printing Office, (Statens Offentliga Utredningar 1959: 35).

Peck, H. I. (1963). "An Evaluation of Topics in Modern Mathematics", *The Arithmetic Teacher*, May, 277–279.

Peterson, H., et al. (1963). "Determination of 'Structure of Intellect' Abilities involved in 9th Grade Algebra and General Mathematics", report from the Psychological Laboratory, No. 31, University of Southern California.

Pidgeon, D. A. (1958). "A Comparative Study of Basic Achievements", *Educational Research*, Vol. I, No. 1, 50–68.

Postlethwaite, T. N. (in press). *School Organization and Student Achievement*, Stockholm, Almqvist & Wiksell.

Ruddell, A. K. (1962). "The Results of a Modern Mathematics Program", *The Arithmetic Teacher*, October, 330–335.

Scottish Council for Research in Education (1963). *The Scottish Scholastic Survey, 1953*, University of London Press.

Sewell, W. H. (1963). "The Educational and Occupational Perspectives of Rural Youth. Proceedings of the National Conference on Problems of Rural Youth in a Changing Environment. Washington D.C.: The National Committee for children and Youth. 18.

Sheehan, T. J. (1965). "The Relationship Between Students' Degree of Freedom and Success in Higher Mental Process Learning", unpublished Ph.D. dissertation, Department of Education, University of Chicago.

Shepard, S. (1963). "A Program to raise the Standard of School Achievement" in *Programs for the Educationally Disadvantaged*, USOE, Washington, Government Printing Office.

Smilansky, M. (1964). *Progress Report on a Program to Demonstrate Ways of Using a Year of Kindergarten to Promote Cognitive Abilities, Impart Basic Information, and Modify Attitudes Which are Essential for Scholastic Succes of Culturally Deprived Children in Their First Two Years of School*, Henrietta Szold Institute, Jerusalem.

Spearman, C. (1927). *The Abilities of Man: Their Nature and Measurement*, London, Macmillan.

Svensson, N. E. (1962). *Ability Grouping and Scholastic Achievement: Report on a 5 Year Follow-Up Study in Stockholm*, Stockholm, Almqvist and Wiksell.

Thurstone, L. L. (1938). *Primary Mental Ability*, University of Chicago.

Tredway, D. C., and Pollister, G. E. (1963). "An Experimental Study of Two Approaches to Teaching Percentage", *The Arithmetic Teacher*, December, 491–495.

Trow, M. (1961). "The Second Transformation of American Secondary Education", *International Journal of Comparative Sociology*, 2, 144–166.

Tyler, L. E. (1956). *The Psychology of Human Differences*, 2nd Edition, New York, Appleton, Century.

United Nations Statistical Year Book (1964). Sixteenth Edition. Statistical Office of the United Nations Department of Economic and Social Affairs, New York.

Witkin, H. A., et al. (1962). *Psychological Differentiation: Studies of Development*, New York, Wiley.

Wolf, R. M. (1964). "The Identification and Measurement of Environmental Process Variables Related to Intelligence", unpublished Ph.D. dissertation, Department of Education, University of Chicago.

Wrigley, J. (1958). "The Factorial Nature of Ability in Elementary Mathematics", *British Journal of Educationel Psychology*, 28, 61–78.

Yates, A. (1966). *Grouping in Education*, New York, Wiley.

Subject Index

Age of ceasing school education 58, 68–74, 140, 296
Age of entry to school 58, 61–68, 139, 288, 295
Aspiration and mathematics achievement 204, 250
Attitude scale scores
　difficulty of learning mathematics 44
　man and his environment 45
　nature of mathematics 44, 289–290, 298
　relationship to achievement 48
　role of mathematics 45–48, 290
Attitudes toward
　difficulty of learning mathematics 247
　further education 231
　mathematics 146, 147, 152–157
　school and school learning 45, 290

Case studies 309
Class size 59, 79–85, 140, 297
Coeducation versus single sex schools 305
Cognitive styles 310
Comprehensive systems and schools 59, 87–107
Correlations, median
　between countries 49–55, 289
　within countries 49–55
Curriculum
　conclusions 194
　implications 197

Data bank 306
Differentiation 103–107
"Does more mean worse?" 60, 128–135, 141–142

Educational Motivation 210
Efficiency of school systems 290–291
Equality of opportunity 109
Expectation of students 303–304

Financial support for education 203, 232–233, 257, 301, 304

Further education,
　attitudes toward 146, 147, 152–157, 231
　plans for 209

General school programs 211
Grouping 292

Home environment 200, 253
Homework 182, 211–212

Industrialization 117, 298
Inquiry-centered methods 144–146
Interest in mathematics 146–147, 152–162, 211, 243, 299
　in relation to attitude 152–162
　father's occupation 147, 161
　inquiry 148, 157
　view of mathematics as open system 146, 154
Item analysis
　form of reporting 42–43, 289
　statistics on—see Appendix II 312–359

Level of mathematics instruction 209–210

Mathematical model 61, 135–139
Mathematics test scores
　correlations with other variables 36–41, 290
　increments in 29
　part scores by country 31
　range of 288
　societal factors 301–304
　total scores by country 21–31, 288

National emphasis 157, 163–164, 170
National or regional examinations 231
Need for mathematicians 57, 142
"New Mathematics" 183–184
Non-cognitive outcomes 298–299
Number of hours
　all homework per week 300
　mathematics homework per week 300

366

mathematics instruction per week 300
in the school week 300–301
Number of subjects 59, 86

Occupations 107–115
OECD 56
Opportunity to learn mathematics 162, 167–169, 183–184, 249, 299–300

Parents as teachers 201
Parents
  education 303
  occupation 303
Percentile levels 128–130
Phase II
  objectives 307–308
  subject areas 309
Place of residence 203, 256–257, 304

Regression analysis Chap. 6 passim
Retentivity 61, 71–73, 116–135, 141–142, 288, 293–294
Role and self-image of girls 235

Schools
  organization 57
  programs 90–102
  selective and comprehensive 59, 87–115, 293, 297
  single sex and coeducational 247–250
  size 58, 74–79
  socio-economic variability and mathematics achievement 219–222
  social composition 166, 219–222
  specialized 87–115, 141
  staffing policy 233
  vocational 92
Selection 298
Sex differences 204, 233–250, 239, 304–306
  in mathematics achievement and attitudes toward mathematics 233–250
  in attitude toward difficulty of learning mathematics 233–235
Social selectivity 204
Socio-economic status 60, 66, 107–115, 141, 303
  and selectivity 204
  and interest in mathematics and amount of homework 211–212

and level of instruction and further plans for education 209–210
and mathematics achievement 205–209, 212–219
Specialization 59, 85–87

Teacher competence 299–300
  freedom 164–165, 175, 274, 285
  in-service training 165–166, 177–178
  pre-service training 165–166, 178–181
  qualifications 230
  salaries 232–233
Time spent on
  all homework 183–188
  mathematics 183–188
  mathematics homework 183–188
  schooling 182–183

Urban-rural
  differences in mathematics achievement 223–232
  scale 224–225

Variability 87–102
Variables, parental 260, 265–269
  father's education 265–267, 285
  father's occupation (status) 266
  father's occupation (scientific or non-scientific) 285
  mother's education 265, 285
  place of parents' residence 265
Variables, school 261, 274–281
  cost per student in teachers' salaries 278, 279–280
  educational differentiation 278, 279–280, 281
  number of hours of all homework in the week 275–276, 277
  number of hours of mathematics homework in the week 275–276, 277, 285
  number of hours in the school week 274
  number of weekly hours of mathematics instruction 275–276
  number of subjects taken in grade 8 279–280, 281
  number of subjects taken in grade 12 279–280
  percentage of men teachers on the school staff 278, 279–280, 281, 285
  total roll of school 278, 279–280, 281

Variables, student 261, 281–284
  age of student 265, 281, 282–283
  sex of student 265, 281, 282–283
  student's interest in mathematics 263, 265, 281, 282–283, 284
  student's level of mathematical instruction 262, 263, 282–283, 284
Variables, teacher 261, 272–274
  age of teacher 272
  degree of freedom given to the teacher 270–271, 274
  description of mathematics teaching and school learning 270–271, 274
  experience of teacher 274
  length of training 270–271, 273, 285
  recent in-service mathematical training 270–271, 273, 274, 285
  sex of teacher 270–271, 273
  student's opportunity to learn the test items 263, 265, 270–271, 272, 284
Verbal and computational problem scores 241–243

Yield 293, 294, 308–309

6-26-67 J.V.

OHIO